MATHEMATICS
A Human Endeavor

A Series of Books in Mathematics

Editors: R. A. Rosenbaum and G. Philip Johnson

MATHEMATICS
A Human Endeavor

A Book for Those Who Think They Don't Like the Subject

Harold R. Jacobs

W. H. FREEMAN AND COMPANY
San Francisco

Library of Congress Catalog Card Number: 70-116898
ISBN: 0-7167-0439-0 (cloth)
0-7167-0471-4 (paper)

Printed in the United States of America.

Cloth edition:
Copy of a woodcut by Dutch artist Maurits Escher.
Reproduced by permission of the Escher Foundation,
The Hague, Netherlands.

Paper edition:
Cover drawing courtesy of
California Computer Products, Inc.

20 19 18 17 16 15

From *Medical Radiography and Photography*, published by Radiography Markets Division, Eastman Kodak Company.

Contents

3

FUNCTIONS AND THEIR GRAPHS

4

LARGE NUMBERS AND LOGARITHMS

5

REGULAR POLYGONS

6

MATHEMATICAL CURVES

7

SOME METHODS OF COUNTING

8

THE MATHEMATICS OF CHANCE

9

AN INTRODUCTION TO STATISTICS

10

SOME TOPICS IN TOPOLOGY

APPENDIX. SOME FUNDAMENTAL IDEAS

Foreword

Martin Gardner
Editor of the Mathematical Games Department,
Scientific American

The continuing hullabaloo about the "new math" has given many a parent a false impression. What was formerly a dull way of teaching mathematics by rote, so goes the myth, has suddenly been replaced by a marvelous new technique that is achieving miraculous results throughout the nation's public schools. I wish it were true—even if only to the extent implied by entertainer (and math teacher) Tom Lehrer in his delightfully whimsical recording on "The New Math": ". . . in the new approach, as you know, the important thing is to understand what you're doing, rather than to get the right answer."

Let me say at once that I am not against the new math. By and large its basic ideas and vocabulary are admirable. The chief trouble is that it is being taught mainly in the same dusty-minded way as the "old math" has been and the sad result is that the children are as bored as they were fifty years ago, if not more so.

Indeed, there is something to be said for the old math when taught by a poorly trained teacher. He can, at least, get across the fundamental rules of calculation without too much confusion. The same teacher trying to teach new math is apt to get across nothing at all except some phrases as opaque to the pupils as they are to the pupils' parents. Even textbooks purporting to teach new math can be hilariously dull in spite of their profusion of pictures and color overlays. If you wonder why Johnny can't add, consider this definition of "number" from a recent textbook: "the equivalence class of ordered pairs of the equivalence classes of ordered pairs." A graduate student would understand that, but not Johnny. I recently glanced through a less esoteric textbook and found that its author was almost incapable of asking readers to write the product of two numbers. He wanted them to write "the numeral that names the product." (It was gratifying that he did not ask for the numeral that names the number that names the product.) This is typical of the needless obfuscation that can be introduced by writers and teachers who suppose they are effective merely because they have acquired a new terminology.

Recently the son of a friend, having completed his assignment in a high school algebra study period, began to amuse himself by analyzing the game of ticktacktoe. His teacher caught him at it, snatched away the sheet, tore it up, and said: "When you're in my classroom I expect you to work on mathematics and nothing else." She obviously did not know that ticktacktoe is used in several modern math programs as an excellent way of introducing elementary concepts in such areas as set theory, group symmetry, game theory, probability, n-dimensional geometry, topology, and abstract algebra. Although the boy did not realize it, he was actually doing mathematics on his own, which, with a little expert guidance, might have been much more rewarding than what his teacher was teaching.

Mr. Jacobs is a pedagogue of a different breed. It takes only a few glances at his book to see that he enjoys mathematics hugely, that he appreciates its incredible variety and structural beauty. Above all, he knows how to present mathematics in a way most likely to provide an average student with what it is fashionable to call "motivation." It is unfortunate that there are not many more textbooks having this kind of approach, with the author's sense of humor and with his insight into the minds of the students whom our educational system is supposed to teach.

Preface

A Letter to the Student

Mad magazine recently suggested this as a mathematics problem: "If one handy home owner can repair one stopped-up sink in 25 minutes, how many hours will it take two plumbers charging $6.00 an hour to do the same job?"

Perhaps this problem is actually more "practical" than some of the tiresome "practical" exercises that regularly appear in general mathematics books. You won't find that kind of problem in this book. Neither will you find a lot of drill in arithmetic, which some students call "busy-work." Mark Twain's Huckleberry Finn once said: "I had been to school . . . and could say the multiplication table up to $6 \times 7 = 35$, and I don't reckon I could ever get any further than that if I was to live forever. I don't take no stock in mathematics, anyway." If this is your idea of the subject, then you may agree that it is hard to see why anyone would care very much for it. Many people, perhaps like you, think that mathematics is the art of computing and that mathematicians must spend all their time making very advanced calculations. Such an idea could hardly be farther from the truth.

My goals in writing this book are to introduce you to what mathematics is really like, to reveal its extent and power, and to give you some insight into its historical development. Of course, we cannot go into the subject very deeply in a book of this size, but luckily it is not necessary to learn how to extract square roots, or memorize the quadratic formula, or even prove a lot of geometric facts in order to appreciate many interesting areas of mathematics.

You may come across topics in this book that would seem to be of little use to you in later life, but although mathematics is indispensable to most activities of the modern world, its true significance does not depend upon its practical use. Don't be like the woman flying over the Grand Canyon who asked, "What good is it?" It was many years before some mathematics created in the past was discovered to apply to any useful problem. Some mathematics of today may never have any applications to other subjects. But, like the Grand Canyon, mathematics has its own kind of beauty and appeal to the person who is willing to look.

★

On a recent flight here from Denver, as the pilot announced to passengers that they were about to fly over the Grand Canyon, writer Courtney Anderson, sitting next to the window, offered to change seats with the woman next to him. A woman of about 45, head of an employment bureau in Chicago, she'd told him this was her first trip west.

She looked silently at the great gorge and after a thoughtful moment turned to him and said, "What good is it?" He silently tightened his seat belt and wished he were back home in West L.A.

★

Courtesy of Matt Weinstock, Los Angeles Times.

A great mathematician of our century, G. H. Hardy, said: "A mathematician, like a painter or a poet, is a maker of patterns. If his patterns are more permanent than theirs, it is because they are made with *ideas*."

A Letter to the Teacher

An item in the *Wall Street Journal* tells of a mother discussing her son's slow progress in algebra with his teacher. She says: "It isn't that my son refuses to try—I rather think he just doesn't believe it!"

So much emphasis seems to be placed now on abstract "symbol-shoving" and the rigor of mathematics that many students who have a limited interest in the subject to begin with seem to end up the same way as this skeptical algebra student. George

"I see trouble with algebra."

Boehm in his excellent book, *The New World of Math,* says:

"It is a pity that so few people today are given a chance to appreciate mathematics. Somehow the subject has been lost from the liberal-arts course, where historically it held a central position.

"When mathematics is taught, it is presented mainly as a collection of slightly related techniques and manipulations. The profound, yet simple concepts get little attention. If art appreciation were taught in the same way, it would consist mostly of learning how to chip stone and mix paints."*

Hopefully, this book will succeed in developing more than stone-chippers and paint-mixers. Admittedly it cannot create any Rembrandts either, but, just as important, there may be more Rembrandt appreciators. The basic objective of the book is to develop an interest in mathematics in students who previously have had little or none. Ideally, at the end of the course, every student will be able to say as one did: "After taking this class I feel as though at the start of the semester, I was sort of sheltered. This class opened up a relatively new world to me."

Although a quick glance at the table of contents suggests a rather sophisticated treatment, the topics chosen are presented in a way that requires a minimal mathematical background and maturity. The exercises emphasize inductive thinking and discovery and, because of their strong intellectual interest, are the

**The New World of Math,* by George A. W. Boehm and the editors of *Fortune,* Dial Press, 1959, pp. 8-9.

type that attract students. It is not assumed that students who use this book know how to solve even simple equations or that they have much acquaintance with geometric figures. An appendix is provided for those who are unfamiliar with, or don't remember, several basic topics that were not felt to be a proper part of the main presentation. Reference to these topics is made in the text at the appropriate places.

The omission of any discussion of sets and certain other topics from the "new math" is deliberate, motivated by the belief that these ideas have been overemphasized; by the time they reach this course, many students have already been exposed to these topics a number of times and tend to find them tiresome. A major goal in writing this book has been to present mathematics with a *fresh* approach.

Although some lessons can be covered in a single day, others may take two or even three days. It is recommended that assignments be planned so that there is ordinarily time in class to do most of the work assigned. You will probably find, as I have, that the material in this book so motivates the students that full-period tests are unnecessary. Grading based on selected problems not discussed in class, and on brief quizzes (ordinarily open-book), works very well and tends to promote a more positive attitude toward the subject and a healthier interest in it for its own sake.

The exercises in most lessons are highly sequential so that it is best to assign complete sets, rather than even or odd problems. Exercises are divided into three sets, and although every student should do Sets I and II, Set III may ordinarily be considered to be optional. Most of the Set III exercises are intended to challenge the students and it is not expected that everyone will be able to do them. The answers to selected exercises are given in the back of the book.

An attempt has been made to incorporate a number of experiments into the course, somewhat comparable to what one might expect in a science class. In many cases, these require some advance preparation by the teacher, but I think the effort involved will make the course more worthwhile to the students. Don't be tempted to skip them!

The teacher's guide, available from the publisher, contains many specific and practical ideas for presenting the material in the classroom, a suggested time schedule, and a complete solution key to all of the exercises.

HAROLD R. JACOBS

Photo by The New York Times.

The antenna of the National Radio Observatory at Green Bank, West Virginia. This antenna has been aimed at several stars in hope of receiving signals from another civilization.

Introduction: Mathematics — A Universal Language

IS man the only intelligent being in the universe? If there is life on a planet of another star and the people of that far-off world tried to communicate with us, what kind of message would they send? Suppose we tried to send a message to them. What should we say, and how should we say it?

We could hardly expect that something like Morse code would work or that any earthly language would make sense. How, then, would it be possible to begin a conversation with another world? Scientists agree that the kind of message most likely to be understood would be a mathematical one.

Here are some diagrams of radio signals suggested by an English physicist as a way of starting a conversation. Each line of pulses represents a mathematical statement. Can you figure out what the statements say?

1.

Hint: This message seems to have 5 parts.
What does each part mean?

2.

3.

4.

5.

6.

7.

8.

A problem somewhat like that of understanding a message from outer space is that of making sense out of the messages left behind by an early civilization here on earth. The Babylonians were the first people to create much mathematics, about 4,000 years ago. They left behind thousands of clay tablets, some of which reveal their number system and their discoveries in algebra and geometry.

A Babylonian tablet of about 1800 B.C., dug up in this century. The wedge-shaped writing, made by a stylus on wet clay, is called cuneiform.
Courtesy of University Museum, University of Pennsylvania

On the next page is a copy of the front and back of a tablet. Can you translate the groups of wedge-shaped symbols into familiar symbols, and explain what the table is about? Hint:

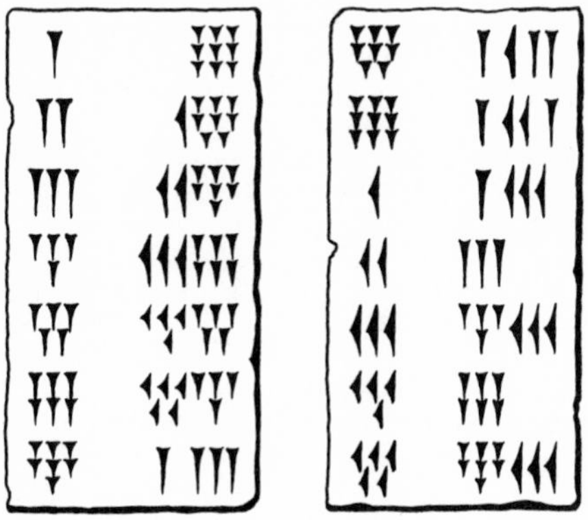

Figure out what all the symbols in the left-hand columns mean before working on the right-hand columns.

INTERESTING READING

We Are Not Alone, by Walter Sullivan, McGraw-Hill, 1966: Chapter 18, "Celestial Syntax."

Is Anyone There?, by Isaac Asimov, Doubleday, 1967: Chapter 22, "Is Anyone There?"

The Pantheon Story of Mathematics, by James T. Rogers, Pantheon, 1966: Chapter 2, "Out of the Mists."

Episodes from the Early History of Mathematics, by Asger Aaboe, Random House, 1964: Chapter 1, "Babylonian Mathematics."

MATHEMATICS
A Human Endeavor

Chapter 1

THE MATHEMATICAL WAY OF THINKING

"For a minute I thought we had him stymied."

Lesson 1

The Path of a Billiard Ball

AN expert billiard player's ability to control the path of a ball seems almost miraculous. Mathematicians like the game of billiards because the paths of the balls can be precisely calculated by mathematical methods. Lewis Carroll, the author of *Alice in Wonderland* and a mathematics teacher at Oxford University as well, liked to play billiards. He even invented a version of the game to be played on a circular table!

A skilled billiard player can picture the path of a ball before he hits it. The path is determined by how the ball is hit, by the *shape* of the table, and by the positions of the other balls. The ordinary billiard table is about twice as long as it is wide (approximately

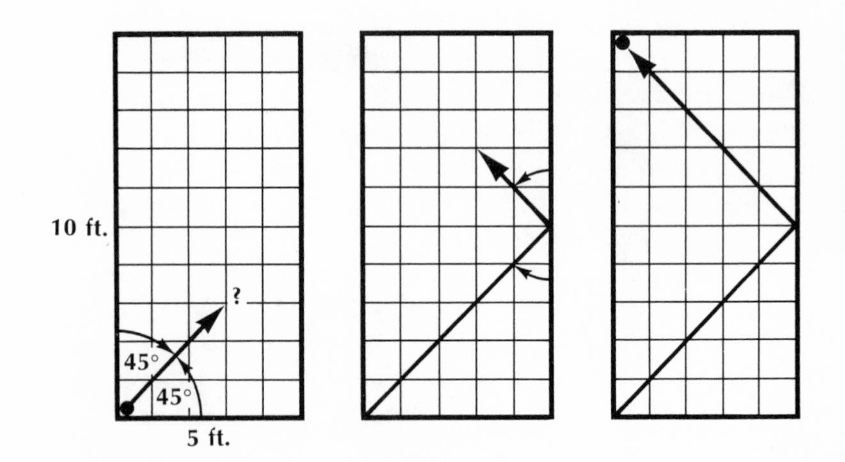

10 feet by 5 feet). Suppose a ball is hit from one corner so that it travels at a 45° angle with the sides of the table.* If it is the only ball on the table, where will it go? The first diagram (on the facing page) shows us the direction of the ball as it is hit from the corner. The second diagram shows that the ball hits the midpoint of the longer side of the table. (A billiard table, unlike a pool table, does not have any pockets.) When the ball strikes the cushion, it rebounds from it at the same angle. The angles of hitting and rebounding have been marked with curved lines to show that they are equal. The third diagram shows that the ball goes to the corner at the upper left, and we will assume that the ball always stops when it comes to a corner.

Let's try this again on a table with a different shape. Suppose the table is 6 feet by 10 feet.

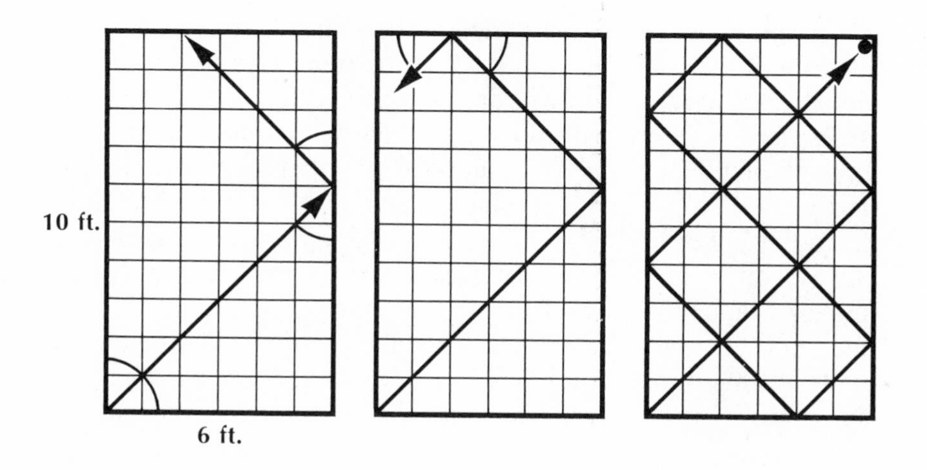

Again the ball is hit from the lower left-hand corner at a 45° angle with the sides. This time, after the first rebound, it misses the other corner and hits the top side as shown in the second diagram. It rebounds from the side in a new direction so that the angles of hitting and rebounding are again equal. When we follow the path of the ball on the rest of its journey, we see that it rebounds several times more before finally coming to rest in a corner. This time, however, it is the corner in the upper right.

These two tables suggest all kinds of questions about tables of other shapes. Will the ball always end up in one of the table's corners, or, ignoring friction, could it go on rebounding from the walls forever? Can it ever come back to the original corner? If

*If this isn't clear to you, look on page 494.

the ball ends up in a corner, and you know the length and width of the table, is it possible to predict which corner without drawing a diagram? You can probably think of other questions as well. We are faced with quite a puzzle. A 20th century American mathematician has said:

"Puzzles are made of the things that the mathematician, no less than the child, plays with, and dreams and wonders about, for they are made of the things and circumstances of the world he lives in."*

EXERCISES

Set I

On graph paper (4 units per inch is convenient), make a diagram of each of the following tables. Use the *same number of units of length and width* as shown, and write the dimensions along the sides, as has been done for table 1. Now continue drawing the path of each ball as far as it can go. (We will assume that the ball comes to a stop when it reaches a corner.) Always start the path from the lower left-hand corner and notice that since the ball is always hit at a 45° angle with each side of the table, it always moves 1 unit up or down for 1 unit left or right. If the ball ends up in a corner, mark the corner with a large dot. (Please do not write in this book. You *won't* learn anything extra by doing so, and you *will* spoil the problems for the next student who uses it.)

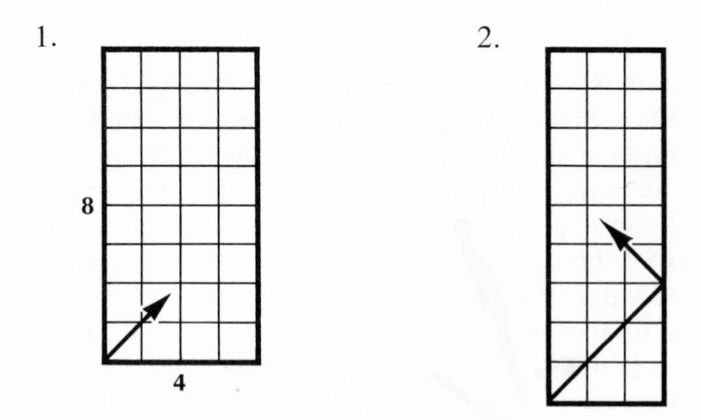

*Edward Kasner, in his book, *Mathematics and the Imagination,* Simon and Schuster, 1940, pp. 188–189.

3.

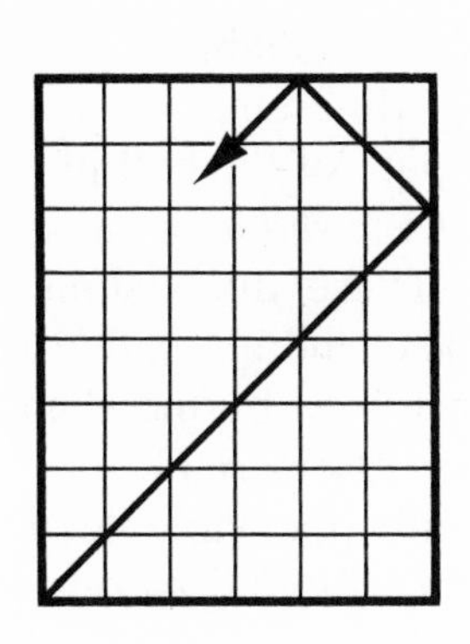

4.

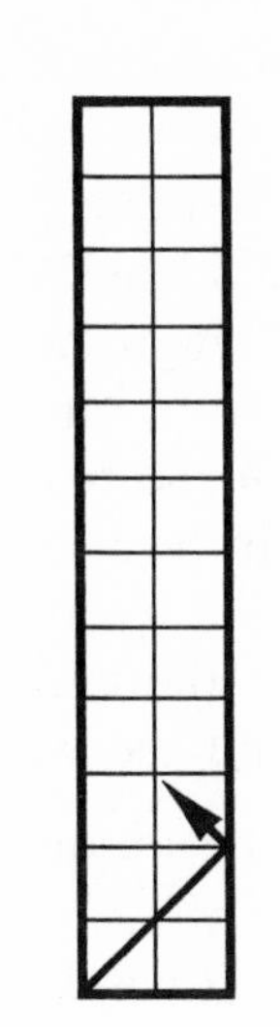

5.

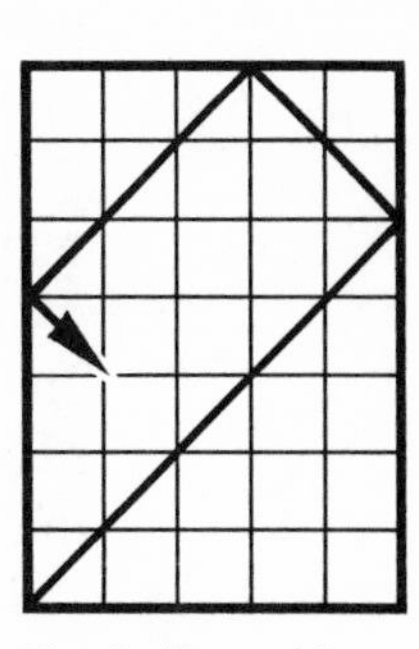

The ball on this table really travels!

6. 7. 8.

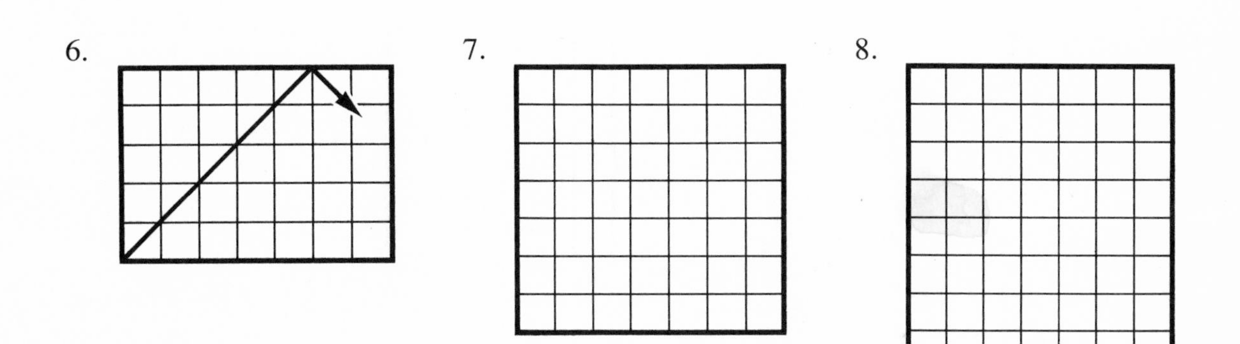

9. On which table does the ball have the simplest path? Can you explain why?
10. What do you notice about the paths on tables 5 and 6? Can you explain?
11. Do you think the ball will always end up in a corner?
12. If the ball starts from the lower left-hand corner, do you think it can end up in any of the four corners?

Set II

The paths we have drawn so far are wildly unpredictable. A slight change in the shape of the table can make a tremendous difference in where the ball will go. Compare your drawings of the last two tables. Table 8 is the same width as table 7 and only one unit longer, yet the paths are entirely different.

The shape of the table determines the path of the ball in some way that is not yet clear. What determines the *shape of the table?* Two things: its length and its width. These dimensions can change, or vary, from one table to the next and are called *variables.* The path of the ball, then, is determined by the shape of the table which, in turn, is determined by two variables.

It would be simpler if there were only one variable. Let's keep the length of the table the same (we will hold it *constant*), vary the width, and see what happens.

1-6. Draw a set of six billiard tables with lengths of 6 units and with widths of 1, 2, 3, 4, 5, and 6 units. Start the ball from the lower left-hand corner as before and mark where the ball ends up with a large dot.

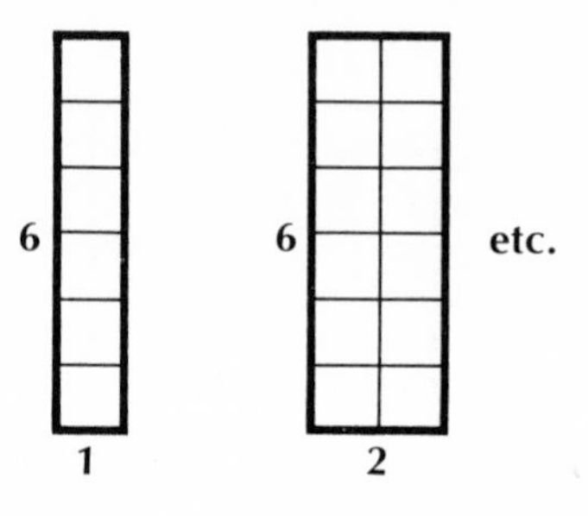

7. Does the result for any of these tables surprise you? Which one and why?

8. What are the two dimensions (length and width) of the table with the *simplest* path?

9. What are the dimensions of the table with the most *complicated* path?

10. If a giant billiard table had a length of 100 feet, what width should it have for the ball to travel the simplest possible path?

11. What width should it have for a very complicated path?

Set III

Draw a set of seven billiard tables with lengths of 7 units and widths from 1 unit through 7. When you have done this, you will see that the results are very different from the previous set with length 6.

On most of the tables the ball travels over every square. Can you think of a rule for telling in advance which tables these are without drawing them? (Hint: The rule has to do with factors of numbers. A factor of a number is a whole number by which it can be divided without leaving a remainder. For example, the factors of 10 are 1, 2, 5, and 10.)

Look back at the tables you drew in Sets I and II to see if your rule always works.

Lesson 2

More Billiard Ball Mathematics

DO you think that our study of the path of a billiard ball on a table is mathematics? If you do, then you disagree with the idea that mathematics must mean just making calculations with numbers. If you don't, then you should know that problems like this are of great interest to mathematicians and that such problems have sometimes led to the development of new and often very practical areas of mathematics.

So far, in every diagram we have drawn, the ball, after starting at the lower left-hand corner, has ended up at one of the other three corners of the table. We haven't yet solved the problem of predicting *which* corner on the basis of the length and width of the table.

Since the path of the ball depends upon the shape of the table, it would seem that on tables whose shapes were the same, the paths would be identical, meaning that the ball would end up in the same corner. Let's see if this is true. Five tables (top of the facing page) are twice as long as they are wide. The paths are the same. What is the same about each table? The *ratios** of

*A ratio is a fraction, a way of comparing two numbers by division. To find the simplest ratio between two numbers, we divide both by their largest common factor. In the case of the ratio 6/3, the largest common factor of 6 and 3 is 3. (The factors of 6 are 1, 2, 3, and 6; the factors of 3 are 1 and 3.) Dividing both numbers of the ratio 6/3 by 3, we get 2/1, or 2.

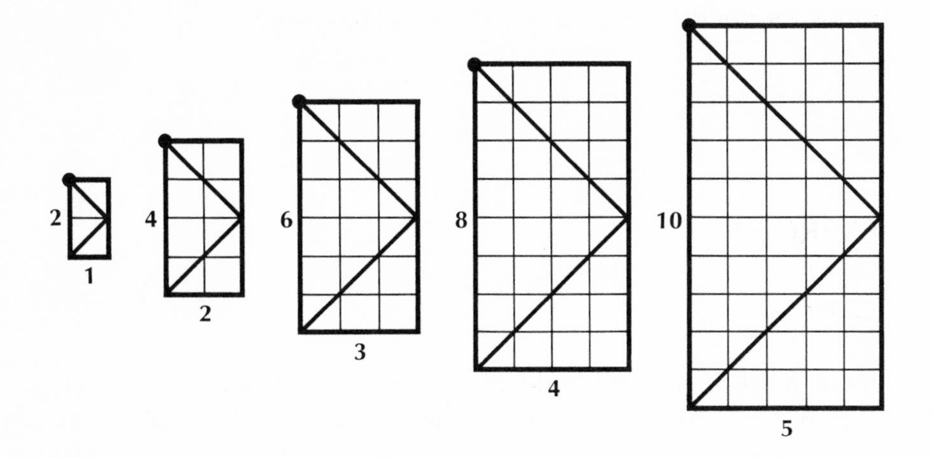

length to width are the same.

$$\frac{2}{1} = \frac{4}{2} = \frac{6}{3} = \frac{8}{4} = \frac{10}{5} = 2$$

We have been representing the top of each table with a rectangle. Rectangles that have the same shape are called *similar;* the ratios of the lengths and widths of similar rectangles are the same.

Below is a rectangle whose dimensions are 6 feet and 4 feet. If we increase its length and width by the same number, say 2, will the new table have the same shape? In other words, will the two rectangles be similar?

The path of the ball on each table shows that the answer is no.

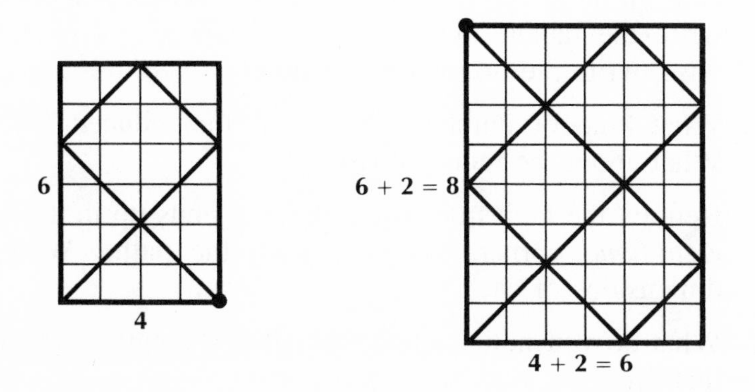

The ratio of the length and width of the first table is 6/4 or 3/2; the ratio of the length and width of the second table is 8/6 or 4/3.

After writing each ratio in simplest form, we can see that the ratios are not the same. In mathematical language,

$$\frac{3}{2} \neq \frac{4}{3}.$$

EXERCISES

SET I

Draw a set of eight billiard tables with length 10 and widths from 1 through 8. Write the two dimensions of each table along its sides. Show the path of the ball starting from the lower left-hand corner, traveling at an angle of 45° with each side of the table, and mark the corner where the ball ends up each time with a large dot.

5

1

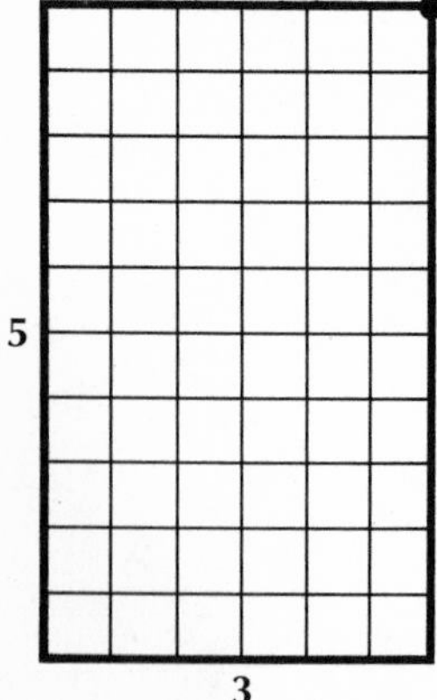

Set II

On two of the tables you have drawn, the ball ends up in the *upper right-hand* corner. They are shown (left) with the paths omitted. The two dimensions of each have been written along the sides in "reduced" form to show the simplest ratio between them. Notice that the "reduced" lengths and widths of these tables are *odd* numbers.

1. Omitting the paths, redraw the tables on which the ball ends up in the *upper left-hand* corner. Where it is possible, write the tables' dimensions in "reduced" form. Otherwise, write the original dimensions.
2. What kind of numbers are all of the reduced *lengths?* What about the reduced *widths?*
3. Redraw the tables on which the ball ends up in the *lower right-hand* corner, again omitting the paths. Write the dimensions of each in "reduced" form.
4. What do you notice about the reduced lengths and widths this time?
5. Can you write a rule for telling in advance in which corner the ball will end up?

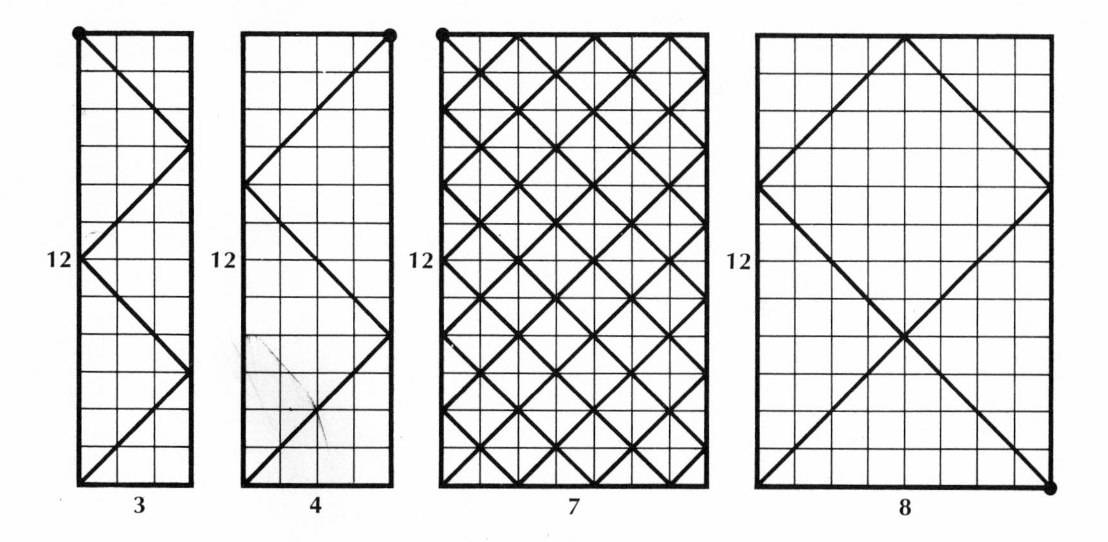

6. Above are some tables with lengths of 12. Test your rule on each of them. Does it always work?
   Explain in each case.

Set III

If you have figured out a rule that predicts the corner correctly, you shouldn't have any trouble with the following question. If you know the dimensions of a table, is it possible to predict how many times the ball will strike the cushions during its trip around the table?

Let's count the original and final corner positions as "hits" so that, for example, on a 4 ft. by 2 ft. table there are three hits. How many hits on a 4 ft. by 3 ft. table?

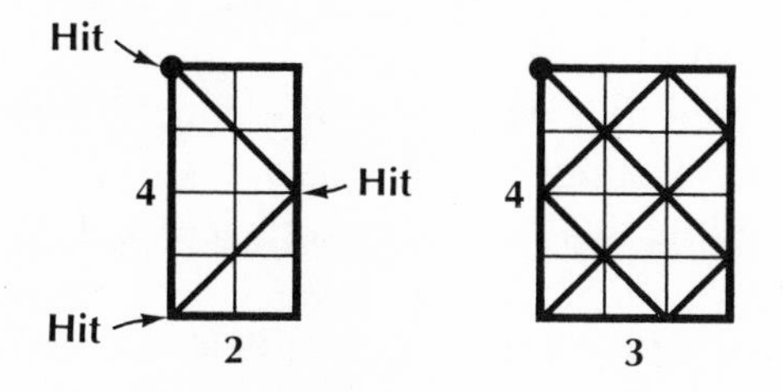

If you agree that there are seven, go back and count the number of hits on each table you drew in Set I and see if you can finish this rule: "To predict the number of times the ball will hit the sides of the table. . . ."

Lesson 3

The Nature of Inductive Reasoning

WE have had quite a bit of practice in *collecting evidence* (drawing billiard tables of various shapes), *noticing patterns,* and *forming conclusions* on the basis of these patterns (making up rules about where the ball will end up and how many times it will hit the cushions). The method of reasoning we have been using is that of the scientist when he makes observations and discovers regularities from which he proposes general laws of nature. In science, this is often called the experimental or scientific method. In mathematics, it is referred to as reasoning **inductively.**

Although inductive reasoning is of tremendous importance in developing mathematical ideas, it has a basic weakness. Since we draw our conclusions from only the evidence we have collected, the possibility always exists that some more evidence may be discovered that will prove the conclusions incorrect. We may end up as wrong as the cave man who concluded, on the basis of boiling water, snow, and ice, that everything boils down to nothing.

Here is an example of how easy it is to draw an incorrect conclusion when reasoning inductively. In studying the mathematics of billiard ball behavior, suppose we had drawn a set of tables with a constant length of 8. In every one of our diagrams the ball ends up in the same place, even after taking widely varying paths. A reasonable conclusion seems to be: "If the length of the table is 8, the ball will always end in the upper left-hand corner." What

"Water boils down to nothing . . . snow boils down to nothing . . . ice boils down to nothing . . . everything boils down to nothing."

do you think? Does it seem a safe bet to claim this on the basis of these seven cases?

What about a square table with length and width both 8?

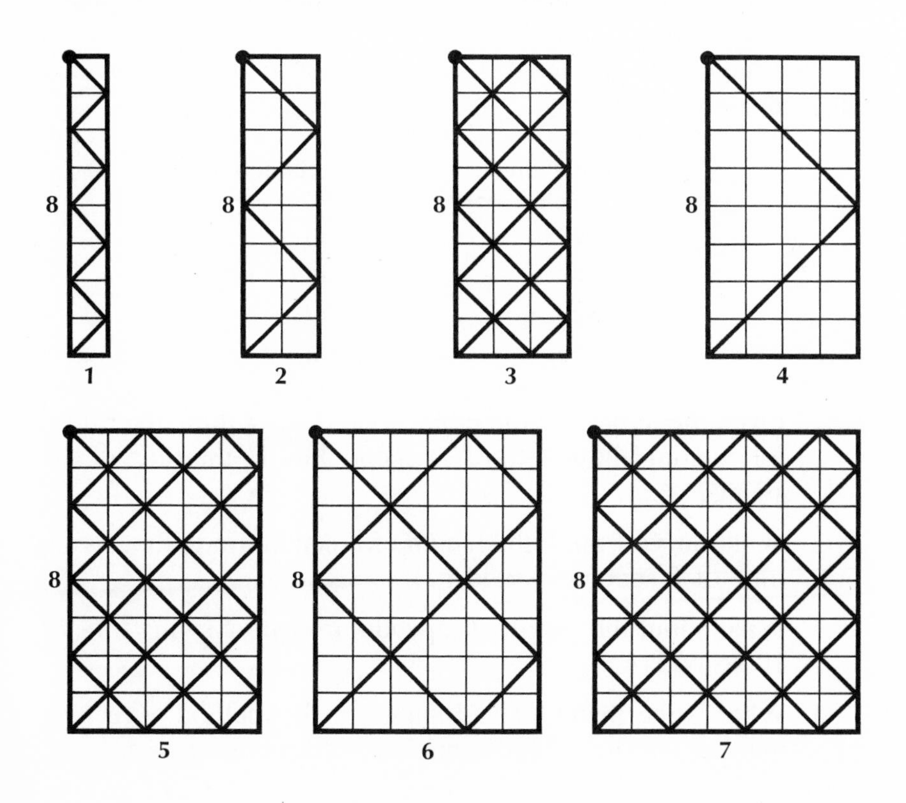

EXERCISES

Galileo Galilei

Set I

The great Italian scientist Galileo made a number of discoveries about the behavior of swinging weights – discoveries that led to the invention of the pendulum clock. One of these discoveries was that the time required for one swing is the same whether the weight moves through a long distance or a short one. Another discovery was of a relationship between the *time* of the swing and the *length* of the pendulum.

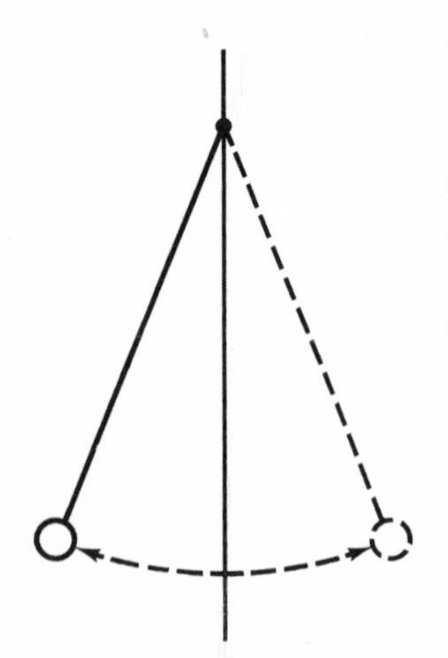

This table lists the lengths of a series of pendulums having different swing times.

Time of swing	Length of pendulum
1 sec.	1 unit
2 sec.	4 units
3 sec.	9 units
4 sec.	16 units

1. From the pattern in the table, what do you think the rule is for relating the length of the pendulum to the time of the swing?
2. What do you think the length of a pendulum with a swing time of 5 seconds would be?
3. What about the length of a pendulum with a swing time of 10 seconds?
4. What is the mathematical name for the method of reasoning that you have been using in answering these questions?

In 1772, a German astronomer named Bode discovered a pattern in the distances of the planets from the sun. At that time only six planets were known. His table of numbers and the actual relative distances from the sun looked like this:

Planet	Bode's numbers	Actual distance
Mercury	0 + 4 = 4	3.9
Venus	3 + 4 = 7	7.2
Earth	6 + 4 = 10	10.0
Mars	12 + 4 = 16	15.2
Jupiter	48 + 4 = 52	52.0
Saturn	96 + 4 = 100	95.4

5. You can see that there is a gap in the table between Mars and Jupiter. What number do you think Bode decided belonged there?
6. What do you suppose Bode thought this gap meant astronomically?
7. William Herschel discovered Uranus, the next planet beyond Saturn, in 1781. What number do you think Bode calculated for Uranus? (The actual distance of Uranus is 192 units.)
8. What distance does Bode's pattern predict for the next planet after Uranus? (That planet, Neptune, was discovered in 1846. Its actual distance is 301 units.)

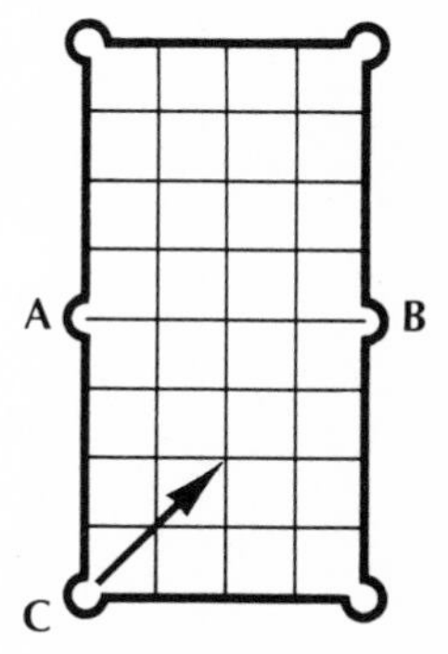

Set II

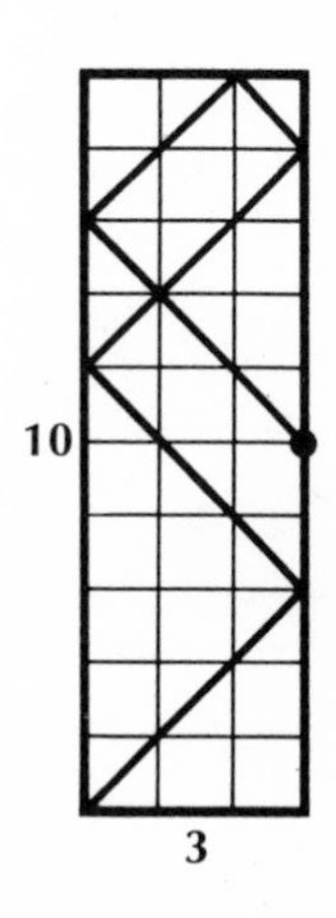

A pool table has six pockets: one at each corner and two at the midpoints of the longer sides. In the diagram, top right, the two middle pockets have been labeled A and B. If a ball is hit from corner C at a 45° angle as shown, is it possible for it to end up in the pocket labeled A? For example, on a table 10 ft. by 3 ft., the path looks like the diagram on the bottom right.

1. Draw some tables with different shapes to see if you can arrive at any conclusion about whether the ball can end up in pocket A.
2. Can the ball ever end up in a pocket at an upper corner?

Set III

Here is a set of three circles.

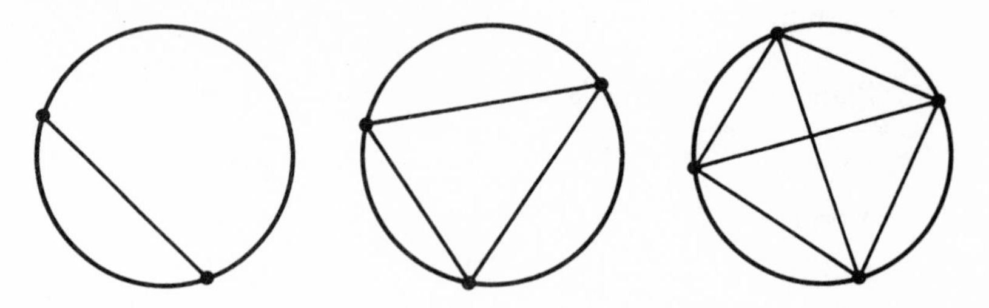

Two points have been chosen on the first circle and a straight line segment drawn between them. The circle is separated into two regions as a result. Three points were chosen on the second circle and connected with three line segments, forming four regions. Four points were chosen on the third circle and after being connected in all possible ways, eight regions resulted.

Here is a table that includes the results so far:

Number of points connected	2	3	4	5	6
Number of regions formed	2	4	8	▯	▯

Two cases have been added.

1. Guess from the pattern in the second line of numbers what the missing numbers are.
2. Draw a couple of large circles and choose five points on one circle and six on the other. Join the points of each in every possible way.
3. How many regions are formed in each?
4. Do both results agree with your guesses? (Remember that the cave man also reasoned from three cases when he concluded that everything boils down to nothing.)

Lesson 4

Mathematical Illusions

PETER is upset because he "knows" that parallel lines never meet and yet, from wherever he looks, they do. His difficulty comes from confusing a mathematical *idea* with a *representation* of it. What he and Thor have drawn on the ground is a crude picture of the idea they are arguing about. Peter has been fooled by an illusion.

The modern way of mathematical thinking began in Greece in about the sixth century B.C. The ancient Egyptians and Babylo-

Photograph from Moody Institute of Science.

The Parthenon

nians had previously made many discoveries in mathematics by experimenting as we have been doing, and reasoning inductively from their observations. The Greeks developed another way of reasoning called the *deductive method.* Mathematicians ever since have used this method when they want to be sure that their conclusions are reliable. We will become acquainted with the deductive method in the lessons following this one.

Another important contribution of the Greeks was that they made mathematics *abstract.* Ask an Egyptian what straight lines are, for instance, and he would say that they are what Peter and

Thor have drawn on the ground. Ask a Greek mathematician the same question and he would say that no one can really draw a straight line because it is merely an idea. Since it is an idea, it is perfect and everlasting. Remember the statement of the mathematician Hardy quoted in the Preface to this book that said, in effect, that mathematics is permanent because it is made with *ideas*.

Our senses can often fool us and our intuition may be completely wrong. The architects of ancient Greece understood the nature of illusions in geometric figures and prepared for them in their work. The Parthenon at Athens, built in the fifth century B.C., contains rows of columns which seem to be perfectly straight. The Greeks knew that straight lines in perspective may appear to be curved and so they cleverly bent the lines of the columns and tilted their directions just enough to compensate. Being aware of how misleading our senses and intuition can be, the Greeks also knew that these are shaky foundations on which to build a solid structure of mathematics.

This remarkable illusion is by the Dutch artist Maurits Escher.

Courtesy of G. W. Breughel, Zwolle, Netherlands.

Courtesy of C. F. Cochran.

Would you believe this is a photograph of a wooden crate used for shipping optical illusions?

EXERCISES

Set I

These diagrams contain illusions. For each diagram, tell what kind of geometric figures you see and what the illusion is.

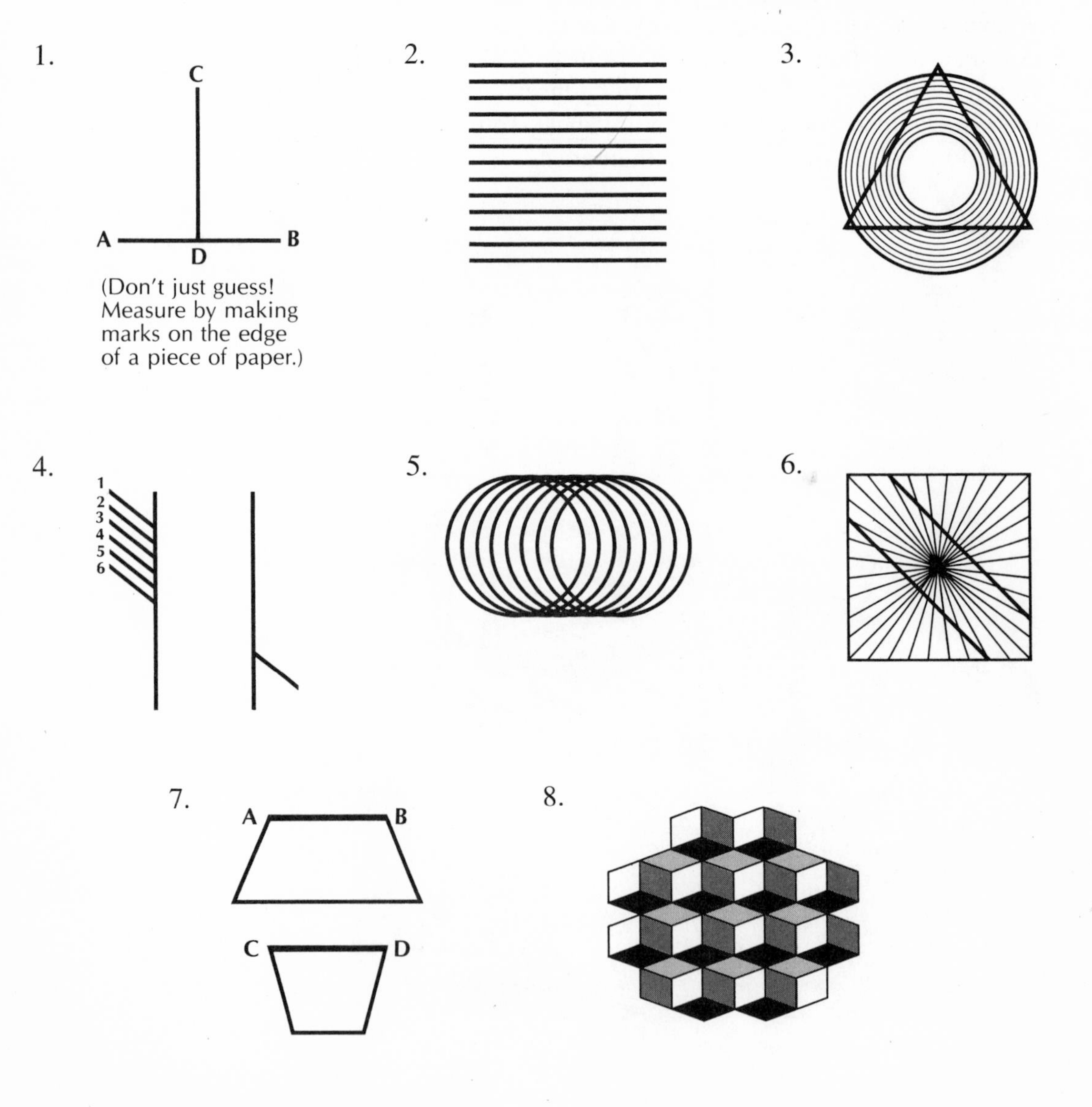

Set II

Experiment. An Area Paradox (or, How to Get Something from Nothing)*

Make a neat drawing of this figure on half of a sheet of graph paper ruled 4 units per inch. Cut out the six pieces and throw the two shaded rectangular pieces away.

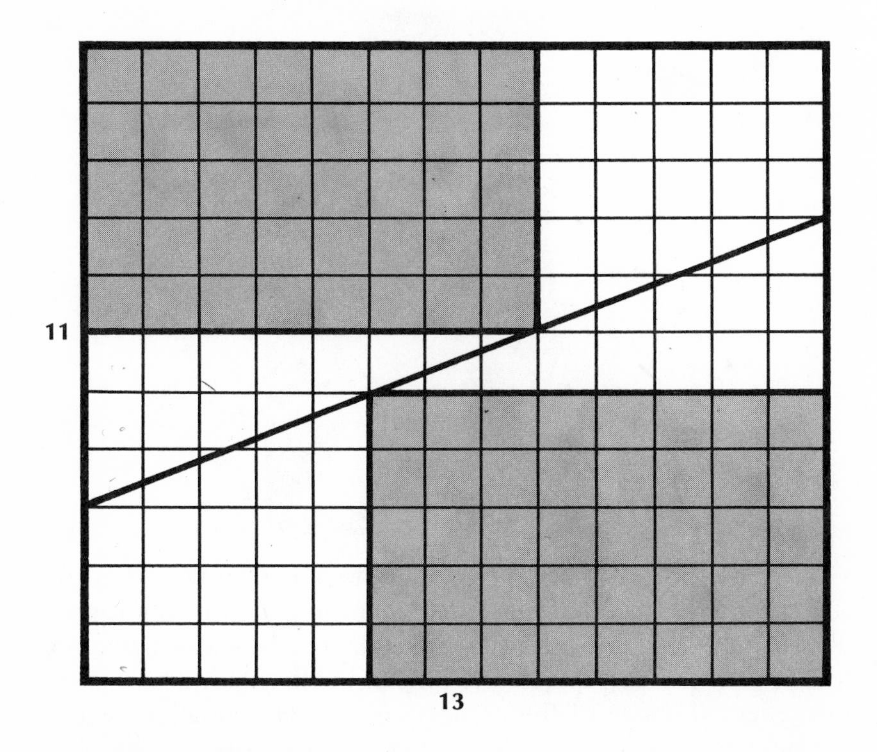

As you can see from the figure, the area of the four remaining pieces is 63 square units. (Check this.)

1. Now rearrange the four pieces to form a square. Make a drawing of the result on the other half of your sheet of graph paper.
2. What is the area of this figure?
3. Rearrange the pieces again to form a long rectangle with a width of 5 units. Make a drawing of the result.
4. What is the area of this figure?
5. Can you explain how this can be?

---

*A paradox is something that seems to be correct but which contradicts our common sense.

Set III

A stereoscope is an instrument that seems to change a pair of flat, two-dimensional, pictures into a solid, three-dimensional, one. It does this by enabling the viewer to merge the two pictures, one of which is seen only with the left eye and the other only with the right, into one image.

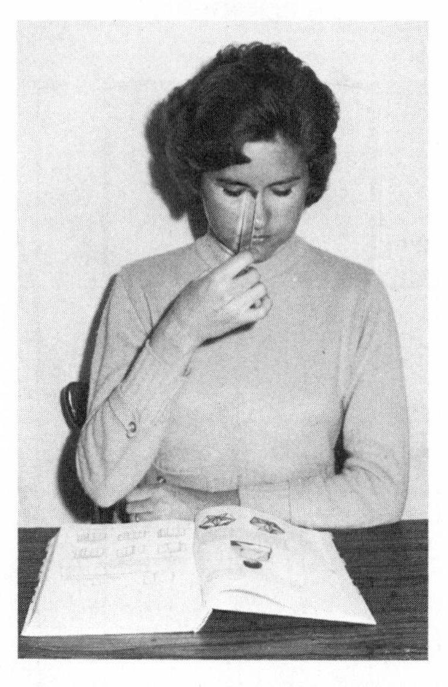

To get an idea of how this works, hold a small mirror against the right side of your nose as illustrated in the photograph above and look down at the two pictures below. Focus your left eye on the picture on the left and move the mirror so that the reflected image of the picture on the right coincides with it.

Can you explain why the picture seems to be three-dimensional when the images are merged? If you took your own photograph of an object and reversed it to produce a second image, do you think the two pictures would give a three-dimensional illusion?

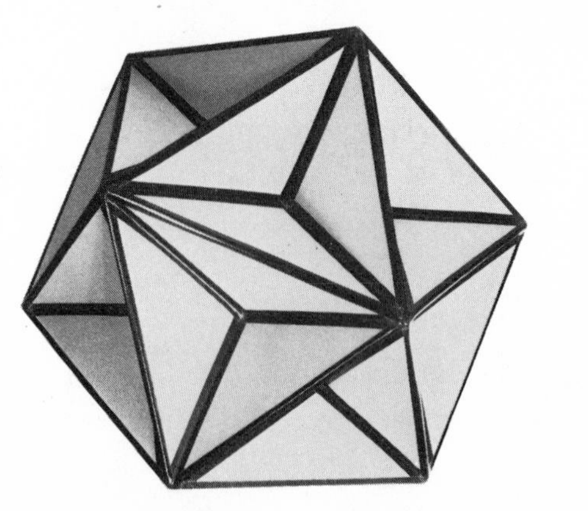

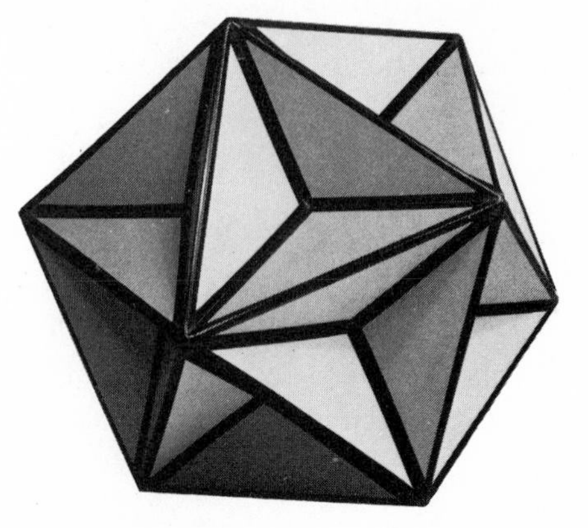

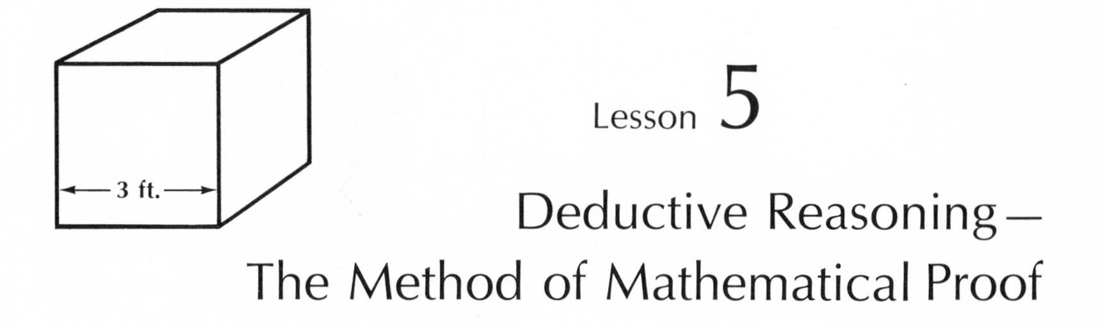

Lesson 5

Deductive Reasoning— The Method of Mathematical Proof

HERE is a block of marble in the shape of a cube that measures 3 feet on an edge. An artist wants to cut it up into twenty-seven 1-foot cubes. One way to do this is to make a series of six cuts through the cube while keeping it together in one block. The cuts are indicated by the arrows in the diagrams below.

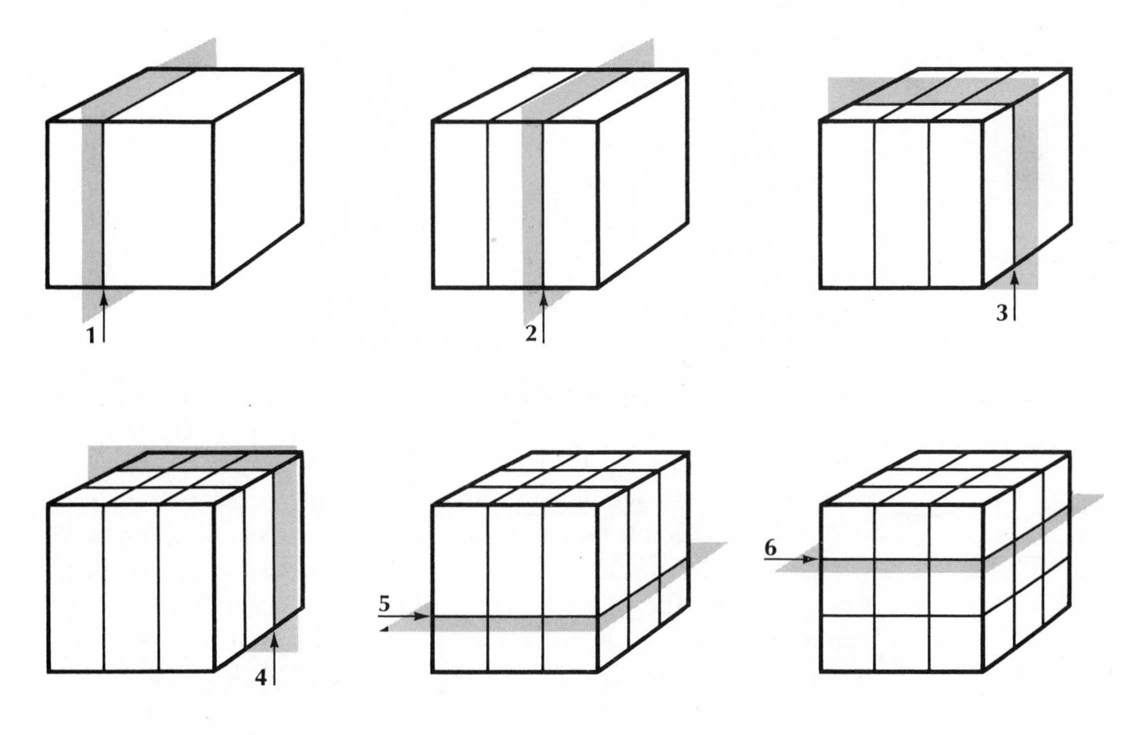

Is it possible to do the job with *less* than six cuts if the pieces are rearranged between each cut? The next two diagrams show one way of rearranging the first two pieces. Notice that the second cut will now cut through more marble than it would have if the piece at the left had not been moved to the bottom.

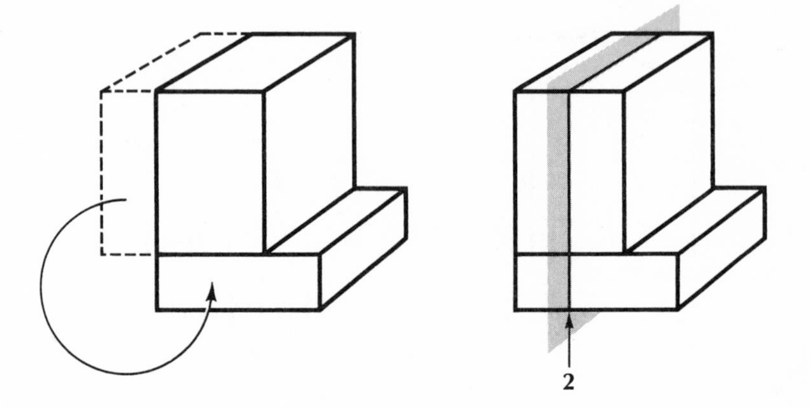

This seems like a very difficult problem, since the number of pieces increases with each cut and there are so many ways of rearranging them. As the blocks are heavy and inconvenient to move around, the artist doesn't feel like trying out all of the possibilities.

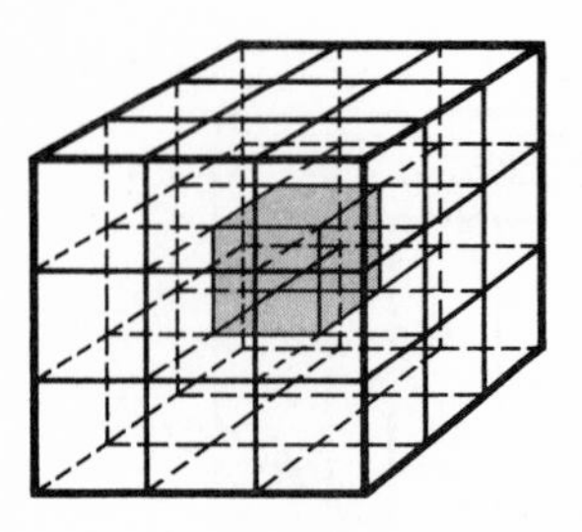

Instead of reasoning inductively by testing many different cases, he could use another method. Every cube *except one* will have at least one face (surface) that was originally part of the surface of the original block. The one exception is the cube in the *center,* which has every one of its faces formed by cuts. Since this cube has *six faces, six cuts* are necessary to form it, and that solves the problem. We have proved that rearranging the pieces will not help at all. The job can not be done with less than six cuts.

The method of reasoning we have used in arriving at this conclusion is called **deductive** reasoning. In contrast to *inductive* reasoning, this method guarantees its results when used correctly. Deductive reasoning is the method of drawing conclusions by combining in a logical way other facts that we accept as true.

A mathematician generally uses deductive reasoning when he wants to prove something true. Deductive reasoning has been used for more than twenty-five centuries, from the time that the mathematicians of ancient Greece first began to prove statements about geometric figures. In 300 B.C., Euclid, in his book the *Elements* (the most famous textbook ever written), used deductive reasoning exclusively in dealing with mathematical ideas. Euclid's method was so successful that it is still followed today in the study of geometry.

EXERCISES

Set I

First, here are the names of different parts of a cube, which you should understand.

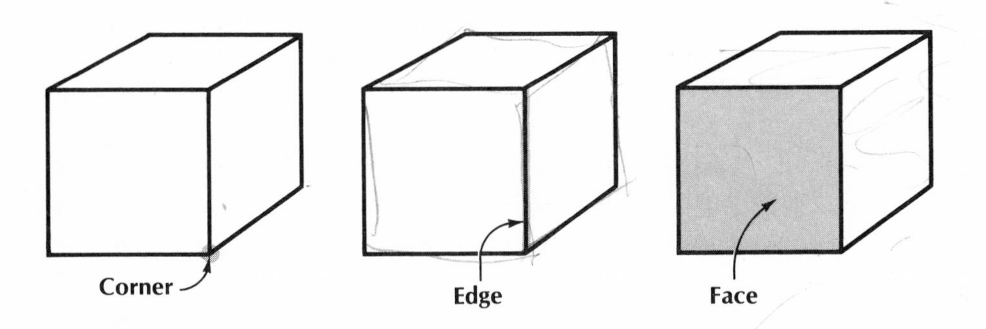

1. How many corners does a cube have?
2. How many edges?
3. How many faces?

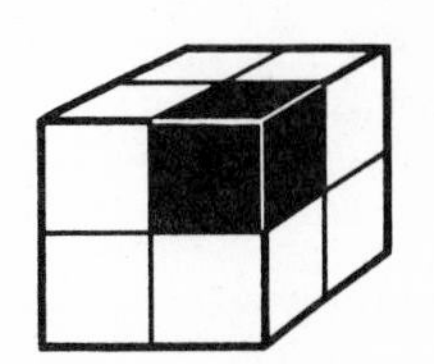

Suppose we have a small wooden cube, which is painted black, and we cut it up into 8 smaller cubes. Notice that each of these cubes comes from a corner of the larger cube.

4. How many faces of each of the smaller cubes will be painted black?

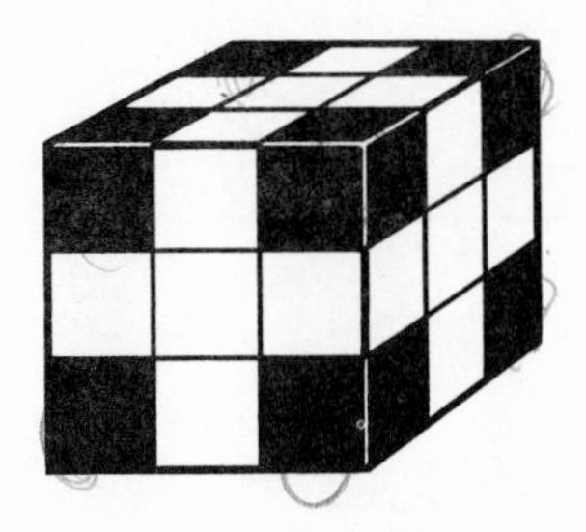

Now suppose we have another wooden cube, which is also painted black, and we cut it up into 27 smaller cubes. This time the new cubes will not all look the same. Some of the cubes will again have 3 faces painted black.

5. From where on the original cube do these cubes come?
6. How many of these cubes are there?

Other cubes will have exactly 2 faces painted black.

7. Where are these cubes located on the original cube?
8. How many are there?

There are several cubes that have exactly 1 painted face.

9. Where are these cubes?
10. How many are there?
11. Are there any cubes that have *no* painted faces?
12. Check to see whether you have accounted for all 27 cubes. (The sum of your answers to problems 6, 8, 10, and 11 should be 27.)

Suppose a large wooden cube, which has been painted black, is cut up into 64 smaller cubes.

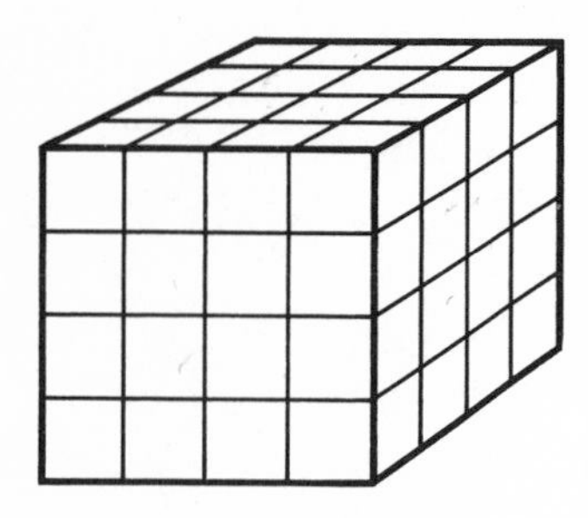

13. How many of the smaller cubes have 3 faces painted black?
14. How many have 2 painted faces?
15. How many have 1 painted face?
16. How many have no painted faces?
17. Check to see that you have accounted for all 64 cubes.
18. (Optional.) Can you figure out how many cubes there would be with each different possible number of painted faces, if a very large cube is cut up into 1,000 smaller cubes?

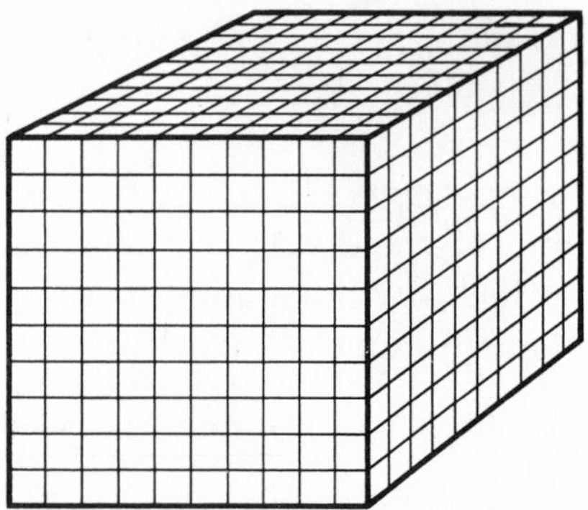

Set II

Here is a puzzle in deductive reasoning about a checkerboard and some dominoes. The board originally had 64 squares, but two at a pair of opposite corners have been cut off, leaving 62 squares. Each domino is just large enough to cover two adjacent squares of the board. Since each domino covers two squares, it will take 31 dominoes to cover the board. How should they be arranged?

1. Let's try this out on a smaller board. Draw several boards identical to Figure A on graph paper and sketch in dominoes in different positions to try to cover the board. (Do not draw any dominoes "at a slant.") One way that doesn't work is shown in Figure B.

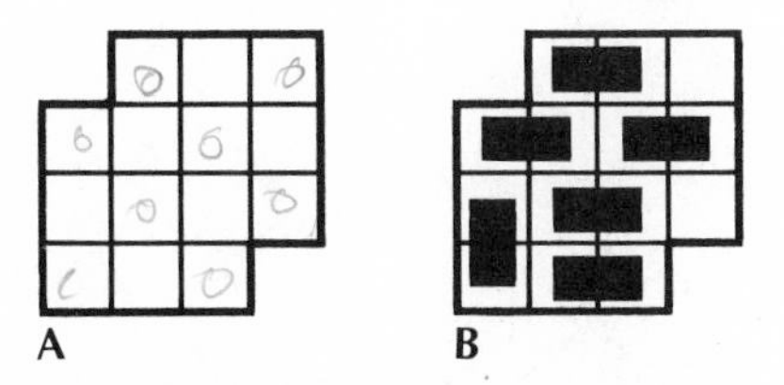

2. Now, instead of experimenting with the large board, let's try to reason deductively about it. The squares of the large board are colored black and white. Each black square is surrounded by white squares and vice versa. Look at the board again. What color were the two squares that were cut off?
3. How many black squares are on the board?
4. How many white squares?
5. Can a domino cover two squares of the same color?
6. After 30 dominoes have been put on the board, how many black squares and how many white squares will be covered?
7. What color will the two remaining squares be?

8. A domino has to cover two adjacent squares. Can these two remaining squares be adjacent?
9. Is it possible to solve the puzzle?
10. It is impossible to cover the board with dominoes long enough to cover three squares each. Considering that there are 62 squares on the board, prove why this can't be done if each domino must cover exactly three squares.

Set III

A recent postage stamp from Greece which illustrates the Pythagorean Theorem.

Mathematical statements that are proved true by using deductive reasoning are called **theorems.** The most famous theorem of all time is the one named for Pythagoras, a Greek mathematician of about 500 B.C.

The Pythagorean Theorem states that if squares are drawn on the three sides of a right triangle,* the square on the longest side of the triangle (called the hypotenuse) will be equal in size to the other two squares put together.

For example, look at this right triangle with sides of lengths 3, 4, and 5 units. Notice how the diagram "proves" the theorem for this "3-4-5" triangle.

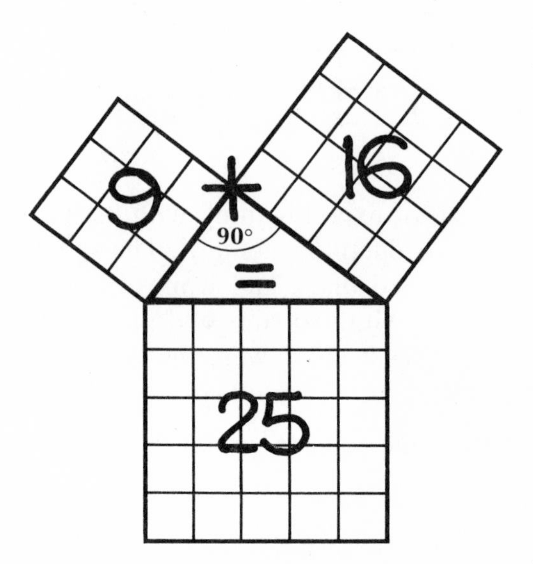

*A right triangle is a triangle that has a right (90 °) angle.

Right triangles come in many different shapes. Another of these is the isosceles right triangle, which has two equal sides.

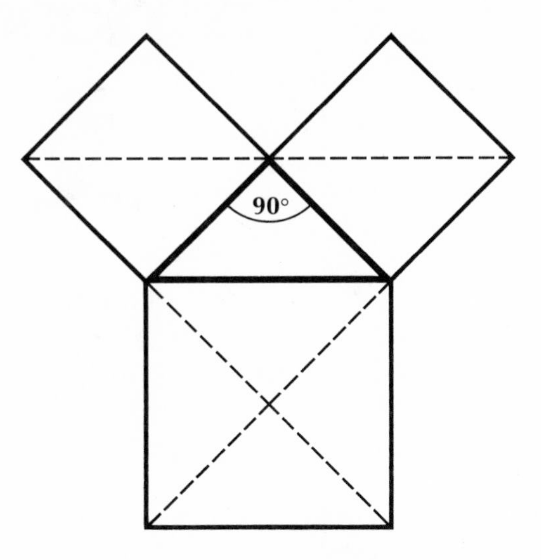

Can you explain how the diagram shown here "proves" the Pythagorean Theorem for isosceles right triangles? (Hint: It's as easy as 2 + 2.)

Lesson 6

Some Number Tricks and Deductive Proofs

Courtesy of Eastern Airlines.

WHEN a mathematician writes a deductive proof, he may list his statements in a column to show how they fit logically together. Eastern Airlines recently put out an ad that illustrates this method very well. The ad refers to the company's shuttle service between several eastern cities. The first sentence, "If we served you drinks to Boston or Washington, we couldn't guarantee you a seat when you want to go," is the statement to be proved, and is called a *theorem* in mathematics. The rest of the statements make up the proof of the theorem; notice how each follows in order from the last.

Have you ever had someone ask you to think of a number and tell you several things to do with it? Then, without knowing your original number, the person is somehow able to tell you what number you ended up with.

Here is an example: "Choose a number. Add five. Double the result. Subtract four. Divide by two. Subtract the number you started with. Your result is three."

Is it always three, no matter what the original number was? How does this trick work? Here is a table showing the steps for four numbers chosen at random:

Choose a number.	4	7	12	35
Add five.	9	12	17	40
Double the result.	18	24	34	80
Subtract four.	14	20	30	76

Divide by two.	7	10	15	38
Subtract the number you started with.	4	7	12	35
Your result is three.	3	3	3	3

And three it is in all four cases. But that doesn't prove that the trick will always work.

We are reasoning inductively, and inductive reasoning cannot prove anything. For a proof, we need to use the deductive method. Let's go through the trick again, but this time with one slight change. Instead of choosing a particular number, we will use a black box to represent all possible numbers. When we think of it as standing for a certain number, however, it stands for that number for the rest of the trick. To represent numbers we know, we will use dots.

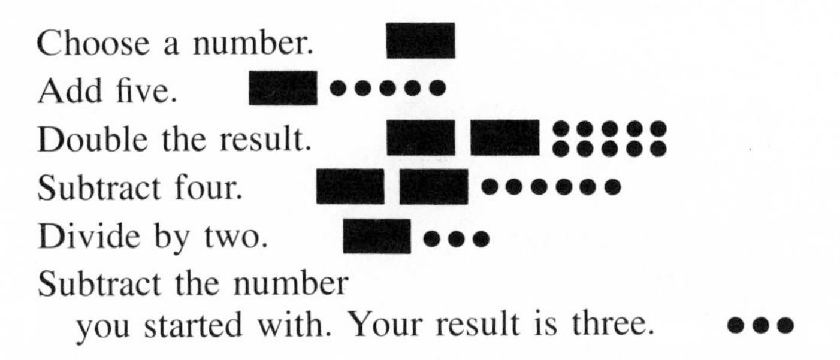

Here we have a proof that the result is *always* three! Now it is easy to see why the original number doesn't make any difference in the final result.

Black boxes and dots are somewhat clumsy and so a mathematician prefers instead to use a letter of the alphabet to represent the original number and ordinary numerals for the other numbers. A proof using the symbols of algebra looks like this:

Choose a number.	n
Add five.	$n + 5$
Double the result.*	$2n + 10$
Subtract four.	$2n + 6$
Divide by two.	$n + 3$
Subtract the number you started with.	
Your result is three.	3

---

*The principle being used here is called the "distributive law" by mathematicians. Look on page 497 if you are not familiar with this idea.

EXERCISES

Set I

Copy each of the following number tricks, writing each step on a separate line. Try each trick out with at least two different numbers. Then prove, using first black squares and dots and then algebraic symbols, that each trick will always work.

1. Choose a number.
   Add three.
   Multiply by two.
   Add four.
   Divide by two.
   Subtract your original number.
   Your result is five.

2. Choose a number.
   Double it.
   Add nine.
   Add your original number.
   Divide by three.
   Add four.
   Subtract your original number.
   Your result is seven.

3. Choose a number.
   Add the next larger number.*
   Add seven.
   Divide by two.
   Subtract your original number.
   Your result is four.

4. Choose a number.
   Triple it.
   Add the number one larger than your original number.
   Add eleven.
   Divide by four.
   Subtract three.
   The result is your original number.

---

* If the original number is represented by ■, the next larger number, being one more, would be shown as ■ •, and their sum would be shown as ■ ■ •.

Set II

Here is a number trick that is much more surprising than the ones you have just done.

1. Write down any three digits and then write them again in the same order to make a six-digit number. For example, if you wrote 123, your number would be 123,123.
2. Seven is considered by many people to be a lucky number; try dividing your number by 7 and you will find that it divides exactly, without leaving a remainder! (A difficulty with number tricks is that people frequently make mistakes in doing the required arithmetic. If you get a remainder, do it again more carefully.)
3. Eleven is another "lucky" number. Take your *result* and divide it by 11; again there won't be any remainder!
4. Even "unlucky" 13 will work! Divide your last result by 13. Does the result look familiar?
5. Start all over again with another number and carry out the same steps.

How does this trick work? To answer this question, let's go back to the beginning. We wrote down a three-digit number and then *repeated it* to make a six-digit number.

6. To see what we were *actually doing* to the three-digit number, multiply it by 1,001. What is the result?
7. When you make a six-digit number out of a three-digit number by repeating the digits, what are you actually doing to the number?

If we represent the original three-digit number by the letter n, then the six-digit number is $1{,}001n$. Dividing this number by 7, we get

$$\frac{1{,}001n}{7} = 143n.$$

8. Divide $143n$ by 11. What is the result?
9. What is the result of dividing this result by 13?
10. What is the product of 7, 11, and 13? (In other words, what is $7 \times 11 \times 13$?)

Set III

Now that you know how the "7-11-13" number trick works, try this trick out.

1. Write down any two digits and then repeat them two more times to make a six-digit number. For example, if you wrote 47, your number would be 474,747.
2. Now divide your number by 3, the result by 7, the new result by 13, and the last result by 37. All of the divisions will come out even!
3. What is the final result?
4. See if you can prove that this trick will always work.

Chapter 1 / Summary and Review

IN this chapter we have been introduced to:

Inductive reasoning *(Lessons 1, 2, & 3)* The method of making observations, noticing patterns, and forming conclusions. Such conclusions may seem very probable but cannot be proved true by this method.

Some mathematical illusions *(Lesson 4)* The method of guessing based on appearances or what we feel "must be true," sometimes called "intuition," may lead to false conclusions.

Deductive reasoning *(Lessons 5 & 6)* The method of drawing conclusions from other ideas we accept as true by using logic. Statements that are proved true by using deductive reasoning are called *theorems.*

EXERCISES

Set I

Albrecht Dürer, a German artist of the 16th century, made a famous engraving titled *Melancholy,* which contains an interesting square of numbers. Two of the numbers in the square have been blacked out in this copy of that engraving.

Melancholy

16	3	2	13
5	10	11	8
9	6		12
4	15	14	

1. Can you guess what the number in the corner is? The number inside?
2. When you *assume that there is a rule* for figuring out these numbers, what kind of reasoning are you using?
3. When you *use the rule* you have figured out to tell what the missing numbers are, what kind of reasoning are you using?
4. What kind of reasoning did the early Greek mathematicians use to prove facts about geometric figures?
5. We have done a lot of experimenting with the paths of billiard balls on rectangular tables of different shapes. What kind of reasoning did you use in making up a rule for predicting the corner in which each ball will end up?

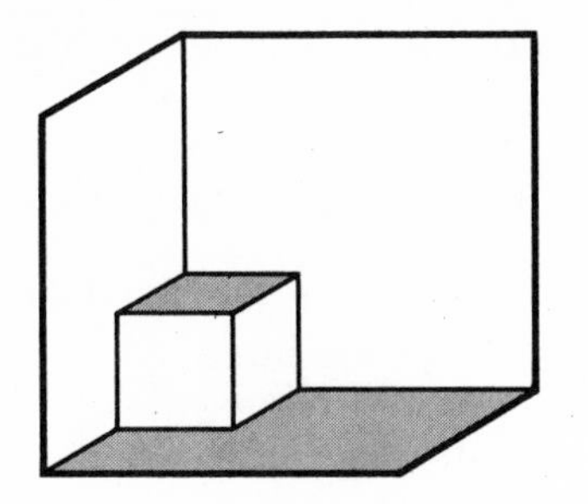

6. This diagram of a solid figure contains an unusual optical illusion because it is possible to interpret it in *three* different ways. Study it closely and try to describe in three sentences what these interpretations are.

7. Three golfers named Tom, Dick, and Harry are walking to the clubhouse. Tom, the best golfer of the three, always tells the truth. Dick sometimes tells the truth, while Harry, the worst golfer, never does.
   Use deductive reasoning to figure out who is who and explain how you know. (Hint: First, figure out which one is Tom.)

Set II

Here is an interesting number trick.

1. Write down a three-digit number in which the difference between the first and last digits is more than one.
2. Write the same three digits in reverse to form another three-digit number.
3. Subtract the smaller of your two numbers from the larger and circle your answer.

4. Reverse the digits in the number you circled to form another number and circle it.
5. Add the two circled numbers together.
6. Start all over again with another three-digit number and carry out the same procedure.
7. Try the trick a third time with another three-digit number.
8. What do you conclude about the result of this number trick?
9. If you try out the trick one hundred times, would this prove that your conclusion is correct?
10. The first step in the trick was to choose a number in which the first and last digits differ by *more* than one. Suppose the first and last digits differ by *exactly* one. Would the original trick still work?
11. Try this several times. Would part of the trick work in another way?
12. What would happen if you started with a number in which the first and last digits were the same?

Set III

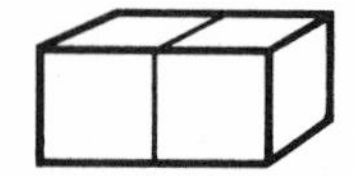

Here is a three-dimensional version of the checkerboard puzzle.

Suppose we have a wooden cube, which we cut up into 8 small cubes. If we glue pairs of the small cubes together to form 4 "bricks," it is easy to put the 4 bricks back together to form the original cube.

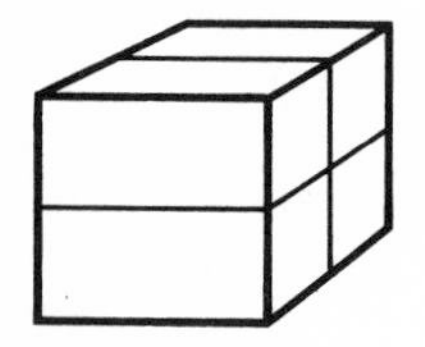

Now suppose we have a larger cube, which we cut up into 27 small cubes. If we glue these together to form 13 two-cube bricks, there will be one cube left over. Suppose we throw this extra cube away. Is it possible to put the 13 bricks back together to form the original cube with a hole in the center?

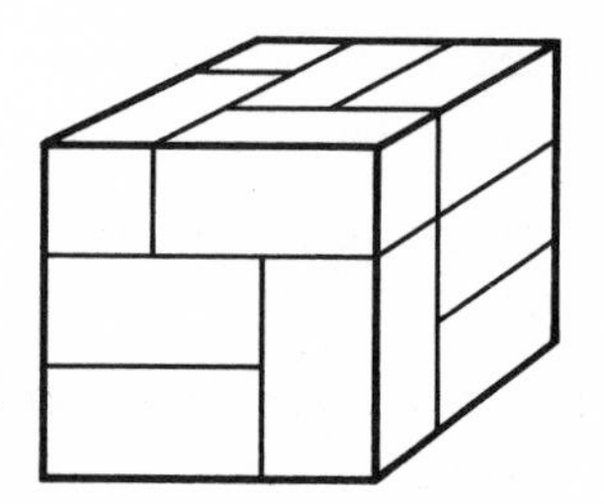

To answer this without trying it out, let's pretend the problem has already been solved and that the 13 bricks have been put together. Suppose we paint the 26 cubes, which make up the bricks, alternately black and white like this:

1. How many white cubes would there be altogether? (An easy way to count them is layer by layer. How many in the top layer? the middle? the bottom?)
2. How many black cubes would there be altogether? (Remember that there is no cube in the center.)
3. Notice that each brick is made of two cubes that are different colors. How many cubes of each color would be necessary to form 13 bricks?
4. How many of each color did you say were in the large cube with the hole in the middle?
5. Is it possible to solve the puzzle if you are very patient and keep trying out different arrangements of putting the bricks together?

Chapter 2

NUMBER SEQUENCES

Lesson 1

Arithmetic Sequences—Growth at a Constant Rate

IN an ancient Chinese book on numbers written about 1100 B.C., the following number sequences appear:

1 3 5 7 9 11 . . .

and

2 4 6 8 10 12 . . . .

You recognize these number sequences as the odd numbers and the even numbers.

► A **number sequence** is an arrangement of numbers in which each number follows the last according to a uniform rule.

The rule for each sequence above happens to be the same and is illustrated by the diagrams below.

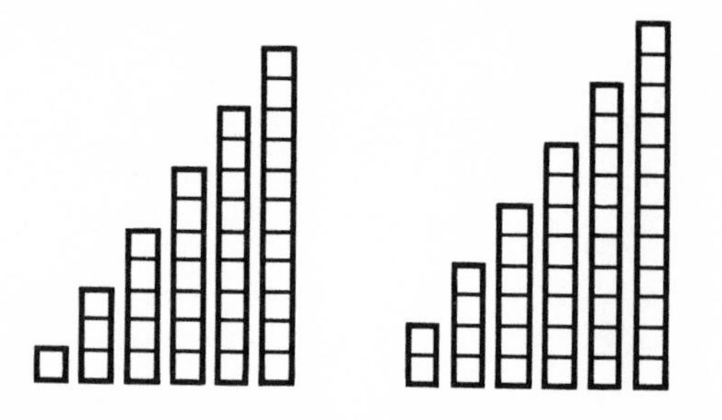

The rule is: Add two to each number (or term) to get the next one. In symbols, if we represent any term as n, the next term is $n + 2$.

By permission of Johnny Hart and Field Enterprises, Inc.

A number sequence that is built by always *adding the same number* is called an **arithmetic sequence.**

The counting numbers form the simplest arithmetic sequence:

1 2 3 4 5 6 7 8 9 10 11 . . . .

Here the rule is: add one; and two successive terms may be represented as n and $n + 1$. The diagram runs off the edge of the page to show that a number sequence can be continued as far as you want. A mathematician once said:

> "A number is like a bus; nobody ever doubts that there is always room for one more."

A number sequence can start with any number. For example, in giving the dates of the Saturdays in a certain month, you might say

3 10 17 24 31.

What is the number added here?

When a parachutist jumps from an airplane, the distances in feet he falls in successive seconds before he pulls the rip cord are

16 48 80 112 144 . . . .

What is the number added here?

EXERCISES

The exercises in this book are intended to make you think! They are not necessarily arranged in order of increasing difficulty and every once in a while you will come to one that is rather tricky.

Set I

1. Write the first five terms (numbers) of the number sequence suggested by this figure.

Copy the following arithmetic sequences, writing in the missing numbers.

2.	1	4	7	10		
3.	5	13	21			
4.	11	15		23		
5.	3		17		31	
6.			23	32	41	
7.	6				14	
8.	7					

9. Each number shown in this sequence is a prime number.*

   5 11 17 23 29 . . .

   What are the next five numbers in the sequence? Are they all prime numbers?

Are the following number sequences examples of arithmetic

*A prime number is a number that has no factors (whole numbers that will divide it without leaving a remainder) except itself and 1.

sequences? (Read the definition of an arithmetic sequence again.) Explain each answer with a sentence.

10. 3 3 3 3 3 . . .

11. 2 4 8 16 32 . . .

12. 10 100 1,000 10,000 . . .

13. 10 9 8 7 6 . . .

14. 1/6 1/3 1/2 2/3 5/6 . . .

15. If each term in an arithmetic sequence is doubled, is the resulting sequence also arithmetic? Try it out and see.

Set II

What is the tenth term of the sequence

1 2 3 4 5 . . . ?

It is easy to see that it is 10. How about the 100th term of the sequence

2 5 8 11 14 . . . ?

Instead of counting that far, there is an easier way to find out. The same sequence written a different way looks like this:

1st term	2nd term	3rd term	4th term	5th term
2	$2 + 1 \cdot 3$	$2 + 2 \cdot 3$	$2 + 3 \cdot 3$	$2 + 4 \cdot 3$ . . . .

So the 100th term would be

$$2 + 99 \cdot 3 = 2 + 297 = 299.$$

Use this method to figure out the terms in each of the following:

1. The 10th term of

   4 9 14 19 . . .

   (Hint: $4 \quad 4 + 1 \cdot 5 \quad 4 + 2 \cdot 5 \quad 4 + 3 \cdot 5$ . . .)

2. The 20th term of

   3 7 11 15 . . .

3. The 50th term of

   7 9 11 13 . . .

4. The 30th term of

$$100 \quad 97 \quad 94 \quad 91 \ldots$$

5. The 111th term of

$$24 \quad 35 \quad 46 \quad 57 \ldots$$

(The digits of the answer form an arithmetic sequence.)

Set III

What is the sum of the odd numbers from 1 through 15? One way to find out would be to add them in order:

$$1 + 3 + 5 + 7 + 9 + 11 + 13 + 15.$$

An easier way is this:

$$\begin{array}{r} 1 + 3 + 5 + 7 + \\ \underline{15 + 13 + 11 + 9} \\ 16 + 16 + 16 + 16 = 4 \cdot 16 = 64. \end{array}$$

Use this method to figure out the sum of the terms shown in each of the following sequences.

1. 2 7 12 17 22 27

2. 2 4 6 8 . . . 100

   (The three dots represent all the terms between 8 and 100. It isn't necessary to write out all the terms in order to figure out their sum. How many terms are there? How many *pairs* of terms?)

3. Anyone who likes to bowl knows that the pins are set up in four rows with one pin in the front row, two pins in the second row, three pins in the third, and four in the last. Suppose a gigantic set of pins was set up with 20 rows, the first row having one pin and each succeeding row having one more, so that the last row has 20 pins. How many pins would you have to knock over to make a strike?

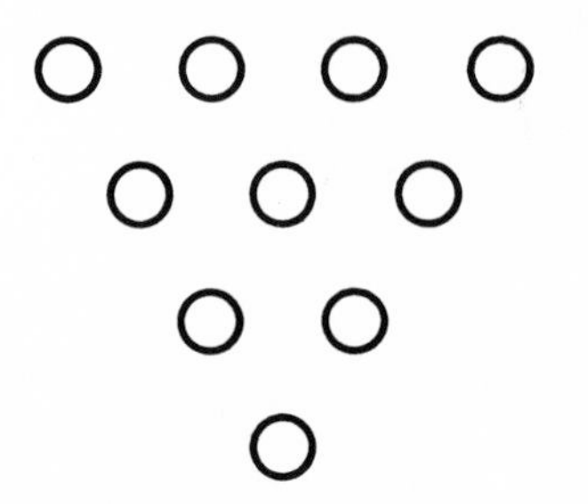

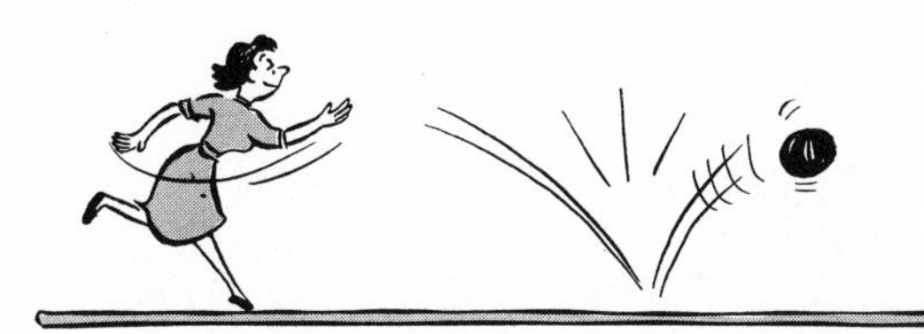

Courtesy of Scott Val Taber;

4. The Boodle Noodle Company is running a contest in which they will send the winner some money every day for one month. The amounts are $100 the first day, $200 the second day, $300 the third day, and so on. If the month has 30 days and you win the contest, how much money will the Boodle Noodle Company send you altogether?

IN THIS PAPER YOU WILL FIND DAILY A. P. WIREPHOTOS AND THE NEWS SERVICES OF THE ASSOCIATED PRESS, NEW YORK TIMES, CHICAGO TRIBUNE, UNIVERSAL, INTERNATIONAL NEWS AND UNITED PRESS. NO OTHER NEWSPAPER IN THE WORLD HAS ALL OF THESE GREAT WIRE SERVICES

DIME-A-DAY CHAIN LETTERS STILL FLOOD MAILS DESPITE GOVERNMENT WARNING

The Post Telephone MAin 2121

The Paid Circulation of THE DENVER POST Yesterday Was 167,129

Average Paid Sunday Circulation for March, 319,019

The Best Newspaper in the U. S. A.

THE DENVER POST

3c by Newsboys in Denver 5c on Trains and Outside of Denver—Sunday 10c

NRA MEMBER U.S. WE DO OUR PART

HOME EDITION | DENVER, COLO., SATURDAY, APRIL 20, 1935 | 20 PAGES VOL. 43—NO. 261

THE WEATHER

'Tis a Privilege to Live in Colorado.

Saturday—Sun rose in Denver at 5 15 a. m. Sun sets in Denver at 6 43 p. m.

There are 13 hours and 28 minutes of sunlight in Denver Saturday, two minutes more than on Friday.

Highest temperature in Denver on Friday, 67; lowest temperature on Friday night, 44.

Denver and Vicinity (radius 20 miles) — Partly cloudy Saturday night and Sunday; little change in temperature; lowest Saturday night about 45 degrees.

Shippers' Forecast (radius 250 miles)—Protect shipments next 36 hours from temperatures as follows: North, 35 degrees; east, 40 degrees; south, 40 degrees; west, 40 degrees; mountain passes, 25 degrees.

CLOUDY

Colorado—Generally fair east, unsettled west portion Saturday night and Sunday; warmer southwest portion Saturday night.

Wyoming—Partly cloudy Saturday night and Sunday; warmer extreme southwest portion Saturday night.

Nebraska—Generally fair Saturday night and Sunday; not much change in temperature.

Kansas—Generally fair Saturday night and Sunday; slightly warmer in east and south portions Saturday night.

ADDITIONAL WEATHER REPORTS ON PAGE 19, COLUMN 6.

TODAY

(By Arthur Brisbane.)

(Copyright, 1935, King Features Syndicate, Inc. International Copyright and All Other Rights Reserved.)

WITH Easter Sunday comes wide rejoicing. The sorrow of Good Friday, its memory of the long hours of agony are passed. The promised resurrection day has come. This country on Easter, 1935, sees a Roosevelt rainbow in the sky, a 4-billion 800-million-dollar rainbow, of hope and promise.

During Easter services at St. Peter's, Rome, Pope Pius, standing on the balcony of the basilica, will, by radio, "impart his blessing to all mankind."

That the head of a great church should be able, literally, to speak to all the earth is a sufficient miracle. The Christians of earlier times would have said it could not happen until the day of judgment, when the world would hear Gabriel's horn. But that is far away, if science is accurate in its statement that the earth, and men on it, will last a million, million years more—1,000,000,000,000 years.

Miss May Robson, well known on the stage fifty years ago, acting in "movies" now, celebrating her seventieth birthday, says, "It's grand."

Miss Robson takes the proper view of a seventieth birthday. She had a birthday cake nearly as tall as herself, at the top a real canary singing in a cage of candy. Miss Robson likes to go with her companion, Miss Harner, to sit in the patio and talk until 2 or 3 o'clock in the morning because "That is peace."

"It was a fine birthday," said she, "but wait until you see what I do when I am 80."

Young women of 70 and over have this advantage over old men of the

(Turn to Page 7—Col. 1.)

SCHECHTEL SAYS HE'S GUILTY OF $500,000 FRAUDS

Sentence of Broker Deferred; Others Plead Not Guilty

(CRIME NEVER PAYS)

Obviously under a great strain, Harry Schechtel of 801 Race street, widely known promoter, stood Saturday before Judge J. Foster Symes in the United States district court and confessed he was guilty of participating in the operation of a fraudulent brokerage enterprise in which investors in at least seventeen western states are alleged to have been swindled of approximately a half million dollars in two years.

Schechtel pleaded guilty to all three indictments returned against him and twenty-three others by a federal grand jury Feb. 4. Two of the indictments charged violation of the national securities act in addition to fraudulent use of the mails and one charged mail fraud and conspiracy.

Judge Symes continued Schechtel's bond and postponed his sentence thirty days.

FOURTEEN DEFENDANTS PLEAD NOT GUILTY.

Fourteen codefendants, including Miss F. Edythe Adams of 1050 Washington street, and Edward Schulman, who formerly lived at 400 South Marion street, denied they are guilty of the charges.

Their trials were set for May 20 at the request of Ivor O. Wingren, assistant United States district attorney, who represented the government at the arraignments.

Wingren said the trials, which are expected to be long and tedious, should be held as soon as possible in view of the fact that Schechtel has pleaded guilty and for the additional reason that so many defendants are involved.

In this contention Wingren was opposed by Ralph L. Carr, former United States district attorney and now counsel for Schulman, who asked that the trial be deferred until the September term of court.

Speaking for Schulman, who was in court, Carr told Judge Symes he wanted to preserve his client's right to change his plea from not guilty to guilty if it develops later that he so desires.

SCHECHTEL GIVES NO EXPLANATION.

Schechtel was pale and haggard, but he entered his plea in a clear, firm voice, as if he was struggling to control his emotions. He did not indulge in any explanation of the plea, and said he had no statement to make.

The federal indictments against him and the codefendants were based on the alleged fraudulent activities of three brokerage concerns. They were Schechtel & Co., formerly at 714 First National Bank building; W. J. Bona & Co., formerly at 732 United States National Bank building, and

(Turn to Page 3—Col. 6.)

Admits Charges Of Fraud

HARRY SCHECHTEL, Widely known Denver promoter, who pleaded guilty Saturday to three federal indictments charging violation of the national securities act, fraudulent use of the mails and conspiracy. His sentence was deferred thirty days.

Car Failed to Make Turn

JULIUS F. EARNEST

C. U. STUDENT DIES IN CRASH NEAR SPRINGS

Auto Driven by Son of Denver Druggist Hits Filling Station.

Julius F. Earnest Jr., 18, a student of the University of Colorado and son of a Denver druggist, was fatally injured shortly before 1 o'clock Saturday morning when his automobile struck a filling station concrete island, six miles south of Colorado Springs.

Earnest had been to Pueblo in connection with the rush activities of the Chi Psi fraternity, of which he was a member, and was returning to Boulder. He was driving alone, according to authorities, and his car failed to make the turn where the filling station is located. He was taken to the Beth-El hospital at Colorado Springs and died shortly after being admitted.

E. C. Lipscomb, proprietor of the

(Turn to Page 3—Col. 8.)

MILK TRUCK KNOCKS MAN OUT OF BATHTUB AND INTO KITCHEN

Babylon, N. Y., April 20.—(A. P.)—A heavily-loaded five-ton milk truck, traveling at high speed, got out of control here early Saturday, plowed into the house of Augustus Haff, drenched the living room furnishings with milk and went on thru another wall to knock Haff out of the bath tub into the kitchen.

The truck had imbedded itself so completely in the dwelling that it was necessary to call upon a house-wrecking firm to get it out.

Neither Haff nor the driver, Frank Albino, were seriously injured.

COLORADO MAN SLAIN IN MEXICO WITH PAYROLL

Brother of Denver Radio Executive Shot Near Matamoras.

Raymond Bengson, 26, a native of Golden and the brother of Elmer L. Bengston, production manager of the KLZ radio station, was found shot to death near Matamoros, Mexico, Friday night, according to press dispatches.

He is said to have left Matamoros with an $800 payroll for the Titania Oil company, by whom he was employed as geophysical camp chief. The oil camp is fifty miles from the city. When his body was found, pierced by four bullets, the money was missing.

The payroll had been obtained in Brownsville, Tex., just across the Mexican border, the dispatches state. It is believed that recent trouble among the oil workers may be con-

(Turn to Page 2—Col. 6.)

THIRTEEN INJURED WHEN DRIVER FAINTS

Los Angeles, April 20.—(A. P.)—Thirteen pedestrians were injured, none seriously, when Duncan M. Conley fainted at the wheel of his automobile Saturday and it plowed into a lane of people on a downtown street.

Most painfully hurt was Mrs. Eveline Gaudet, radio singer and daughter of Madame Constance Balfour, well known New York soprano, who was with her daughter.

Conley's driver's license was revoked.

Sunday's the Day For Post's Great Easter Egg Party

(By A. G. BIRCH.)

The biggest Easter party for the children ever held in the west will be staged by THE DENVER POST at Cheesman park at 2 o'clock Sunday (tomorrow) afternoon. The whole thing is FREE, and all the youngsters are invited.

The chances are this party will be even larger than the famous "Egg-Rolling" at the White house in Washington. Last year's POST party in Denver was just about as big as the White house affair, and far more children are bound to attend this year.

The weather man promises to co-operate with THE POST. He says Easter Sunday in Denver will be only "partly cloudy," and that there is little likelihood of a storm or chilly temperature.

In the big candy plant of Baur's, 1512 Curtis street,

(Turn to Page 3—Col. 5.)

TWO FOUND GUILTY OF SELLING DOPE

Frank Marino Given Five-Year Term in Leavenworth

(CRIME NEVER PAYS)

George (Chubby) Aiello, widely-known Denver night club operator, and Frank Marino, an associate, were found guilty of selling heroin in violation of the federal narcotic laws, in a sealed verdict read Saturday morning before Judge J. Foster Symes in the federal district court.

Two other defendants, Ernest Gollo of Denver and Frank Carramusa, Kansas City importer, were found not guilty. A fifth defendant, Lawrence Cipolla of Kansas City, had received a directed verdict of not guilty.

MARINO ONCE GIVEN TERM FOR STICKUP.

Marino was sentenced immediately by Judge Symes to five years in Leavenworth penitentiary and fined $500 on one count in the indictment, and sentenced to three years in Leavenworth on each of the other two counts, to be served concurrently.

Deputy United States District Attorney David H. Morris, who prosecuted the case, said the additional sentences would bar Marino from parole and necessitate his serving a

(Turn to Page 7—Col. 3.)

GEORGE AIELLO (top) and FRANK MARINO, Who were found guilty by a United States district court jury of violation of the federal narcotics law. Marino was given a long penitentiary term and a heavy fine on one count. Aiello indicated he will seek a new trial.

4TH PAIR OF TWINS BORN TO COUPLE

Ypsilanti, Mich., April 20.—(I. N. S.)—Mr. and Mrs. Joseph Shock were caring for their fourth pair of twins Saturday—a boy and a girl, as yet unnamed.

The first pair of twins was born to the couple twenty-one years ago. One died in infancy. The second pair are now 13, and the third pair 7.

U.S. Cites Lottery Statute

Chain Letter Illegal Under Law, Says Inspector.

Decision by Postal Inspector Roy E. Nelson that the present widespread chain letter circulation in Denver is illegal was based by Nelson on two sections of the postal law, one prohibiting lotteries and the other use of the mail for purposes of fraud.

Nelson Saturday cited sections 336 and 338 of 18 United States criminal code.

The first section reads in part as follows:

"No letter, package, postal card or circular concerning any lottery, gift enterprise or similar scheme offering prizes dependent wholly or in part on lot or chance . . . shall be deposited in or carried by the mails or be delivered by any postmaster or letter carrier."

Penalty provided is a fine of not more than $1,000 and imprisonment for not more than two years on first offense and imprisonment for five years on second offense.

Nelson said that the "send-a-dime" chain letters are a lottery in that the amount which might

(Turn to Page 2—Col. 4.)

EX-KAISER REPORTED ILL FROM INFLUENZA

Amsterdam, April 20.—(A. P.)—The newspaper Het Volk Saturday published a report that former Kaiser Wilhelm of Germany is suffering from influenza.

Rumors that he was seriously ill were denied at Doorn castle, but it was said he had not been seen outside the castle for several days and physicians have called there frequently.

DENVER'S POSTOFFICE STAFF TAKES QUESTION UP WITH WASHINGTON

Nearly Every Home in Denver Believed to Have Been Solicited on Scheme to Make 10 Cents Grow to $1,562

Send-a-dime chain letters continued to flit Saturday thru the Denver postoffice in a perfect blizzard of mail which threatened to blanket Denver.

Despite the announcement of Roy E. Nelson, postoffice inspector, that circulation of the chain letters was a violation of two sections of the postal laws, persons stimulated by the hope of getting $1,562.50 in exchange for a dime continued to circulate them.

It was estimated that in the three or four weeks since the first letters were started, nearly every family in Denver had received one or more letters.

The matter had reached a position of such importance in Nelson's opinion that he made it the subject of a report to Washington.

Nelson pronounced the ingenious scheme a lottery and also a violation of the section of the statutes prohibiting the use of the mails for purposes of fraud.

Every persons who sends one of the letters is guilty of these violations, Nelson said.

He was ready to admit, however, that it would hardly be feasible to proceed against half of the adult population of Denver, and said he was concentrating on catching the

(Turn to Page 2—Col. 3.)

Clerks Were Swamped at the Denver postoffice Saturday with an unprecedented volume of mail attributed to the chain letter enterprise. The clerks took it cheerfully. Vera Haines and Frank J. Dioguardi are shown putting some of the mountain of mail thru the stamp-canceling machines.

ONE CHAIN LETTER COULD GROW INTO OVER 300 MILLION

The chain-letter progression, which has Denver up in the air, soon runs into astronomical figures, Postal Inspector Roy E. Nelson discovered Saturday as he started to calculate what would happen if a chain, starting from a single letter, went unbroken thru twelve phases.

He found that in the twelve progressions, more than 305 million persons would receive letters in the unbroken chain. This is two and one-half times the population of the United States.

Nelson also calculated that of this number 3,911 would get $6,103,515.50 in dimes and the rest would get nothing.

HERE'S HOW YOUR DIME IS SUPPOSED TO BECOME $1,562

If the Chain Isn't Broken You're a Cinch to Cash In, But in Dealing With Human Beings There's Always That IF.

The mathematical theory of the chain letter plan in which each individual who invests one dime cash and six postage stamps stands to reap a harvest of $1,562.50 is what is known in arithmetic as "raising five to the fifth power."

The chain letter carries a list of six names. Suppose, for example, that your name is Joe Gilch and that the name of John Doe is at the top of the list of six names. Here is the way this chain letter system operates:

You send John Doe a dime and then cross off his name from the top of the list. You add your name at the bottom of the list. Next, you make five copies of the letter and send one to each of five persons. Each recipient sends a dime to the person whose name is at the head of the list, then crosses off that name and adds his or her own name at the bottom.

Each of the five persons to whom you have sent the chain letter sends it to five. That makes twenty-five persons who have the letter, and your name has advanced from sixth place on the list to fifth place.

Each of the twenty-five send the letter to five. That makes a total of 125, and your name has advanced from fifth place to fourth place.

Each of the 125 send the letter to five. That makes a total of 625, and your name now is in third place.

Each of the 625 send the letter to five. That makes a total of 3,125, and your name has been raised to second place.

Each of the 3,125 sends the letter to five. That makes a total of 15,625 who receive the letter with your name at the top of the list of six names. Each of the 15,625 is supposed to send you a dime, so you stand to cash in to the extent of $1,562.50 after having spent only a dime in cash and six postage stamps. This of course, is on the theory that the chain is never broken.

The larger the number of letters sent out, the larger, of course, the financial reward which is possible. The calculation given here presupposes that each recipient of the letter sends out only five copies.

WRECKERS OF TRAIN CONVICTED BY JURY

Pittsburgh, April 20.—(A. P.)—Two men accused of wrecking a Pennsylvania railroad passenger train and causing the death of the engineer were convicted of second-degree murder Saturday by a jury in criminal court.

The defendants, Robert Edwards, 43, and Alfred Van Shuren, 21, face ten to twenty-year terms in prison under the verdict.

The Pennsylvania's American flyer was wrecked at nearby Sturgeon last October. Engineer Robert Black was killed and a dozen persons injured.

TWO QUINTUPLETS CONTRACT HEAD COLDS BUT ARE BETTER NOW

Callander, Ont., April 20.—(U. P.)—Marie and Emilie, ailing members of the Dionne quintuplets, were reported Saturday by Dr. Allan Roy Dafoe as showing slight improvement. Both are suffering from head colds, and are being kept apart from their sisters.

Emilie, who has been fighting her cold for four days, was almost well, but Marie, who developed the infection Friday, had a temperature that reached 100 degrees.

PRINCE NOT BOTHERED BY BARBARA'S DIVORCE

Rome, April 20.—(A. P.)—Prince Alexis Mdivani laughed Saturday when an interviewer told him "You don't look very unhappy over Barbara's divorce."

"It's all in a lifetime," grinned the prince.

Accompanied by his sister, he had just entered a big downtown hotel, smiling and dapper, after his arrival from Spain by automobile.

He declined to reveal whether he had heard recently from Barbara Hutton Mdivani, heiress to the Woolworth millions, who is now in Reno

The Denver Post.

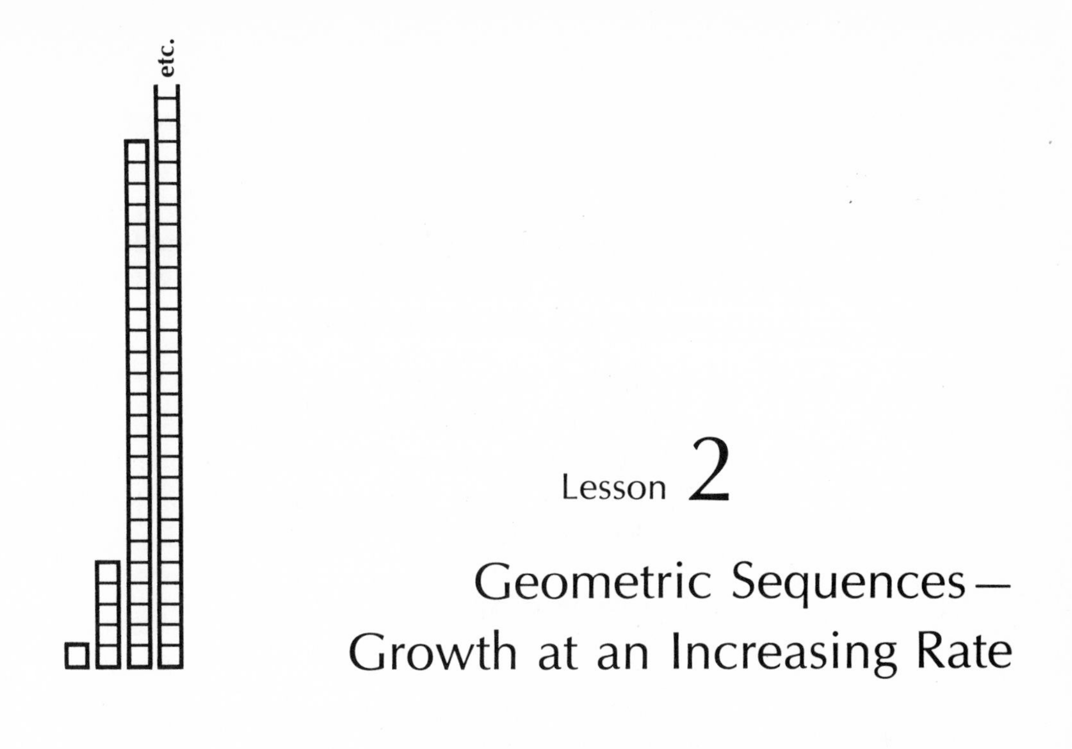

Lesson 2

Geometric Sequences—Growth at an Increasing Rate

HAVE you ever gotten a chain letter? In 1935 a chain letter craze started in Denver and swept across the country. It worked like this. You receive a letter with a list of five names. You send a dime to the person named at the top, cross the name out, and add your own name at the bottom. Then you send out five copies of the letter to your friends with instructions to do the same. When your five friends send out five letters each, there will be 25 in all. If none of the 25 persons getting these letters breaks the chain, 125 more letters will be sent, and so on.

Let's list these numbers in order, starting with the one letter you receive.

$$1 \quad 5 \quad 25 \quad 125 \ldots$$

This is a number sequence because each number follows the last according to a uniform rule. The rule however is to *multiply* by 5 (*not* add 5), so this is not an arithmetic sequence. It is called a geometric sequence and the diagram at the top of this page shows that the numbers are growing faster and faster.

► A number sequence that is built by always *multiplying by the same number* is called a **geometric sequence.**

If we represent any term in the "chain letter" sequence by n, the next term is $5n$.

In his *Poor Richard's Almanack* for the year 1751, Benjamin Franklin calculated the number of a person's ancestors 30 generations back, or approximately 1,000 years ago. To quote his words:

> "It is an amusing Speculation to look back, and compute what Numbers of Men and Women among the Ancients, clubb'd their Endeavours to the Production of a single Modern"!

Franklin's list begins like this:

A present Man's Father and Mother were	2
His Grandfathers and Grandmothers	4
His Great Grandfathers and Great Grandmothers	8
and, supposing no Intermarriages among Relations, the next Predecessors will be	16

This number sequence of ancestors,

2 4 8 16 . . .,

is another example of a geometric sequence. What is the multiplying number here?

EXERCISES

Set I

Copy the following geometric sequences, writing in the missing numbers.

1. 1 5 25 ____ ____
2. 4 12 36 ____ ____
3. 2 14 ____ 686
4. ____ ____ 18 54 162
5. 5 ____ 20 ____ 80
6. 1 ____ ____ 27 81
7. 0 ____ ____ ____

8. The higher the temperature, the faster most chemical reactions take place. In fact, the speed is approximately doubled when the temperature is raised 10°C. Write a geometric sequence to show how many times faster paper will

burn if the temperature is raised 10°C, 20°C, 30°C, and 40°C.

Are the following number sequences examples of geometric sequences? Explain each answer with a sentence.

9. 5 10 15 20 25 . . .

10. 7 7 7 7 7 . . .

11. 1 4 9 16 25 . . .

12. 16 24 36 54 81 . . .

13. 1 1/2 1/4 1/8 1/16 . . .

14. One generation back, you have two ancestors (your parents). How many ancestors do you have 20 generations back?

15. If one is added to each term of a geometric sequence, is the resulting sequence also geometric? Try it out with one and see.

Set II

1. Examine the United States currency notes shown on the next page (p. 52), and then list all of the geometric sequences containing at least three terms that you can find in the denominations of these notes.

2. One of the sequences you probably found is

   5 10 20.

   Is the tenth term of this sequence also the denomination of a currency note?

3. Another way to represent the terms of this sequence is:

1st term	2nd term	3rd term	4th term	5th term
5	$5 \cdot 2$	$5 \cdot 2 \cdot 2$	$5 \cdot 2 \cdot 2 \cdot 2$	$5 \cdot 2 \cdot 2 \cdot 2 \cdot 2$

   Rewrite the 3rd, 4th, and 5th terms, using exponents. (An exponent is a number written at the upper right of another number to indicate how many times that number occurs as a factor. For example, $5 \cdot 2 \cdot 2 \cdot 2$ can be written as $5 \cdot 2^3$.)

From *The Official Guide of United States Paper Money* by Theodore Kemm, © 1968 H. C. Publishers, Inc., New York City. Courtesy of Hal Cohen.

4. Notice the relationship between the exponent and the number of the term. Write the 10th term of this sequence, using an exponent.
5. Write the 100th term, using an exponent.

Set III

1. If you received a chain letter of the type described in this lesson, mailed 5 copies of it to your friends, and no one broke the chain, how much money could you expect to receive?
2. Suppose the letter has a list of 7 names instead of 5, and each person sends a dollar instead of a dime. If everyone cooperated, how much money could you expect from this chain? (This seems like an easy way to get something for nothing, but it is against the law to send chain letters containing money in the mail.)

INTERESTING READING

Fads, Follies and Delusions of the American People, by Paul Sann, Crown, 1967: Chapter 15, "The Chain Letter Craze."

Lesson 3

The Binary Sequence

THERE is a legend that the king of Persia offered the inventor of the game of chess anything he wanted as a reward. What the inventor requested didn't seem like much. He asked that one grain of wheat be placed on the first square of the chessboard, two grains on the second square, four grains on the third and so on, each square having twice the number of grains as the square before. The king thought this was a reasonable request so he sent a servant for a sack of wheat. Can you guess what happened?

The numbers of grains on the successive squares form a geometric sequence, which grows at a rapidly increasing rate. The last (64th) square would have a number of grains equal to the 64th term of the sequence, a number more than 9 quintillion!* This together with all of the wheat on the other 63 squares would be enough to cover the entire country of Persia with a layer more than 3 inches deep.** Needless to say, the king wasn't very happy when he found this out and the inventor may have been rewarded with something other than what he had bargained for.

► The sequence

$$1 \quad 2 \quad 4 \quad 8 \quad 16 \ldots$$

is called the **binary sequence.**

---

*9,000,000,000,000,000,000

**Persia is four times the size of California, so in California the wheat would be more than a foot deep!

The binary sequence is remarkable because each whole number from 1 up can be formed by adding certain terms of the sequence without using any term more than once. For example,

$$3 = 1 + 2$$
$$5 = 1 + 4$$
$$6 = 2 + 4$$
$$7 = 1 + 2 + 4$$
$$9 = 1 + 8, \text{ etc.}$$

Because of this, the binary system of recording numbers is the system electronic computers use. In a computer, a tube or transistor for each term in the sequence may be on or off to show whether that term is counted or not.

Let's reverse the sequence from right to left

. . . 16 8 4 2 1

and think of each term as corresponding to a tube. To represent the number ten, the tubes would look like this:

16	[8]	4	[2]	1
OFF	ON	OFF	ON	OFF.

Representing OFF by 0 and ON by 1, we have

0 1 0 1 0,

which is the binary numeral for ten. The first 0 can be left out, leaving 1010.

To represent the number twenty-five, the tubes would look like this:

and the binary numeral would be

1 1 0 0 1.

Just as each tube has only two possibilities: it can be either off or on, so binary numerals use only two digits: 0 and 1.

If we need to write larger numbers, we can add more tubes (terms from the binary sequence) to the left.

EXERCISES

"For example, Pop, if we write 210 in base 2 it looks like this—11010010."

Set I

1. Continue this list of binary numerals through the number 31.

Number	16	8	4	2	1
1					1
2				1	0
3				1	1
4			1	0	0
5			1	0	1
6			1	1	0
etc.					

2. Some symbols known for thousands of years throughout the Orient are the Pa-kua, or "eight figures."

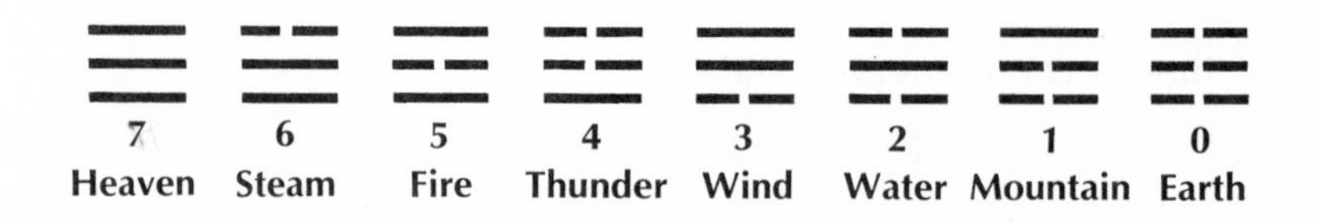

They appear in one of the oldest Chinese books and are printed on compasses and good luck charms. Can you explain the connection between these symbols and binary numerals?

3. Here is a picture showing the pattern in the lengths of the marks along one inch of a ruler. Along the edge, there are 16 small units, which have been numbered. Look just at the increasingly longer marks along the edge. What do you notice about the numbers of these marks?

Set II

Let's take a closer look at the problem of the number of grains of wheat on the chessboard. Exactly how many grains did the inventor request?

We could write the first 64 terms of the binary sequence and add them up,

$$1 + 2 + 4 + 8 + 16 + \ldots$$

but this is no easy job.

When a mathematician is faced with a difficult problem, he sometimes makes up a simpler one to see if it suggests any short cuts. Let's begin with a small chessboard having only four squares. We can write the numbers of grains like this,

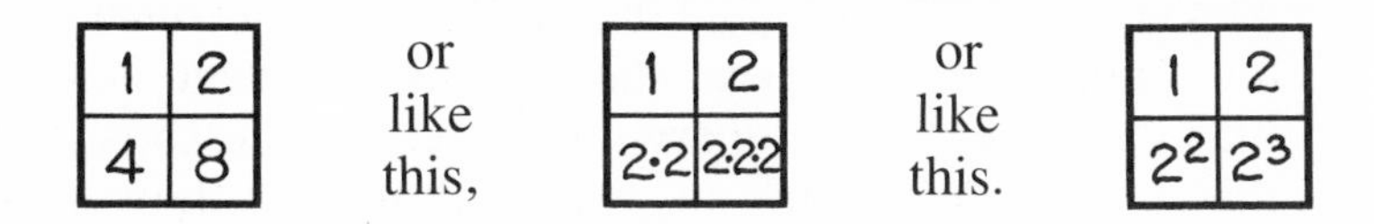

There are 1 + 2 + 4 + 8 or 15 grains altogether.

Let's do the same with a slightly larger board.

1	2	4
8	16	32
64	128	256

1	2	2^2
2^3	2^4	2^5
2^6	2^7	2^8

Adding the numbers here gives 511.

1. The number of grains on the last square of the 4-square board is 2^3. The number of grains on the last square of the 9-square board is 2^8. What do you think would be the number of grains on the last square of a 16-square board?

2. What would be the number of grains on the last square of a full size 64-square board?

The *total* number of grains on all of the squares of the 4-square board is 15, which is 1 less than 16, a number in the binary sequence. The number 15 can be written as $2^4 - 1$. The total number of grains on all of the squares of the 9-square board is 511, which is 1 less than 512, another number in the binary sequence. The number 511 can be written as $2^9 - 1$.

3. What do you think is the total number of grains on the squares of a 16-square board? (Use mathematical shorthand; don't work the answer out.)

4. What is the total number of grains on the squares of a full size chessboard?

5. The largest prime number presently known was discovered in 1963 by a computer at the University of Illinois. The postage meter of the mathematics department prints this number on their mail.

Courtesy of Department of Mathematics, University of Illinois.

The computer used to find the number uses the binary system and, as you can see, the number produced is 1 less than a number of the binary sequence. If a chessboard had this number of grains of wheat altogether, how many squares would it have? (Hint: See the previous two problems.)

Set III

Experiment. A Punched Card Data Sorting System

The binary system can be used to sort cards automatically.

Cut sixteen 3×5 cards in half like this.

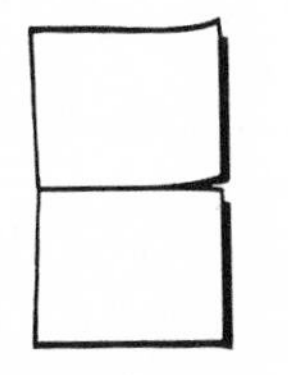

Take one of the 32 cards produced and punch a row of five holes along each of the two longer edges. The holes should be spaced about half an inch apart.

Using this card as a stencil, punch the other 31 cards to match it. Three cards can easily be punched at a time.

Number the cards from 0 through 31. The five holes in each row represent the first five numbers of the binary sequence:

16 8 4 2 1.

Write the number 1 above the appropriate holes along the upper edge and below the corresponding holes along the lower edge of each numbered card to represent the matching binary numeral. For example, on card number 11, write "ones" at the holes representing 1, 2, and 8. (The list of binary numerals that you made in the Set I exercise should be helpful.)

Along the top edge of each card, cut out the space above each hole marked with a "one."

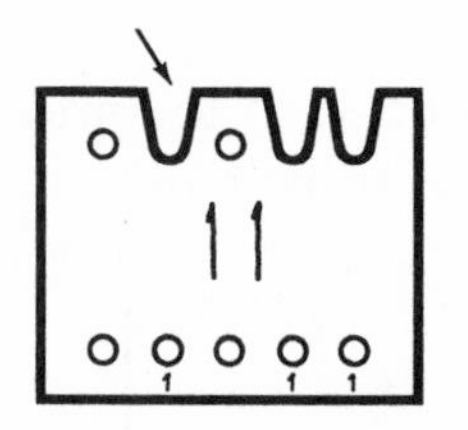

Then do just the opposite along the lower edge, that is, cut out the space below the holes that are *not* marked with a "one."

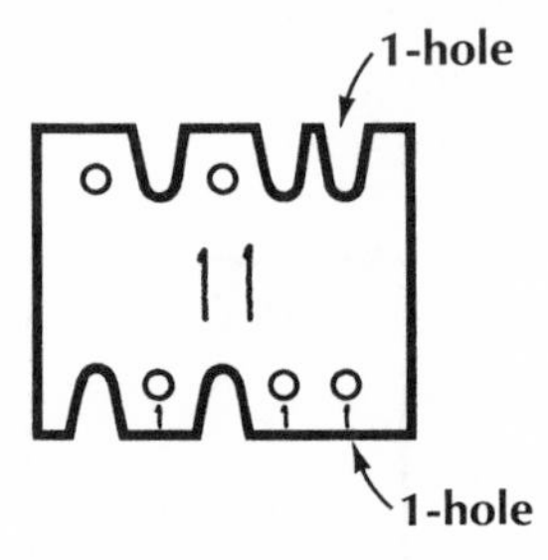

The cards are now ready to use. Shuffle them up, being careful that none get turned upside-down.

Make two hooks something like this, out of paper clips.

Stick the hooks through the 1-holes (on the upper and lower right) and lift up slowly. Half of the cards will be pulled up by one hook and the other half will be held back by the other.

Slide the cards that came up off the hook. Put these cards on top of the other cards that had remained behind. Now stick the two hooks through the 2-holes and carry out the same procedure. Repeat with the remaining three pairs of holes (going from right to left) and you should end up with the cards in correct order from 0 through 31.

1. It takes 5 steps to sort 32, or 2^5, cards. How many steps would it take to sort 64, or 2^6, cards?

2. Can you figure out how many steps would be needed to sort 1,000 cards?

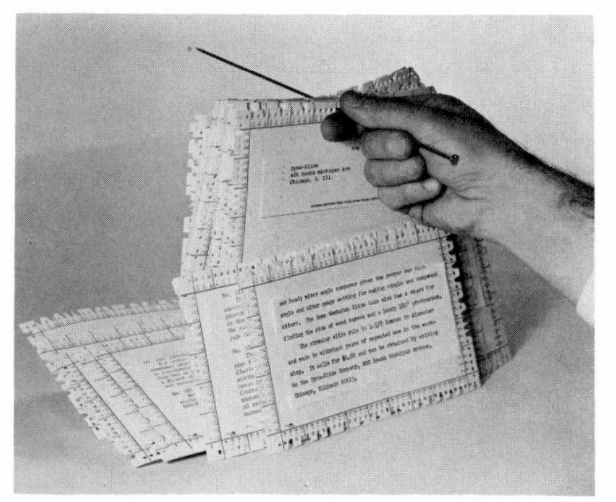

Courtesy of Science Related Materials, Inc., Evanston, Illinois

Only 20 steps would be necessary to sort a million cards!

Data sorting systems using punched cards are available commercially. The cards shown in this photograph have 120 holes spaced along their edges.

Lesson 4

The Sequence of Squares

IN thinking about the problem of communicating with other worlds at the beginning of this course, we learned that many clay tablets made by the Babylonians about 4,000 years ago have been discovered. Some of the mathematical tablets contain number sequences, both arithmetic and geometric. One tablet shows the **sequence of squares** of the numbers from 1 to 60:

1 4 9 16 25 36 49 64 81 . . . .

We can also write the sequence like this:

$1^2 \quad 2^2 \quad 3^2 \quad 4^2 \quad 5^2 \quad 6^2 \quad 7^2 \quad 8^2 \quad 9^2 \quad \ldots$

The Greek mathematicians much later in history (about 600 B.C.) often represented numbers with dots in geometric shapes. They would have shown the sequence of square numbers like this:

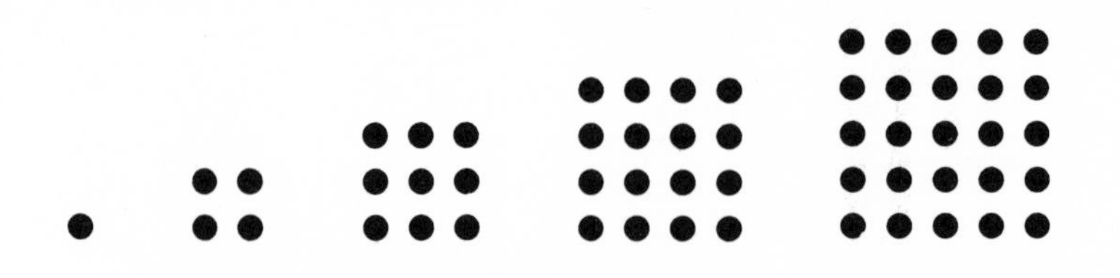

It is easy to see how the square numbers got their name. If you could ask a Greek of that time what the square of five is, he would think: "How many dots are in a square with a side of five?" Now when we talk about the square of five, we think: "What number is five multiplied by five?"

The sequence of square numbers has a special relationship to a certain arithmetic sequence—the sequence of odd numbers:

$$1 \quad 3 \quad 5 \quad 7 \quad 9 \quad 11 \ldots.$$

If we add consecutive odd numbers starting with one, a square number always results:

$$
\begin{aligned}
&1 && = 1^2 \\
&1 + 3 &= 4 & = 2^2 \\
&1 + 3 + 5 &= 9 & = 3^2 \\
&1 + 3 + 5 + 7 &= 16 & = 4^2 \\
&1 + 3 + 5 + 7 + 9 &= 25 & = 5^2
\end{aligned}
$$

and so forth.

The Greeks showed the relationship between square and odd numbers like this:

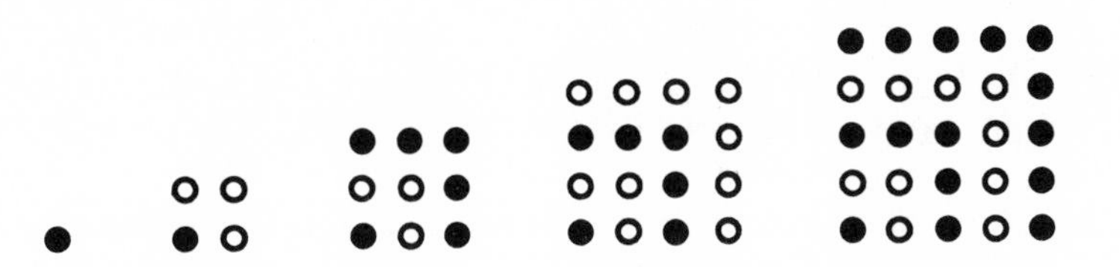

EXERCISES

Set I

Make a table of the squares of the numbers 1 through 30 *to keep* in your notebook for reference. Arrange the numbers in two columns as shown here. (See the hint on the next page.)

Number	Square
1	1
2	4
3	9
⋮	⋮
30	900

Hint: Instead of multiplying, you can use a shortcut. We have learned that the square numbers are also sums of consecutive odd numbers, so a method that may seem clumsy at the beginning, but which makes the work much easier later on, is this:

Number	Square
1	1 + 3
2	4 ← + 5
3	9 ← + 7
4	16 ← + 9
5	25 ← +11
⋮	

Use your table of square numbers to answer the following questions. Give examples where possible.

1. Can a square number of two or more digits have all *even* digits?

2. Do you think that a square number of two or more digits can have all *odd* digits?

3. Look at the *last digit* of each square number. Do you notice a pattern? Write the pattern down.

4. What digits can a square number end in?

The **digital root** of a number is found by adding its digits until only *one* digit is left. The digital root of 16 is $1 + 6 = 7$. To find the digital root of 49 we first get $4 + 9 = 13$; adding the digits in 13, we get 4 for the digital root.

5. Show that the digital root of 256 is also 4.

6. Find the digital root of each square number in your table. (The digital root of a one-digit number is the number itself.) Do you notice a pattern?

7. If a number is a square, what numbers can be its digital root?

8. The square of 111,111,111 is 12,345,678,987,654,321. Is the number 98,765,432,123,456,789 also a square? (Check its digital root.)

Set II

1. Copy this pattern and write in the missing numbers.

$$3^2 - 1 = \underline{\qquad}$$
$$5^2 - 1 = \underline{\qquad}$$
$$7^2 - 1 = \underline{\qquad}$$
$$9^2 - 1 = \underline{\qquad}$$

2. The four answers in the pattern are all even. What else do all four have in common?
3. Write three more lines of the pattern. What do you notice? Are all your answers still even? Do the new lines still have something else in common with the first four answers?
4. Copy this pattern and write in the missing numbers.

$$1^2 + 2^2 + 2^2 = \underline{\qquad}^2$$
$$2^2 + 3^2 + 6^2 = \underline{\qquad}^2$$
$$3^2 + \underline{\qquad}^2 + 12^2 = 13^2$$
$$4^2 + \underline{\qquad}^2 + 20^2 = \underline{\qquad}^2$$

5. Study the pattern closely and complete two more lines of it. Check carefully to see if the lines you add conform to the pattern.

Set III

Square numbers sometimes appear in science in mysterious ways. An example is the relationship of the elements in chemistry. Chemists know of 105 elements, or kinds of atoms, each with a different number of protons in its nucleus. This number of protons is called the atomic number of the element. For example, an oxygen atom has 8 protons, so its atomic number is 8; a hydrogen atom has 1 proton, so its atomic number is 1.

The elements can be arranged in a table in order of increasing atomic number so that elements with similar properties appear in columns. This table, called the periodic table, is shown on page 66. The sixth row of the table is so long that there is not enough room to print it on one line. The last row has been left out, so that only 86 of the 105 elements are shown.

Row																		
Row 1 →	1 H																	2 He
Row 2 →	3 Li	4 Be											5 B	6 C	7 N	8 O	9 F	10 Ne
Row 3 →	11 Na	12 Mg											13 Al	14 Si	15 P	16 S	17 Cl	18 Ar
Row 4 →	19 K	20 Ca	21 Sc	22 Ti	23 V	24 Cr	25 Mn	26 Fe	27 Co	28 Ni	29 Cu	30 Zn	31 Ga	32 Ge	33 As	34 Se	35 Br	36 Kr
Row 5 →	37 Rb	38 Sr	39 Y	40 Zr	41 Nb	42 Mo	43 Tc	44 Ru	45 Rh	46 Pd	47 Ag	48 Cd	49 In	50 Sn	51 Sb	52 Te	53 I	54 Xe
Row 6 →	55 Cs	56 Ba	57 La	58 Ce	59 Pr	60 Nd	61 Pm	62 Sm	63 Eu	64 Gd	65 Tb	66 Dy	67 Ho	68 Er	69 Tm	70 Yb	71 Lu	72 Hf
		73 Ta	74 W	75 Re	76 Os	77 Ir	78 Pt	79 Au	80 Hg	81 Tl	82 Pb	83 Bi	84 Po	85 At	86 Rn			

1. How many elements are in each row of the periodic table?
2. Divide each of these numbers in half. What do you notice?

Lesson 5

Higher Power Sequences

ALONG with the Babylonian tablet containing the squares of the numbers from 1 to 60, part of another tablet was found. It lists the cubes of the numbers from 1 to 32 and probably originally went to 60 also. Sixty was a key number in the Babylonian numeral system (in the same way that ten is in ours.)

The **sequence of cube numbers** can be written as

$$1^3 \quad 2^3 \quad 3^3 \quad 4^3 \quad 5^3 \ldots$$

or

$$1 \quad 8 \quad 27 \quad 64 \quad 125 \ldots.$$

Cube numbers were imagined by the Greeks to be three-dimensional. Their pictures of the first three looked something like this:

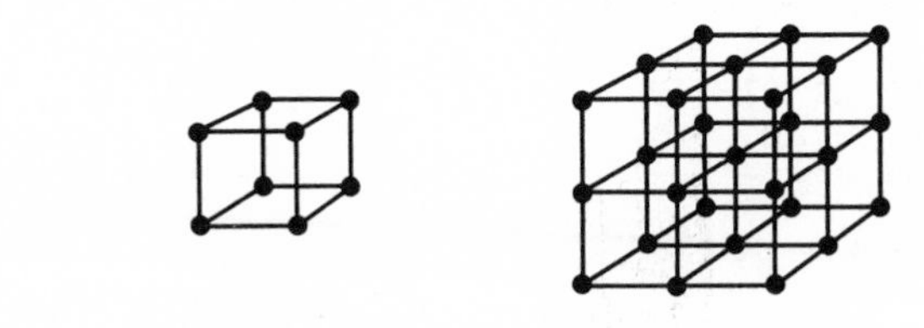

The Greeks thought of the cube of two as: "How many dots are in a cube with a side of two dots?" We think of the cube of two as: "What number is two times two times two?"

Since space as we experience it has no more than three dimensions, it isn't possible to represent sequences of higher powers than the third with dots as the Greeks did with squares and cubes. So there are no special names for them. However, we can get an idea of how the sequences of higher powers grow by listing the first few terms of each. The **sequence of fourth powers,**

$$1^4 \quad 2^4 \quad 3^4 \quad 4^4 \quad 5^4 \quad \ldots,$$

is

$$1 \quad 16 \quad 81 \quad 256 \quad 625 \ldots.$$

Do you recognize that each fourth power number is also a square number? It is the square of a square number.

The higher the power, the more rapidly the sequence grows:

Sequence	First five terms				
Fifth powers	1	32	243	1,024	3,125
Sixth powers	1	64	729	4,096	15,625
Seventh powers	1	128	2,187	16,384	78,125

Notice that one raised to any power is still one. Why?

EXERCISES

Set I

Make a table of the cubes of the numbers 1 through 20 to keep in your notebook for reference. The table you made of square numbers will save some work in making this one. For example,

$$11^3 = 11 \cdot 11^2 = 11 \cdot 121 = 1{,}331$$
$$12^3 = 12 \cdot 12^2 = 12 \cdot 144 = 1{,}728, \text{ and so forth.}$$

Use your table to answer these questions.

1. Can a cube number of more than one digit have all *odd* digits?
2. Is there any pattern in the *last digits* of the cube numbers?
3. Can a cube number end in any digit?

4. Find the digital root of each cube number in your table. What is the pattern?

5. If a number has a digital root of 3, do you think it can be a cube number?

6. Suppose Peter, the inventor of the cubical baseball in the B.C. cartoon, made it with 6,000 windings of string. Is 6,000 a cube number?

Set II

1. Copy this pattern and write in the missing numbers.

$$1^3 = 1^2$$
$$1^3 + 2^3 = \boxed{}^2$$
$$1^3 + 2^3 + 3^3 = \boxed{}^2$$
$$1^3 + 2^3 + 3^3 + \boxed{}^3 = \boxed{}^2$$

2. What is a short cut for finding the last number on each line?

3. Write two more lines of the pattern using the short cut and then check to see if it still gives the right result.

Here is a table of the fourth powers of the numbers 1-10 for reference in exercises 4-7.

Number	Fourth Power
1	1
2	16
3	81
4	256
5	625
6	1,296
7	2,401
8	4,096
9	6,561
10	10,000

4. What digits can a fourth power number end in?

5. What digits can a fifth power number end in? (To answer

this question using the table of fourth powers, you can use the short cut,

$$2^5 = 2 \cdot 2^4 = 2 \cdot 16 = 32$$
$$3^5 = 3 \cdot 3^4 = 3 \cdot 81 = 243, \text{ etc.}$$

Notice also that since we are interested in only the *last* digit of each number, it is not necessary to complete each multiplication.

6. Copy this pattern and write in the missing numbers.

$$3^2 + 4^2 = \square^2$$
$$3^3 + 4^3 + 5^3 = \square^3$$

7. Write what you think is the next line of this pattern and see if it works.

Set III

Areas, "amounts of surface," are measured with numbers of squares. Volumes, "amounts of space," are measured with numbers of cubes. So the sequence of squares and the sequence of cubes are important in determining how a change in size will affect an object.

A model of an airplane and the airplane itself, although they have the same shape, cannot fly the same way. If the plane is 100 times longer than the model, its surface area will be 100^2 or 10,000 times as great. The "lift" of the airplane depends on the surface area of its wings, so the lift of the plane will be 10,000 times the lift of the model.

On the other hand, the volume of the plane will be 100^3 or 1,000,000 times greater than the volume of the model. Since the weight depends upon the volume, the plane may be about 1,000,000 times heavier than the model. How a plane will fly depends on both lift and weight, and so the model cannot behave the same way that the plane will in actual flight.

1. In *Gulliver's Travels* by Jonathan Swift, Gulliver's first voyage took him to Lilliput, a land of people "not six inches high." The king's mathematicians measured Gulliver's height and found that he was 12 times as tall as the Lilliputians. Three hundred tailors were employed to make clothes for Gulliver. If the suits they made were the

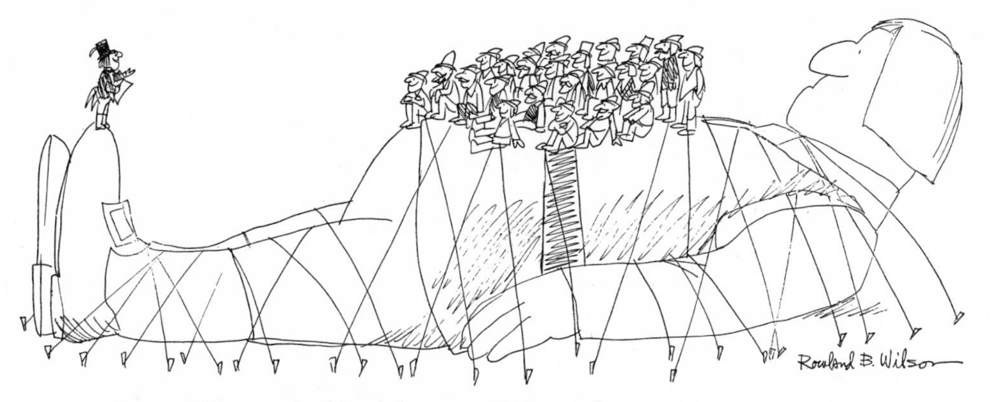

"First of all, let me congratulate everybody on the swell turnout and on the really bang-up job you all did on such short notice."

same thickness as their own, about how many times as much material did they need?

2. The mathematicians calculated how the volume of Gulliver's body compared to their own so that they would know how much to feed him. How many times as much food did they decide he needed?

Lesson 6

The Fibonacci Sequence

Leonardo Fibonacci

Courtesy of Columbia University Library

THE greatest mathematician of the Middle Ages was Leonardo Fibonacci, born in Pisa, Italy. The construction of the famous Leaning Tower of that city was begun during his lifetime, but was not completed for nearly two centuries. In 1202, Fibonacci wrote a book on arithmetic and algebra in which he proposed the following problem:

A pair of rabbits one month old are too young to produce more rabbits, but suppose that in their second month and every month thereafter they produce a new pair. If each new pair of rabbits does the same, and none of the rabbits die, how many pairs of

rabbits will there be at the beginning of each month?

► The numbers that solve this problem form the sequence

$$1 \quad 1 \quad 2 \quad 3 \quad 5 \quad 8 \quad 13 \ldots,$$

which is called the **Fibonacci sequence.**

The first two terms of the sequence are 1 and each succeeding term is the sum of the previous pair of terms.

$$\begin{array}{l} 1 \quad 1 \\ 1 + 1 = 2 \\ \quad 1 + 2 = 3 \\ \quad\quad 2 + 3 = 5 \\ \quad\quad\quad 3 + 5 = 8 \\ \quad\quad\quad\quad 5 + 8 = 13 \ldots \end{array}$$

The Fibonacci number sequence appears in such unrelated topics as the family tree of a male bee and the keyboard of a piano! A male bee has only one parent, his mother, while a female bee has both father and mother. The family tree of a male bee has a strange pattern as a result. If each male is represented by the symbol ♂ and each female by the symbol ♀, the tree looks like this:

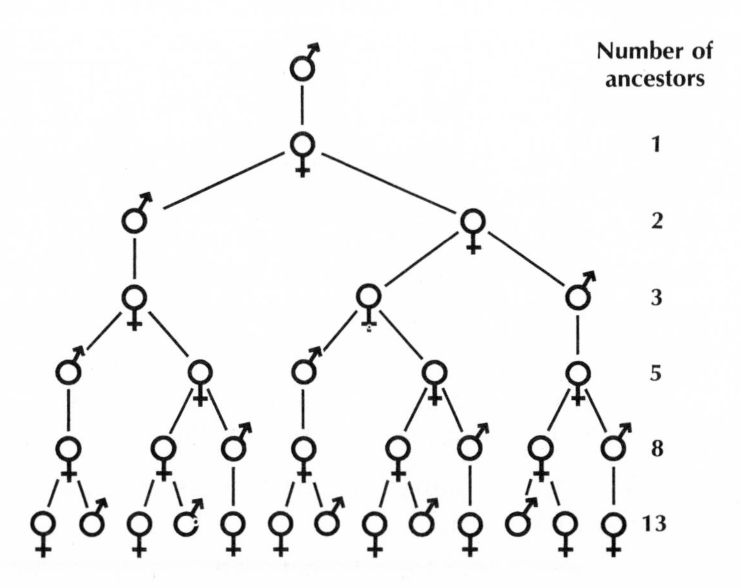

The numbers of ancestors in each successive generation back in time are the numbers of the Fibonacci sequence. The 13 ancestors on the bottom line of the tree have the same relationship as the keys of a piano keyboard! Compare the two closely and you will see what it is.

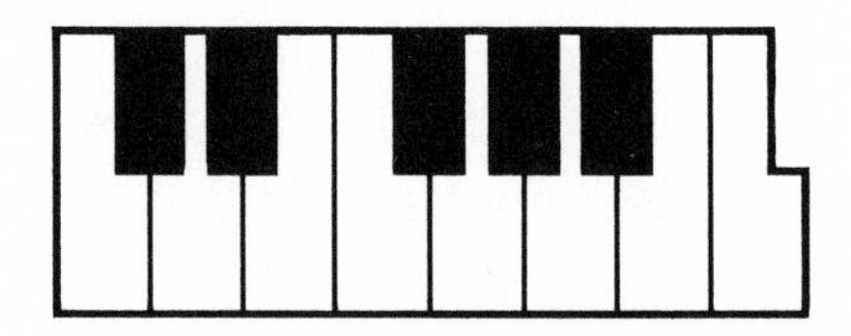

The 13 keys shown are one octave of the "chromatic" scale. The previous ancestor number, 8, corresponds to the number of white keys in an octave, the notes of a "major" scale. The number before that, 5, is the number of black keys in an octave, the notes of the old "pentatonic" scale.

One of the fascinating aspects of mathematics is that an idea that applies to one area of study frequently turns out to be valuable in another apparently entirely unrelated area.

EXERCISES

Set I

1. Write the first 15 terms of the Fibonacci sequence. Remember that the first two terms are each 1; the 15th term is 610.
2. Find the sum of the first *five* terms of the sequence. Notice that the sum is one less than the *seventh* term of the sequence.
3. Find the sum of the first *eight* terms of the sequence. The sum is one less than the *tenth* term of the sequence.
4. Can you guess the sum of the first 13 terms of the sequence without adding them?
5. A great 18th century French mathematician named Joseph Lagrange did not become interested in mathematics until he was 17. He discovered a pattern in the remainders formed by dividing each term of the Fibonacci sequence by 4. The remainders for the first four terms are the terms themselves. Find the remainders for the rest of the terms

in your list. For example, the ninth term, 34, gives a remainder of 2 when divided by 4:

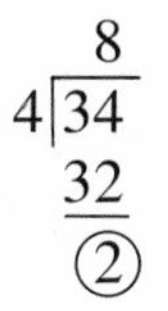

6. What is the pattern in the sequence of remainders?

Set II

1. Find the squares of the first eight terms of the Fibonacci sequence.

(1   1   4   9 . . .)

2. Now add each pair of consecutive squares to make a new sequence:

2   5   13 . . . .

3. What do you notice?

4. Here is a pattern based on the sequence of squares you wrote for exercise 1:

$$1^2 + 1^2 = 1 \cdot 2$$
$$1^2 + 1^2 + 2^2 = 2 \cdot 3$$
$$1^2 + 1^2 + 2^2 + 3^2 = 3 \cdot 5$$
$$1^2 + 1^2 + 2^2 + 3^2 + 5^2 = 5 \cdot 8.$$

Write the next two lines of the pattern and check to see if they are correct.

5. Copy and write in the missing numbers in this pattern based on cubes of consecutive terms of the Fibonacci sequence.

$$2^3 + 3^3 - 1^3 = 34$$
$$3^3 + 5^3 - 2^3 = \text{||||||||}$$
$$5^3 + 8^3 - 3^3 = \text{||||||||}$$

6. What do you notice about each of the three numbers on the right?

Set III

Another area of biology in which Fibonacci numbers appear is plant growth. The numbers of petals of many flowers are Fibonacci numbers. Not every flower is identical, but here are some typical values.*

3 petals	Lilies and irises
5 petals	Buttercups, larkspurs and columbines
8 petals	Some delphiniums
13 petals	Corn marigolds
21 petals	Some asters
34, 55, and 89 petals	Daisies

Fibonacci numbers also appear in the arrangement of leaves on the stems of plants. What Fibonacci numbers can you find in this drawing of a plant?

* *The Language of Mathematics*, by Frank Land Doubleday, 1963, pp. 216–218.

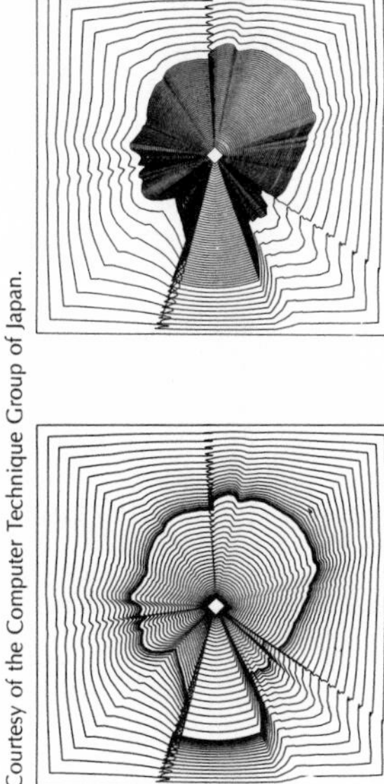

These figures, each showing a square transformed into the profile of a lady and then back into a square, were drawn with the help of a computer. The first was programmed according to an arithmetic sequence and the second according to a geometric sequence.

Courtesy of the Computer Technique Group of Japan.

Chapter 2 / Summary and Review

IN this chapter we have studied four types of number sequences:

Arithmetic sequences *(Lesson 1)* Each term is found by adding a definite number to the previous number.

An example of an arithmetic sequence is

1 4 7 10 13 . . . .

Geometric sequences *(Lessons 2 & 3)* Each term is found by multiplying the previous term by a definite number.

An example of a geometric sequence is

2 6 18 54 162 . . . .

The *binary* sequence is a special geometric sequence:

1 2 4 8 16 . . . .

Power sequences *(Lessons 4 & 5)* Each term is found by raising a counting number to a definite power.

Examples of two power sequences are the *sequence of squares,*

1 4 9 16 25 . . .,

and the *sequence of cubes,*

1 8 27 64 125 . . . .

The Fibonacci sequence *(Lesson 6)* Each term (after the first two) is found by adding the previous pair of terms:

1 1 2 3 5 8 . . . .

EXERCISES

Set I

What kind of sequence is each of the following? Copy the sequences and write in the missing terms.

1. ||||||| 6 10 14 |||||||
2. ||||||| 12 36 108 |||||||
3. ||||||| 8 27 64 |||||||
4. ||||||| 4 8 16 |||||||
5. ||||||| 8 13 21 |||||||

The following sequence can be either arithmetic or geometric, depending upon what numbers are chosen for the missing terms.

10 ||||||| 40 ||||||| |||||||

6. What should the missing terms be so that it is arithmetic?
7. What should they be in order for it to be geometric?

The terms in a number sequence can become smaller instead of larger.

8. How is this possible in an arithmetic sequence?
9. How is this possible in a geometric sequence?

Set II

1. The three chemical elements, lithium, sodium, and potassium, are very much alike. They are soft, light metals that will burn your fingers if you touch them. Their atomic weights are 7, 23, and 39. What kind of number sequence is this?

2. The Ahmes Papyrus, written in Egypt about 1550 B.C., contains this list:

Household	7
Cats	49
Mice	343
Barley	2,301
Hekats	16,807

What kind of number sequence does this seem to be? The sequence as it appears in the papyrus has a mistake. What is it?

3. The number of days in a year, 365, is a unique number because it is the sum of three consecutive square numbers and also the sum of the next two square numbers. Use your list of square numbers to find out what these two sets of numbers are.

4. Among the terms used by the book industry to indicate the size of book pages are "folio," "quarto," and "octavo." These words refer to the number of times the large printer's sheets are folded to make the pages of the book.

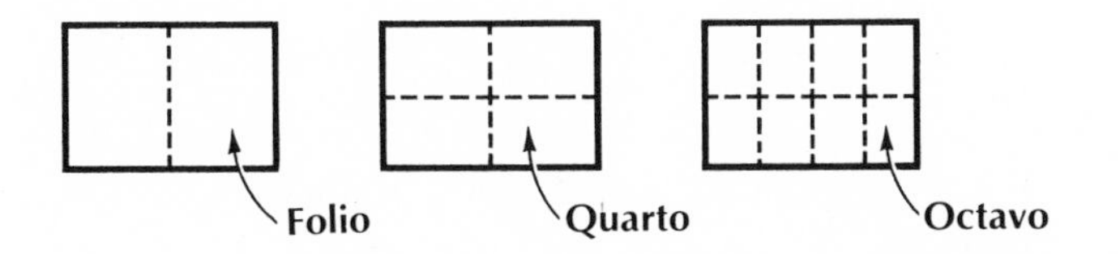

Smaller sheets are referred to as "16mo," "32mo," and so on. What kind of number sequence do these terms suggest?

5. A brilliant Indian mathematician named Ramanujan (1887-1920) specialized in the study of numbers and knew their characteristics in the same way that a baseball fan might know a vast number of statistics about the game. One time a friend went to visit him in a taxi having the number 1729. When the friend mentioned this number, Ramanujan immediately replied: "1729 is a very interesting number; it is the smallest number expressible as the sum of two cubes in two *different* ways."

   Use your list of cube numbers to find out what these *two pairs* of cubes are.

6. One number in the Fibonacci sequence is 89. Find the

reciprocal* of 89 to six decimal places. (The digit in the first decimal place is 0.) What do you notice about the result?

7. As the earth's population increases, so does the need for food. In 1798, in his *Essay on the Principle of Population,* the English economist Thomas Malthus claimed that the number of people in the world increases in a geometric sequence while the amount of available food increases in an arithmetic sequence. He assumed that the population, when unchecked, doubles every 25 years, and said:

   "Supposing the present population equal to a thousand millions, the human species would increase as the numbers, 1, 2, 4, 8, 16, 32, 64, 128, 256, and subsistence as 1, 2, 3, 4, 5, 6, 7, 8, 9. In two centuries the population would be to the means of subsistence as 256 to 9."

   Malthus goes on to give the number relationship for *three* centuries ahead. What numbers did he give? (Since these numbers represent the situation every 25 years, you will need to write four more numbers in each sequence.)

8. The first five terms of the sequence of sixth powers,

   $1^6 \quad 2^6 \quad 3^6 \quad 4^6 \quad 5^6,$

   are

   1 64 729 4,096 15,625.

   Find the digital root of each of these numbers. What do you think the digital root of 6^6 is?

Set III

Wiley has started a "Pyramid Club." (Read the cartoon at the top of the facing page.) It involves a chain letter that is different from the ones we learned about when studying geometric sequences.

1. Study the directions in the letter carefully. Notice that each person who receives the original letter will send 3 copies to Wiley, so that he will first get 3 letters, then 3

*The reciprocal of a number is 1 divided by that number. For example, the reciprocal of 2 is 1/2, or 0.5, and the reciprocal of 3 is 1/3, or 0.333 . . . .

By permission of Johnny Hart and Field Enterprises, Inc.

more letters, and 3 more letters after that, and so on. Write a number sequence to show the successive *total* numbers of letters that Wiley has as they start coming in.

2. What kind of sequence is it?

Here is a side view of a pyramid somewhat smaller than the one Wiley built. All of the blocks in it are the same size and shape.

The top of each layer has the shape of a square and so the number of blocks in each layer is a square number.

3. Can you figure out how many blocks there are in this pyramid?

Chapter 3

FUNCTIONS AND THEIR GRAPHS

Lesson 1

The Idea of a Function

A Greek philosopher named Zeno in about 450 B.C. made up a famous puzzle about a race between Achilles and a tortoise. Achilles runs ten times faster than the tortoise, so it is given a head start of 1,000 yards. Where will Achilles overtake the tortoise?

If we suppose that the tortoise can run 50 yards a minute (an extraordinarily fast tortoise, to be sure), then Achilles, who is ten times faster than the tortoise, runs at a rate of 500 yards a minute. To see the distance Achilles has covered as the race goes on, we can write two number sequences – one for the time and one for the distance:

time (in minutes)	0	1	2	3	4	. . .
distance (in yards)	0	500	1,000	1,500	2,000	. . . .

The time and distance are continually changing and are called *variables.* They are related and the relationship between them is called a *function.* We say that the distance Achilles has run is a *function* of the time the race has gone on.

► We will think of a **function** as a pairing of two number sequences in which any term in the first sequence determines exactly one term in the second.

When we pair two number sequences in this way, we are writing a **table** for the function. The terms of the first sequence are called values of the *independent variable,* because they are

chosen first, or "independently" of the second sequence. The terms of the second sequence are called values of the *dependent* variable, because each one "depends" on the corresponding term of the first sequence.

A more convenient way to represent a function is with a **formula.** If we let the letter t stand for any number in the first sequence and the letter d stand for the corresponding number in the second sequence, we can write the formula

$$d = 500t$$

to represent the function shown by the table given on the facing page. The formula tells us that the distance Achilles has run can be found by multiplying the time by 500.

Now let's check up on the tortoise. Since he has a head start of 1,000 yards, we will consider this to be his distance at the beginning of the race. The tortoise runs 50 yards a minute and so the table for his times and distances looks like this:

time (in minutes)	0	1	2	3	4	. . .
distance (in yards)	1,000	1,050	1,100	1,150	1,200	. . . .

The distance of the tortoise in the race is also a function of the time.

What does the formula for this function look like? Again using the letters t and d to stand for the time and corresponding distance, we can represent the function with the formula

$$d = 1{,}000 + 50t.$$

This formula tells us that the distance of the tortoise in the race can be found by adding 1,000 (the head start distance) to the time multiplied by 50.

We haven't yet solved Zeno's puzzle of where Achilles will overtake the tortoise. Since drawing a graph for the formulas we have written is an easy way to do it, we will postpone working out the rest of the puzzle until we have learned to draw graphs.

EXERCISES

Set I

If we are told the formula of a function, we can write a table by choosing any numbers we want for the first line and then *substituting* them into the formula to find the corresponding numbers of the second line. It is usually convenient to choose consecutive numbers for the first line (the independent variable). The formula then determines the numbers of the second line (the dependent variable). For example, since the speed of Achilles is ten times the speed of the tortoise, a formula for this function is:

$$y = 10x,$$

where y is the speed of Achilles, and x is the speed of the tortoise. For the first line of the table, we choose some numbers for x:

x	0	1	2	3	4.

To find the numbers on the second line, we substitute these numbers into the formula and get:

y	$10 \cdot 0$	$10 \cdot 1$	$10 \cdot 2$	$10 \cdot 3$	$10 \cdot 4$, or
y	0*	10	20	30	40.

Write a table for each of the following functions. The first line of each table is given and you are to use the formula of the function to find the corresponding numbers of the second line.

1. $y = x + 5$

x	0	1	2	3	4

* Don't forget that if *any* number is multiplied by zero, the result is zero.

Courtesy of Al Capp; © 1967 News Syndicate Co., Inc.

2. $y = 7x$

x	0	1	2	3	4

3. $y = 2x + 1$

x	0	1	2	3	4

4. $y = 3x - 2$

x	1	2	3	4	5

5. $y = x^2$

x	0	1	2	3	4

6. $y = x^2 + 4$

x	0	1	2	3	4

7. $y = x^3$

x	0	1	2	3	4

8. $y = 2^x$

x	2	3	4	5	6

(When $x = 2$, $y = 2^2 = 4$; when $x = 3$, $y = 2^3 = 8$, etc.)

The line of numbers you wrote for y in exercise 2 should be:

y	0	7	14	21	28.

These numbers are part of an arithmetic sequence since each number can be found by adding 7 to the previous number.

9. In which of the other tables you have written are the numbers on the second line in arithmetic sequence?
10. The second line of one of your tables is part of a geometric sequence. Which table is it?
11. In which of your tables are the second lines of numbers power sequences?

Set II

Write a formula for the function represented by each of the following tables. Begin each formula with y, which represents the dependent variable.

1.

x	1	2	3	4	5
y	2	4	6	8	10

2.

x	0	1	2	3	4
y	8	9	10	11	12

3.

x	7	8	9	10	11
y	4	5	6	7	8

4.

x	0	1	2	3	4
y	0	5	10	15	20

5.

x	3	4	5	6	7
y	9	16	25	36	49

6.

x	6	7	8	9	10
y	11	13	15	17	19

7.

x	2	3	4	5	6
y	21	31	41	51	61

8.

x	1	2	3	4	5
y	1	8	27	64	125

The size of a motion picture on the screen is a function of the distance of the projector from the screen.

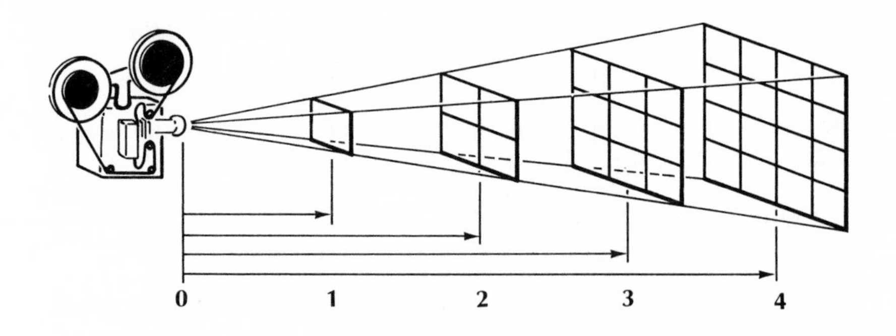

The diagram shows that if the distance between the screen and the projector is 3 units, the size of the picture is 9 squares.

9. Complete the following table suggested by the diagram.

d (distance from screen)	▒	▒	▒	▒
s (size of picture)	▒	▒	▒	▒

10. Write a formula for the function, using the letters s and d.

Set III

You have probably heard stories about people lost on a desert who walk for miles in what they think is a straight line, only to come back to the place from which they started without realizing

Drawing by Jack Tippit;

it. The reason for this is that a person's legs are never exactly the same length and so the steps with his left and right feet are slightly different. An exaggerated diagram of this is shown here. It shows the steps of someone whose step with his left foot is longer than with his right.

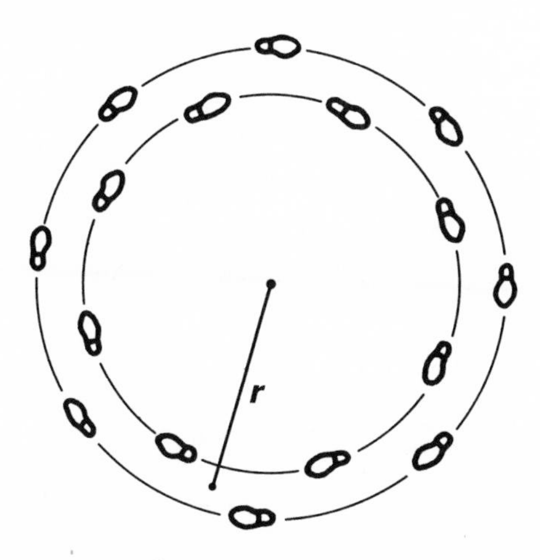

The radius of the circle in which a person walks is a function of the difference between the lengths of his steps. An approximate formula for this function is:

$$r = \frac{24}{d},$$

where r is the radius of the circle in feet and d is the difference between the lengths of the steps in inches.

Suppose the steps you take with your left foot are $1/16$ inch longer than the steps you take with your right foot, and that you are blindfolded and think you are walking straight ahead. What would be the radius of the circle in which you walk?

Lesson 2

Descartes and the Coordinate Graph

A stamp issued by France on the 300th anniversary of Descartes' invention of coordinate geometry.

ONE of the greatest mathematical achievements of all time was the invention of coordinate geometry by René Descartes. It made possible a new method of studying geometric figures and the relationships between them, relationships first proved by the ancient Greeks. The new method involved using algebra, which had been developed many centuries after geometry, and marked the beginning of modern mathematics. With this method, such figures as triangles and circles could be related to equations, so that algebra and geometry were combined into a new subject more powerful than either alone had been before.

The man who developed this method, René Descartes, was born in France at the turn of the 17th century. Europe at this time was in political and religious turmoil, yet the age in which Descartes lived was also a period of great intellectual progress. In England, Shakespeare was writing his plays; great scientific discoveries were being made by Galileo in Italy; and the French mathematicians Fermat and Pascal were developing another new branch of mathematics called probability theory.

As a young boy, Descartes had poor health, and the boarding school where he lived permitted him to stay in bed each morning as late as he pleased. By about the age of 14, he began to doubt the truth of much of what he was being taught. The subject of mathematics, however, appealed to him since its methods of reasoning seemed universal and without fault. His schooling finished, Descartes spent some time leading a wild life in Paris and then, after tiring of this way of living, became a soldier. At the age of 32, he went to Holland where he spent most of the rest of his life studying and writing.

Descartes soon decided that mathematics, to quote his words, "is a more powerful instrument of knowledge than any other that has been bequeathed to us by human agency," and so he tried to apply deductive reasoning, the method of mathematical thinking developed by the Greeks so many centuries before, to other areas of study. In 1637, his book, *A Discourse on the Method of Rightly Conducting the Reason and Seeking Truth in the Sciences,* established Descartes as the "father of modern philosophy." The book ended with a section on coordinate geometry, Descartes' great contribution to the subject of mathematics.

Descartes' invention was clever, yet like many important discoveries in mathematics, very simple. The idea was this: Draw two lines perpendicular to each other. The position of any point can be given by telling its distance from each of these lines. And that's it!

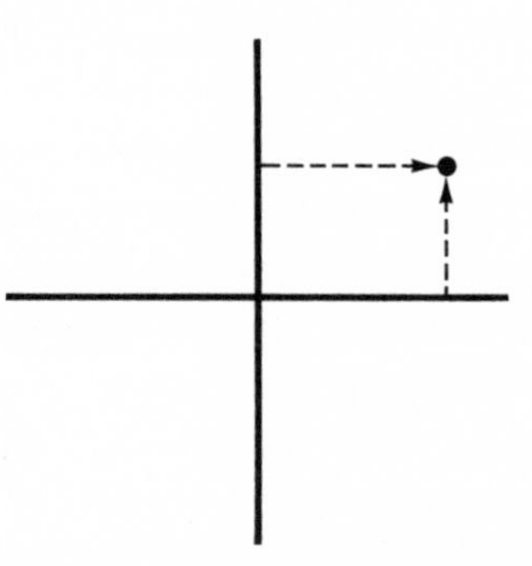

Let's see exactly how this works. We will give the two lines letters for names and call them the *x-axis* and the *y-axis.* The point where they intersect will be called the origin, and labeled

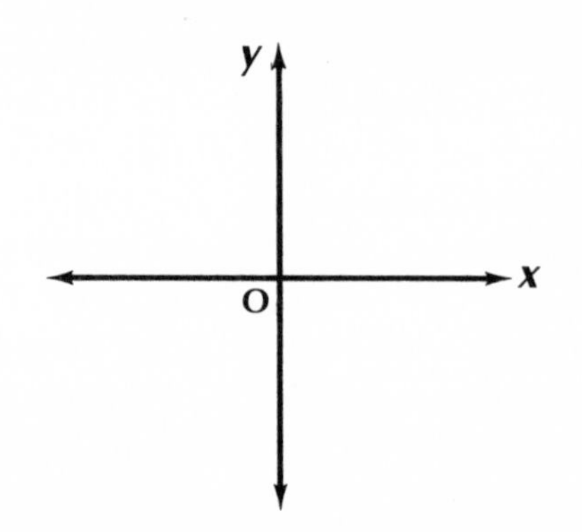

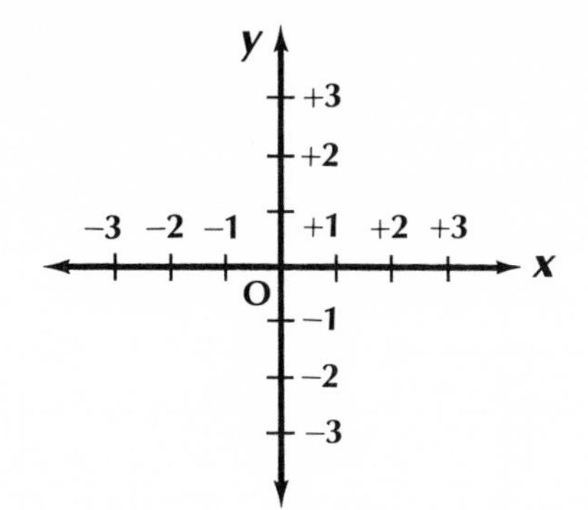

with a capital O. Now the two axes can be numbered like a pair of "double rulers," that is, in each direction from the origin. On the *x*-axis, we use positive numbers from the origin to the right and negative numbers* to the left; on the *y*-axis, positive numbers going up and negative numbers going down.

Now we are ready to locate a point with numbers, namely its two distances from the axes. First we move along the *x*-axis until we are directly below (or above) the point, counting the units as we go. Then we move directly up (or down) to the point itself, again counting the units along the way. These two numbers are called the **coordinates** of the point and are written in parentheses like this: (3, 2). The first number is called the *x-coordinate* and the second number is called the *y-coordinate.* Since we always move along the *x*-axis first, the *x*-coordinate is always given first.

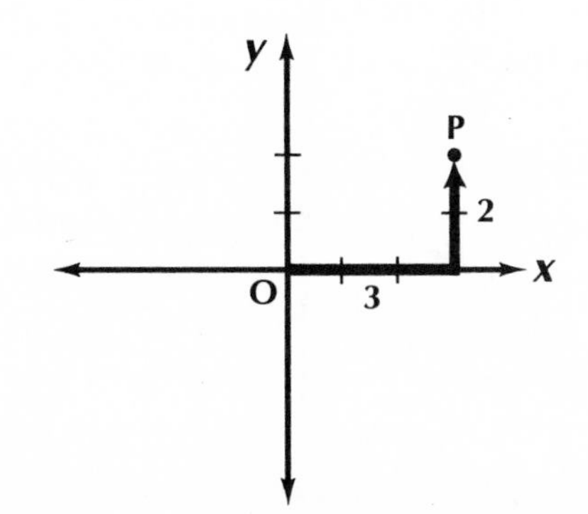

Following are some more examples of how the coordinates of a point are found.

---

*Look on page 498 if you are not familiar with negative numbers.

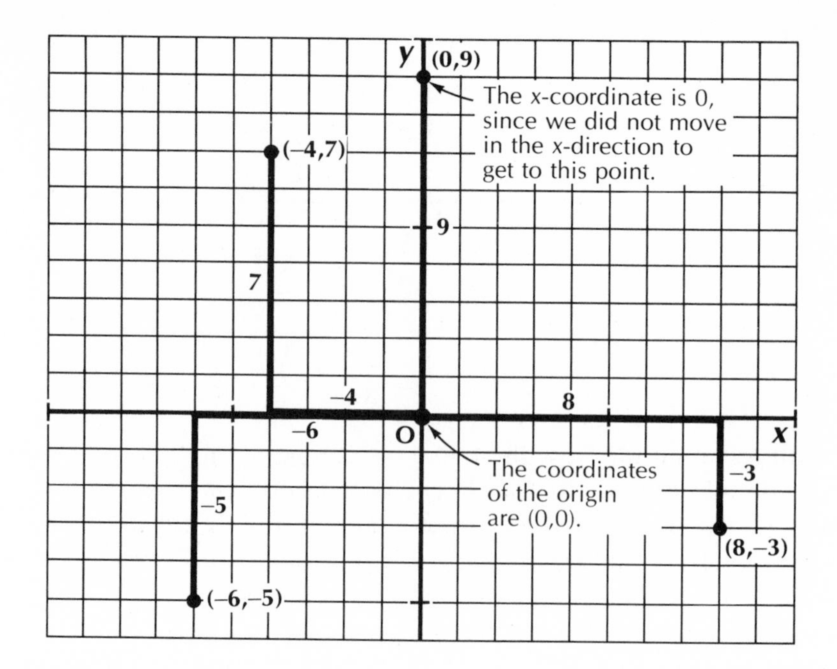

EXERCISES

Set I

The coordinate graph is valuable in the study of geometric figures because each point of a figure can be located with a pair of numbers.

1. In this graph is a circle drawn with its center at the origin. The radius of the circle is 5 units. Write the coordinates of each point on the circle that is named with a letter.

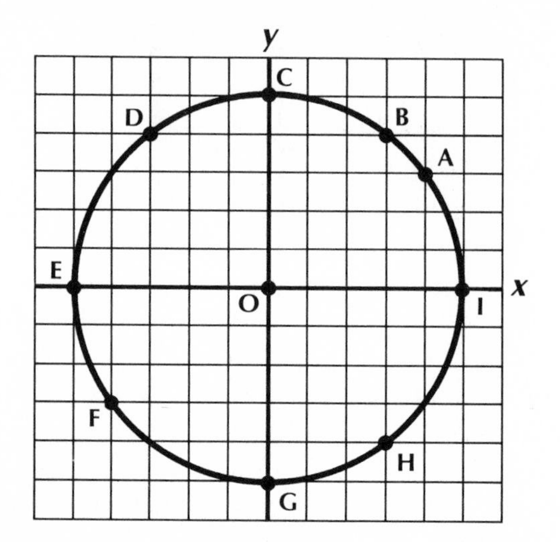

2. The curve in this graph is called a parabola. Write the coordinates of each lettered point on the parabola.

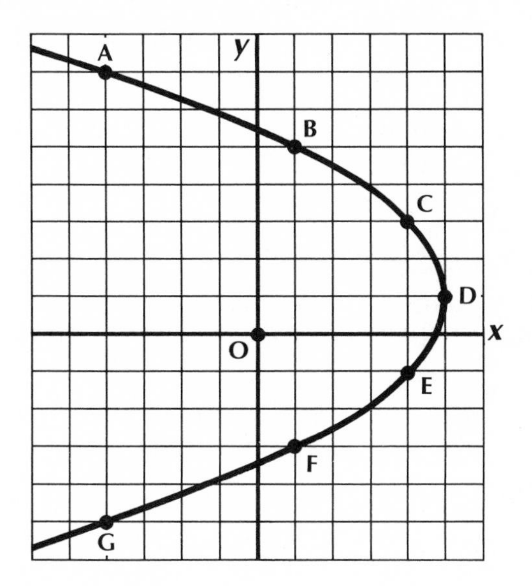

3. This curve is called the "folium of Descartes." Although the coordinates of most of the lettered points on it are not actually whole numbers, write them as if they were.

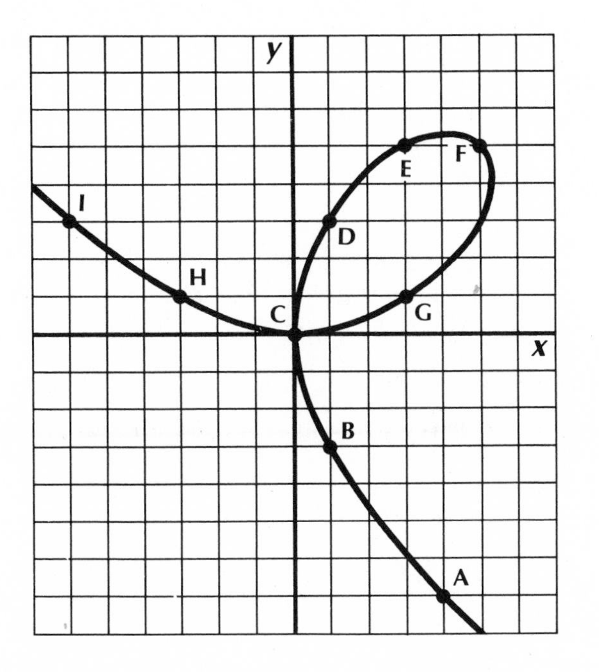

Set II

On graph paper, draw a separate pair of axes extending at least 6 units in each direction from the origin for each of the following exercises. Connect the points in each list in order with straight line segments to form the geometric figure named.

1. Isosceles triangle:
   (2, 5) (5, −2) (−1, −2) (2, 5)
2. Rectangle:
   (4, 2) (4, −4) (−3, −4) (−3, 2) (4, 2)
3. Parallelogram:
   (6, 4) (3, −1) (−5, −1) (−2, 4) (6, 4)
4. Trapezoid:
   (0, 3) (−3, −1) (6, −4) (3, 2) (0, 3)
5. Pentagon:
   (1, 4) (−2, 0) (0, −4) (5, −3) (6, 2) (1, 4)

Set III

Here is an exercise that requires patience but gives an interesting result. Draw a pair of coordinate axes extending from −10 to +10 on the x-axis and from +20 to −10 on the y-axis. Connect the points in each list *in order* with straight line segments and something familiar will appear. (After the points in one list have been connected in order, start all over again with the next list. In other words, *do not connect* the last point in each list to the first point in the next one.)

A. (−3, 10) (−2, 8) (−5, 2) (−7, 1) (−5, 0) (−2, 0) (−2, −8) (−6, −8) (−7, −9) (−4, −9) (−5, −10) (2, −10) (2, −3)

B. (−4, −9) (0, −9) (0, 5) (−2, 1) (0, 1)

C. (2, 4) (2, 0) (3, −1) (3, −2) (2, −3) (0, 0) (−2, 0)

D. (2, 1) (6, 1) (5, 0) (2, 0)

E. (−5, 2) (−4, 1) (1, 7) (1, 8) (3, 9) (4, 9)

F. (1, 7) (5, 1)

G. (2, 13) (4, 14) (5, 13) (4, 12) (6, 11) (6, 10) (3, 9.5)

H. (−7, 12) (−6, 14) (−5, 15) (−1, 15) (0, 16)

I. (0, 15) (1, 16) (2, 16)

J. (−6, 11) (−5, 14)

K. (−5, 10) (−4, 13)

L. (−4, 9) (−4, 11)

M. (−2, 12) (−3, 12) (−3, 11) (−2, 10)

N. (2, 12) (4, 12)

O. (0, 13) (1, 13)

Lesson 3

Graphing Linear Functions

ONE of the problems that interested Descartes, the inventor of coordinate geometry, was Zeno's puzzle about Achilles and the tortoise. You will recall that in this puzzle Achilles runs ten times faster than the tortoise, and so the tortoise is given a head start of 1,000 yards. The question is: where will Achilles pass the tortoise?

If the tortoise runs 50 yards a minute, then Achilles runs 500 yards a minute. The table showing the distance Achilles has run as a function of the time the race has gone on is:

time (in minutes)	0	1	2	3	4	. . .
distance (in yards)	0	500	1,000	1,500	2,000	. . . .

Notice that the table contains *pairs of numbers,* a distance number for each time number. If we think of these pairs of numbers as coordinates of points, we can draw a picture of the function by using the points to make a graph.

We have five points to locate since the table gives the coordinates:

$$(0, 0), (1, 500), (2, 1{,}000), (3, 1{,}500), \text{ and } (4, 2{,}000).$$

We will name the x- and y-axes with the letters t and d instead, since the first coordinate of each point is the time and the second coordinate is the distance. Since the distance coordinates get so

By permission of Johnny Hart and Field Enterprises, Inc.

large, we choose scales on the two axes so that there is room to show all five points.

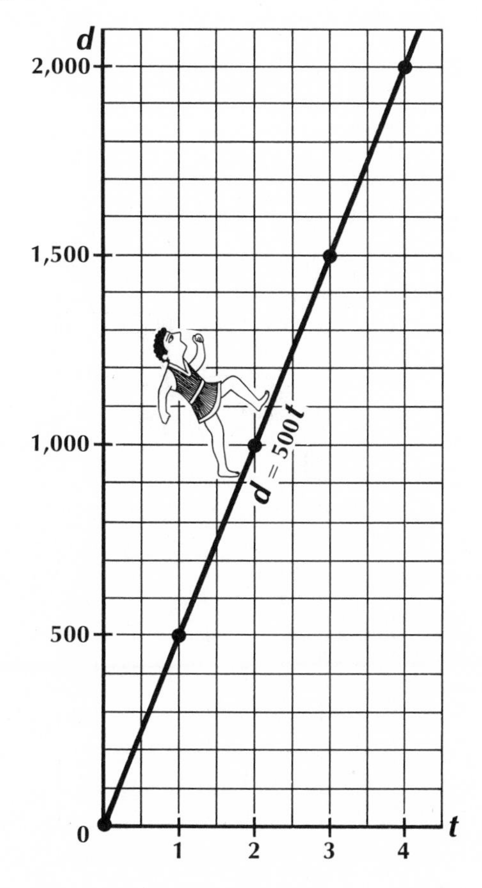

After plotting the points, we see that they lie along a straight line. It makes sense to draw this line, since there are other times and distances between those listed in the table. The formula of the function is

$$d = 500t$$

and the line is called its graph.

Since the tortoise is given a head start of 1,000 yards and runs 50 yards a minute, the table showing its distances in the race as a function of time is:

time (in minutes)	0	1	2	3	4	. . .
distance (in yards)	1,000	1,050	1,100	1,150	1,200	. . . .

This table contains pairs of numbers for five more points. Plotting these points on a new pair of axes, we get the graph below.

Again the points are *collinear* (they lie on a straight line) and

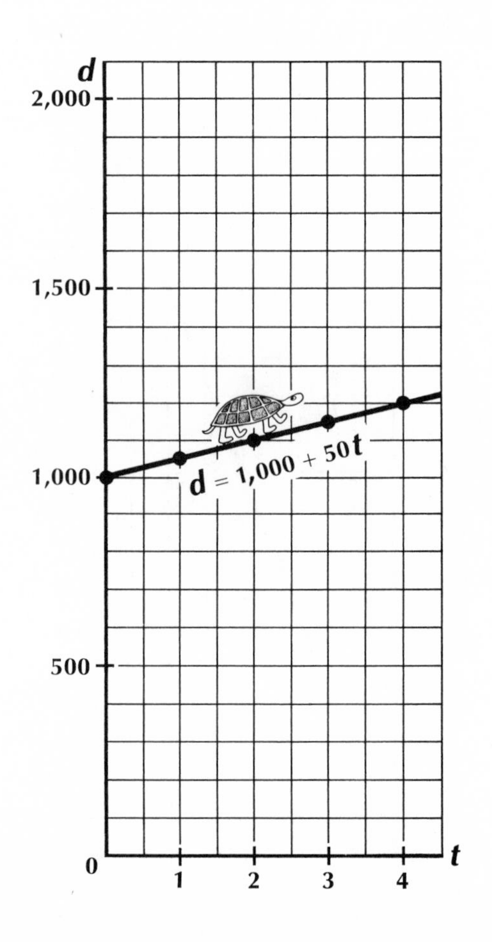

the line can be drawn in. This line is the graph of the formula

$$d = 1{,}000 + 50t.$$

Notice that the line on the Achilles graph is much steeper than the line on the tortoise graph. This is because Achilles is running much faster than the tortoise; the steepness or *slope* of the line shows the *speed*, the ratio of the distance and the time. Notice also how the head start of the tortoise is shown on the second graph.

The puzzle asks where Achilles catches up with the tortoise. Our graphs make finding the answer easy; all we have to do is draw both graphs on the same pair of axes. Where are Achilles and the tortoise at the same place? The point where the lines cross has approximate coordinates: (2.2, 1,110). So we not only have found that Achilles overtakes the tortoise at about 1,110 yards from where he began, but also that it takes him about 2.2

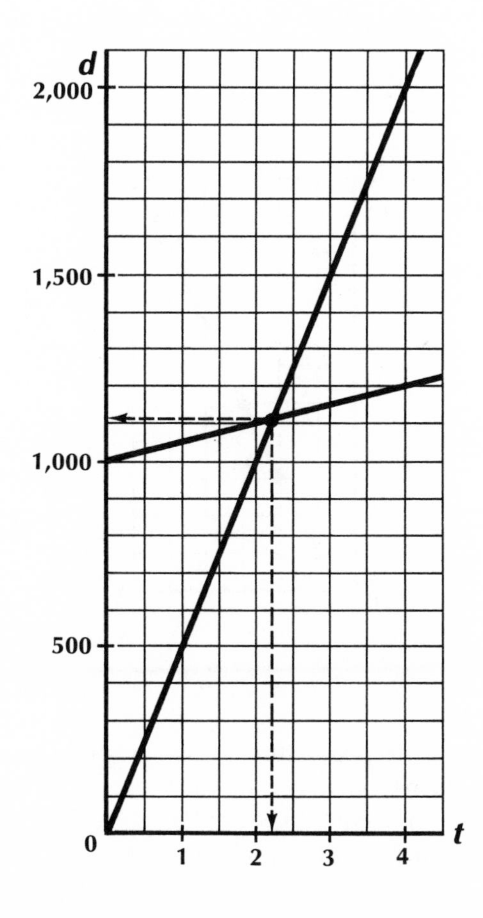

minutes. If we want a more exact answer, we can draw a larger and more detailed graph.

EXERCISES

Set I

To draw the graph of a function for which we are given the formula, it is first necessary to write a table. For example, a table for the function

$$y = 2x - 1$$

is

x	1	2	3	4	5
y	1	3	5	7	9.

This table contains the coordinates of five points; plotting them on a pair of axes and drawing a line through them, we get the graph shown here.

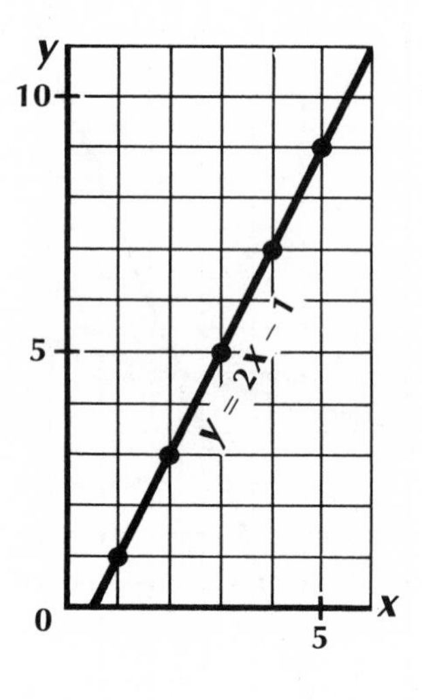

1. Using the same procedure, draw graphs for the following three functions on one pair of axes. (Convenient x numbers to use in your tables are: 0, 1, 2, 3, 4, 5.)
   Write each formula along its line.

$$y = x \qquad y = \frac{1}{2}x \qquad y = 2x$$

2. What do you notice about these three lines?
3. Which line is the steepest?
4. Without drawing graphs of $y = \frac{1}{3}x$ and $y = 3x$, which line do you think would be steeper?
5. Draw graphs for these three functions on one pair of axes. Write the formula for each along its line.

$$y = x + 1 \qquad y = x + 4 \qquad y = x + 5$$

6. What do you notice about these three lines?
7. Write a formula for another line in the same direction.
8. Draw graphs for these two functions on one pair of axes.

$$y = 2x + 1 \qquad y = 2x + 3$$

9. Can you write the formula for the line midway between them?
10. Draw graphs for these two functions on one pair of axes. Part of a table for the first formula is:

x	0	1	2	3	4	5
y	-4	-3				1

$$y = x - 4 \qquad y = 4 - x$$

11. What do you notice about these two lines?

Set II

The distance from you that lightning strikes is a function of the time between the flash and the thunder:

time (in seconds)	1	2	3	4	5
distance (in feet)	1,100	2,200	3,300	4,400	5,500.

1. Draw a graph, letting the x-axis represent time and the y-axis represent distance. It will be convenient to let one unit along the y-axis stand for 1,000 feet.
2. How far away does lightning strike if the time is 0 seconds?

3. Write a formula for this function. Let d represent the distance and t represent the time.

Water pressure in the ocean is a function of the depth:

depth (in miles)	1	2	3	4	5
pressure (in tons per square inch)	1.15	2.30	3.45	4.60	5.75.

4. Draw a graph, letting the x-axis represent the depth and the y-axis represent pressure.
5. In 1960 the U.S. Navy *Trieste* descended to the bottom of the Marianas Trench in the Pacific Ocean, a depth of 6.8 miles. Use your graph to estimate the water pressure at that depth.

Set III

The number of chirps that a cricket makes in a minute is a function of the temperature. As a result, it is possible to tell how warm it is by using a cricket as a thermometer!

A formula for the function is

$$t = \frac{n}{4} + 40,$$

where t represents the temperature in degrees Fahrenheit and n represents the number of cricket chirps in one minute.

1. Copy this table, using the formula to find the missing t numbers.

n	40	60	80	100	120	140
t	50	▒▒	▒▒	▒▒	▒▒	75

2. Draw a pair of axes like the one shown here and plot the six points whose coordinates are given in your table.

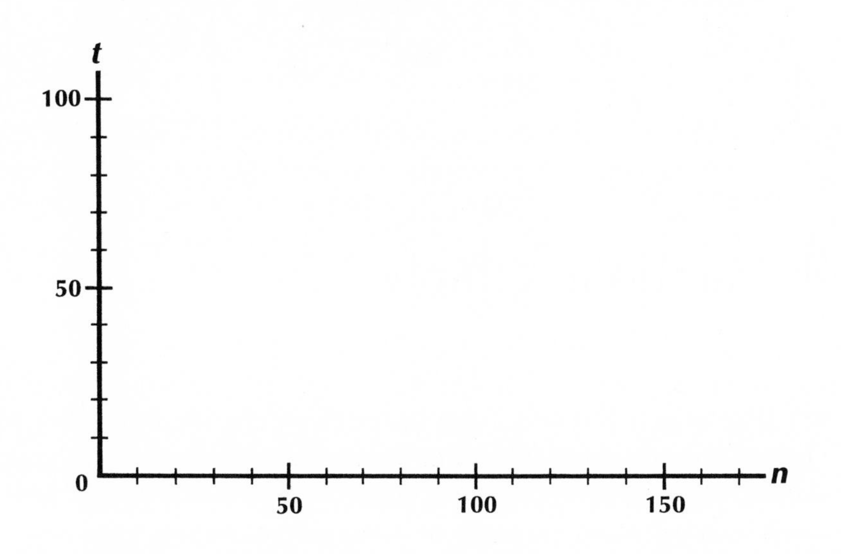

3. Draw a line through the six points and extend it in both directions.

Use your graph to answer these questions:

4. If you hear 170 chirps a minute, what is the approximate temperature?

5. At what temperature do crickets stop chirping?

Lesson 4

Functions with Parabolic Graphs

SO B.C.'s well is 16 seconds deep. Does that mean anything? Doesn't the weight of the stone Peter tossed into the well have something to do with how long it takes the stone to hit the bottom? Suppose B.C. dropped in a large rock; would he perhaps find that the well was 10 seconds deep because the rock fell faster?

The Greeks were so successful in using deductive reasoning in geometry that they decided scientific theories could be worked out in the same way, without the need for experiments to test their ideas. They reasoned about falling objects like this: If one object is heavier than another, it is because it is more strongly attracted to the earth. The stronger the earth attracts an object, the faster it will fall. Therefore, the heavier an object is, the faster it falls. But the Greeks were wrong.

Aristotle used the deductive method and it remained the way of science even through the Renaissance. Finally the great Italian scientist Galileo decided that while *deductive reasoning* is the way to prove ideas in mathematics, *inductive reasoning,* or the experimental method, is the way to make discoveries in science.

One of the first things Galileo experimented with was the study of motion. He discovered that the speed at which an object falls does not depend on its weight. Drop a small stone and a big rock from the same height and they will hit the ground at the same time! Galileo knew that the *distance* an object falls is a

By permission of Johnny Hart and Field Enterprises, Inc.

function of the *time.* Since objects fall so quickly, it is convenient to measure the time in quarter seconds. Here is the table showing the distance covered as a function of time:

time (in quarter seconds)	0	1	2	3	4	5	. . .
distance (in feet)	0	1	4	9	16	25	. . . .

You will recognize that the formula for this function is

$$d = t^2,$$

which no doubt surprised Galileo.*

What does the graph of this function look like? It is shown on page 108; notice the numbering of the two axes. The points are certainly not collinear; no straight line can be drawn through them. However, we can join them with a smooth curve to show the times and distances between those listed in the table.

Although the graph does not extend far enough, we can use

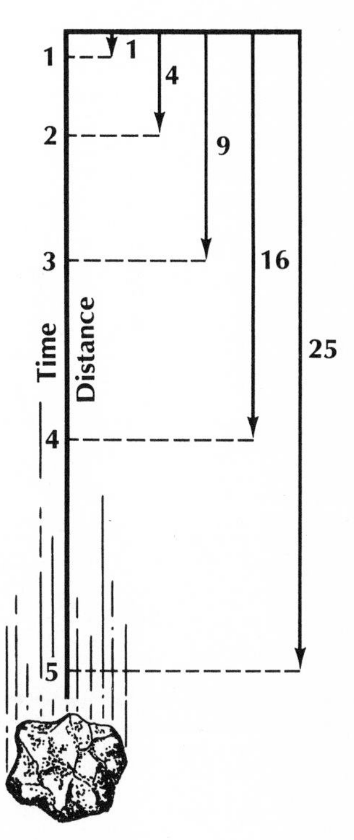

* In 1636, Galileo said: "So far as I know, no one has yet pointed out that the distances traveled during equal intervals of time, by a body falling from rest, stand to one another in the same ratio as the odd numbers beginning with one."

Can you see how the drawing (right) of the falling weight illustrates this? (Compare this drawing to the diagram, on page 63, that shows how the Greeks related square and odd numbers.)

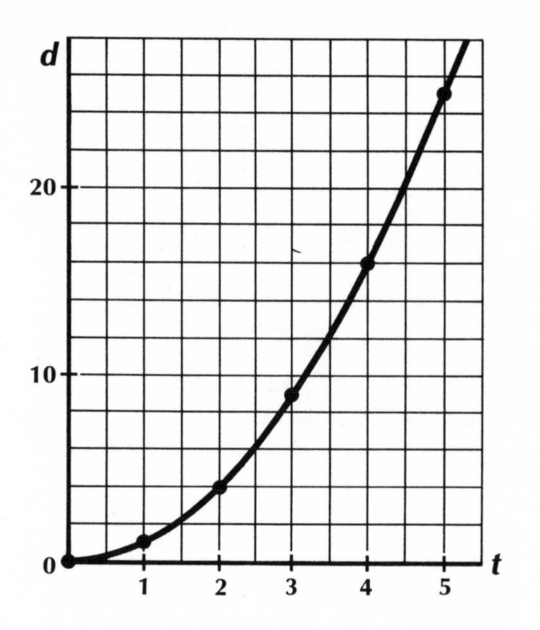

the formula to find how many feet deep B.C.'s well is. Sixteen seconds is the same as 64 quarter seconds. Substituting into the formula,

$$d = 64^2 = 4{,}096 \text{ feet!}$$

Peter and B.C. would be wiser not to stand so close to the edge!

EXERCISES

Set I

Since curved line graphs are continually changing direction, we can get a more complete picture of them by including points with negative coordinates. For example, a table for the function

$$y = x^2$$

could include:

x	−4	−3	−2	−1	0	1	2	3	4
y	16	9	4	1	0	1	4	9	16.

Since the product of any two negative numbers is positive,* the square of a negative number is always positive. Connecting the

*See page 500 if this is not clear.

points in order from left to right with a smooth curve, we get the graph shown below.

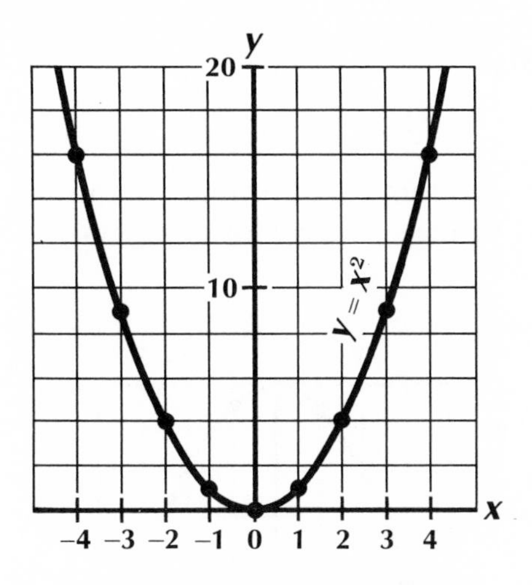

By first making a table with x numbers from −3 to +3, draw graphs of the following functions, each on a separate pair of axes. Use the same scales as shown in the example. Write each formula along its curve.

1. $y = x^2 + 2$
2. $y = x^2 - 2$
3. What do you notice about these two curves?
4. What do you think the formula of this curve is?

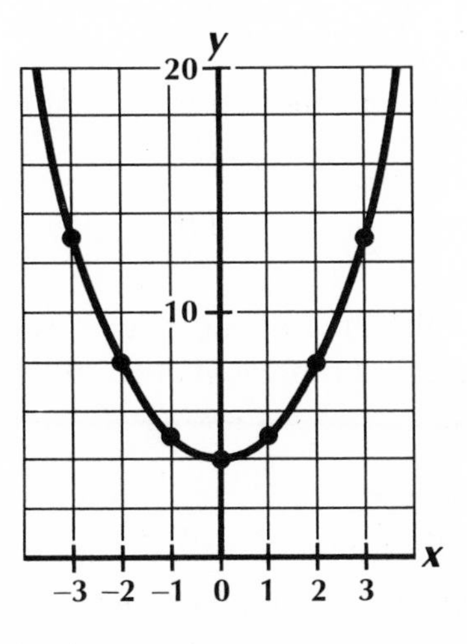

Draw the graphs of the following functions. In each formula, the

x number must be squared *first* and then multiplied by the number in front of the x.

5. $y = 2x^2$
6. $y = -2x^2$
7. What do you notice about these two curves?
8. What do you think the formula of this curve is? (Hint: It may help to make a table of numbers.)

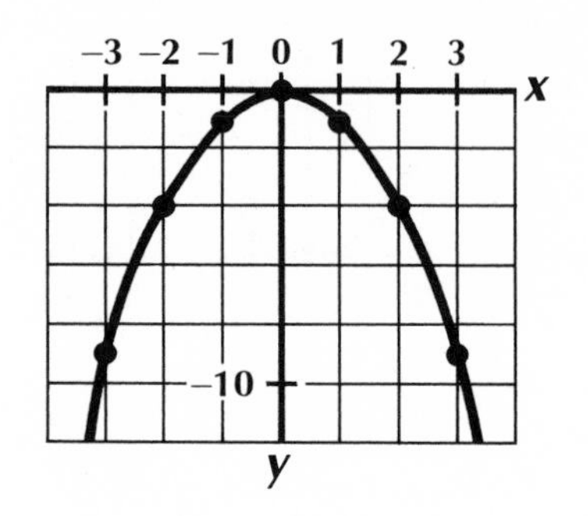

9. What does the graph of this function look like?

$$y = x^2 + x$$

Draw it and see.

Set II

Area

The area of a circle is a function of its diameter. A formula for this function is

$$a = \frac{\pi d^2}{4},$$

where a represents the area of a circle, π represents a number a little larger than 3, and d represents the circle's diameter. A table for this function is:

diameter (in feet)	0	1	2	3	4
area (in square feet)	0	0.8	3.1	7.1	12.6.

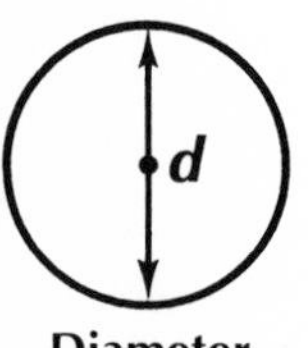

Diameter

1. Draw a graph of this function, letting the x-axis represent diameter and the y-axis represent area. Let 2 units along the x-axis stand for 1 foot.

"Of course you smell pizza pie—
it's right there in front of you!"

2. What would be the approximate diameter of a giant pizza having an area of 10 square feet?

If the earth were flat, then "on a clear day you could see forever." Since the earth is spherical, it is impossible to see past a circle called your "horizon."

On the ocean the height that you are above the water is a function of the distance you can see out to sea: *

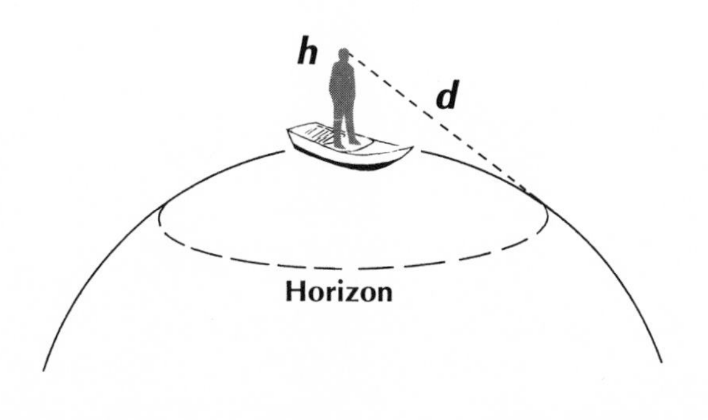

distance out to sea (in miles)	0	3	6	9	12	15
height (in feet)	0	6	24	54	96	150.

* The opposite is also true: the distance you can see out to sea is a function of your height above the water.

3. Draw a graph, letting the x-axis represent distance out to sea and the y-axis represent height. Let 1 unit along the y-axis stand for 10 feet.
4. About how high would you have to be in order to see 10 miles out to sea?
5. Which one of these formulas for this function is correct?

$$h = \frac{2}{3}d \qquad h = \frac{2}{3}d^2$$

Set III

When a ball is thrown straight up in the air, its height is a function of the time. If the ball is thrown with a speed of 40 feet per second, the formula for this function is: $h = 10t - t^2$, where h is the height in feet and t is the time in quarter seconds.

1. Make a table of 11 points, letting $t = 0, 1, 2, \ldots, 10$, and use the formula to find h.
2. Draw a pair of axes like the pair shown here and plot the 11 points from your table.

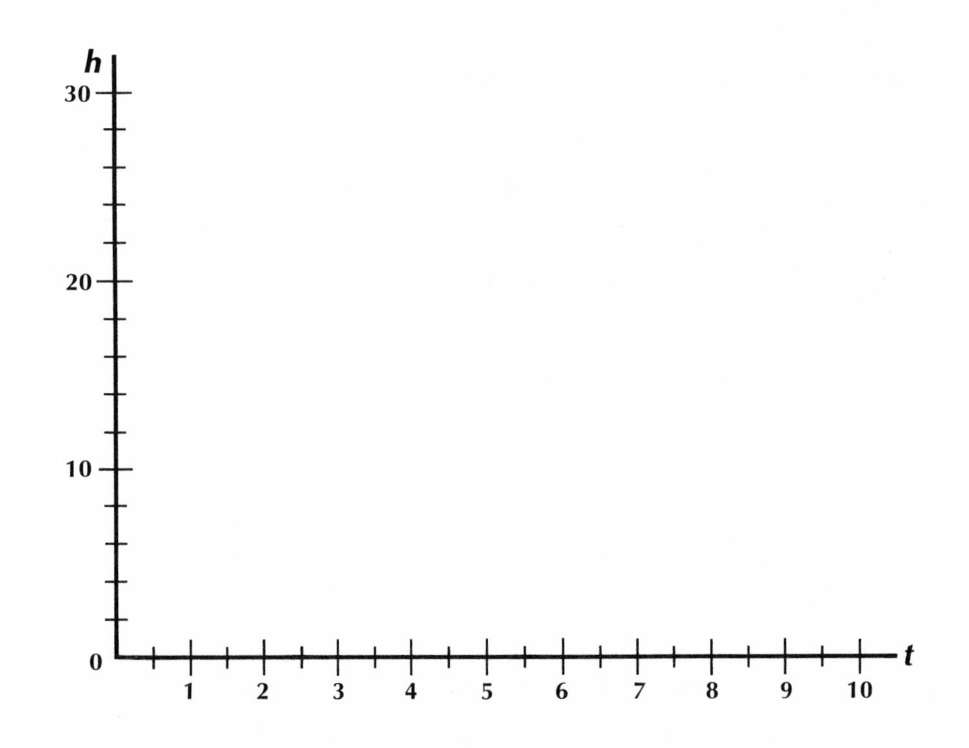

3. Draw a smooth curve through the 11 points.

Use your graph to answer these questions:

4. How high does the ball go?
5. How many seconds is the ball in the air? (Remember that the time is measured in quarter seconds.)
6. How does the time the ball spends going up compare to the time it spends coming down?

Lesson 5

More Functions with Curved Graphs

HOW big is the sun? It appears to be the same size as the moon and during a total eclipse of the sun, the moon is just large enough to cover it completely. The Greeks had some strange ideas about the relative sizes and distances of the sun and moon. Democritus, who lived in the fourth century B.C., thought that the sun was smaller than the earth. In fact, a century before, the people of Athens were surprised by one astronomer's suggestion that the sun might be as large as the country of Greece. We now know that the sun has a diameter of more than 100 times that of the earth; it is about 864,000 miles across. (However, even at that size, the sun is only a medium-sized star compared to others in our galaxy.)

The *apparent* size of the sun is a function of the distance from which we look at it. If the distance from the earth to the sun were half as great, the sun would appear to be twice as large. A formula for this function is

$$w = \frac{1}{d},$$

where w is the apparent width of the sun and d is the relative distance from the sun.

How large would the sun appear to be from some planet other than the earth? From Mercury, the closest planet to the sun, it would seem the largest and from Pluto, the most distant planet,

By permission of Johnny Hart and Field Enterprises, Inc.

the smallest. It would be interesting to know just how large and how small. Here is a table for some of the planets.

	Planet					
	Mercury	Venus	Earth	Mars	Jupiter	Saturn
Relative distance of planet from sun	0.4	0.7	1.0	1.5	5.2	9.5
Apparent width of the sun	2.5	1.4	1.0	0.7	0.2	0.1

In the table, the distance of each planet from the sun is given relative to the earth's distance from the sun, chosen as 1 unit. For example, the distance of Mars from the sun is 1.5 times the earth's distance. The widths of the sun are given relative to its apparent width as seen from the earth. For example, the sun's width as seen from Mercury is 2.5 times its width as we see it.

What does the graph of this "apparent width" function look like? It is shown on the following page.

A set of circles has been added to the graph to represent the apparent size of the sun as seen from each planet. The parts of the curve between the points from the table represent apparent widths of the sun as seen from positions between the planets. Notice that the curve gets closer and closer to the distance axis as we look toward the right. If the graph were extended to in-

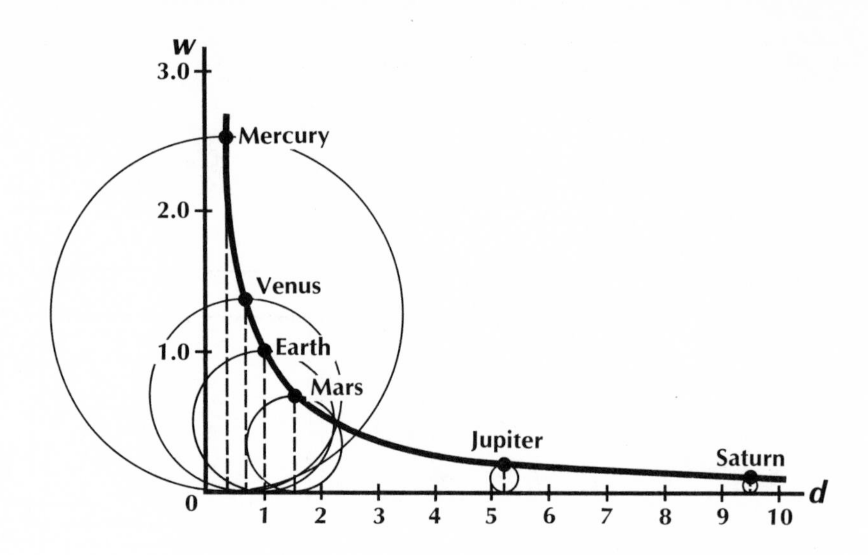

clude the remaining three planets, the curve would be very close to the axis by the time we got to Pluto. The apparent size of the sun as seen from Pluto is smaller than a period on this page!

EXERCISES

Set I

You have learned that in order to get a complete picture of a curved line graph, it is necessary to include points with negative coordinates. For example, a table for the function

$$y = \frac{12}{x}$$

could include:

x	−6	−5	−4	−3	−2	−1	0	1	2	3	4	5	6
y	−2	−2.4	−3	−4	−6	−12	*	12	6	4	3	2.4	2.

When we plot these points to make a graph, a strange thing happens. Examine the graph at the top of the facing page.

As we look at the graph from left to right, we see that there is a break in the curve along the y-axis. The curve approaches the y-axis on both sides, but has no point where $x = 0$.

* No number will fit here. Why not?

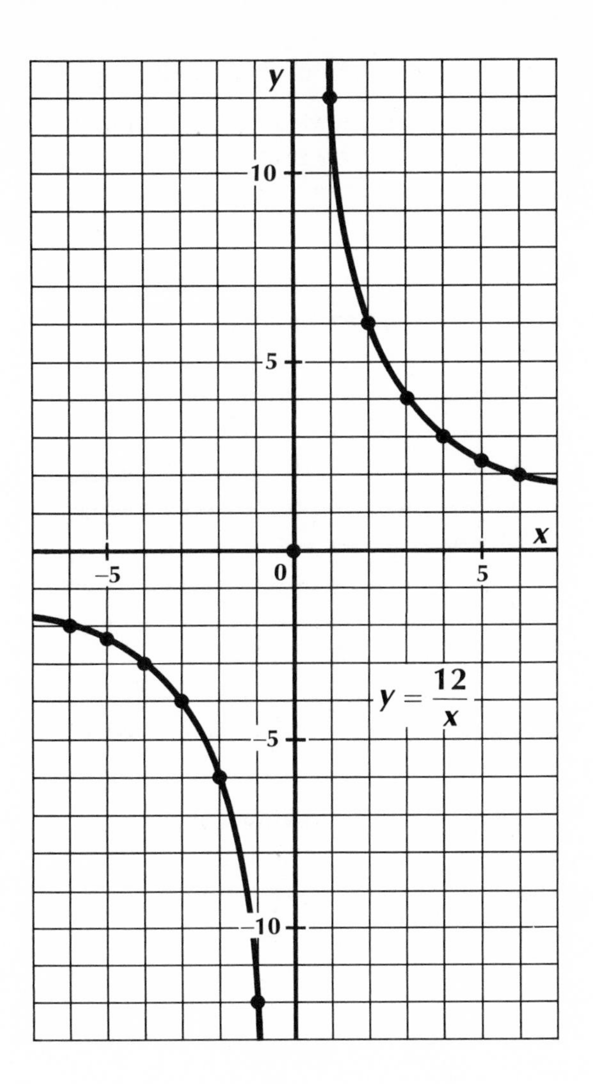

Use the formulas of the following functions to complete the tables. Draw a graph for each and write each formula on its graph.

1. $y = \dfrac{6}{x}$

x	−6	−5	−4	−3	−2	−1	0	1	2	3	4	5	6
y		−1.2					–						

2. $y = \dfrac{10}{x}$

x	−10	−8	−6	−4	−2	0	2	4	6	8	10
y			−1.7			–			1.7		

3. What do you notice about these two graphs?
4. The formula of the curve in the graph below is similar to the formulas for the two graphs you have just drawn. Can you figure out what it is?

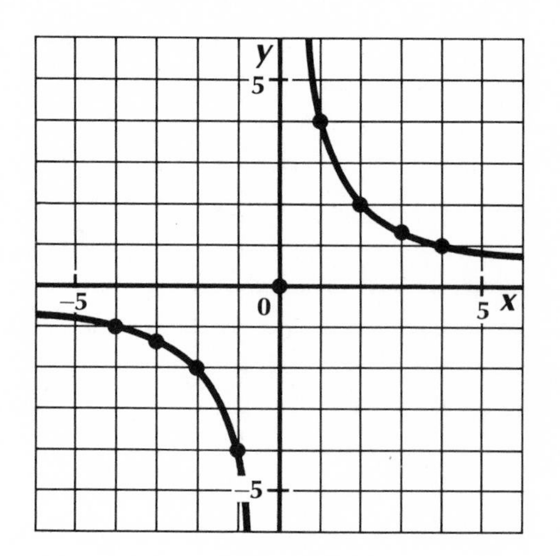

Draw the graphs of the following functions, completing the tables first. In each graph, let 1 unit on the y-axis represent 10. Notice that the cube of a negative number is also negative.

5. $y = x^3$

x	−4	−3	−2	−1	0	1	2	3	4
y	−64				0				

6. $y = x^3 + 10$

x	−4	−3	−2	−1	0	1	2	3	4
y	−54	−17			10				

7. What do you notice about these two graphs?
8. What do you think is the formula of the curve in the graph at the top of the facing page?
9. What does the graph of this function look like? Draw it and see.

$$y = x^4$$

Again let 1 unit on the y-axis represent 10.

x	−3	−2	−1	0	1	2	3
y	81						

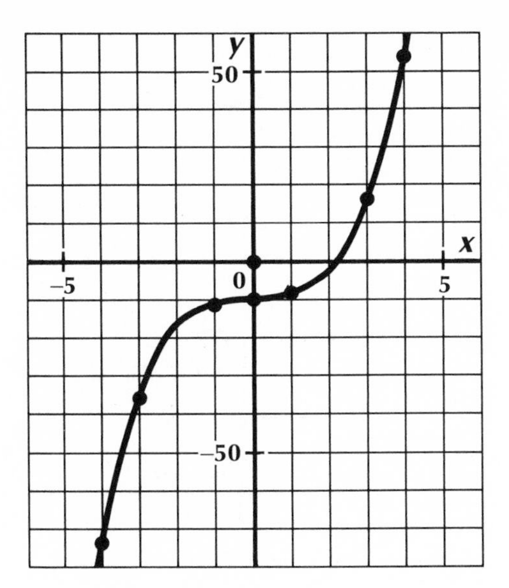

Set II

The wavelength of a sound is a function of its frequency. A high fidelity system can reproduce sounds as low as 15 vibrations per second (with waves more than 70 feet long) and as high as 30,000 vibrations per second (with waves less than half an inch long). The formula for this function is

$$w = \frac{1{,}100}{f},$$

where w is the wavelength in feet and f is the frequency in vibrations per second.

1. Complete the following table for this function.

f	10	20	30	40	50	60	70	80	90	100
w	110		37	28			16	14		

2. Draw a graph for the function, $w = 1{,}100/f$, using your table. Let the x-axis represent frequency and the y-axis represent wave length. Let 1 unit along each axis stand for 10.

3. Some animals can hear much higher sounds than man can hear. Bats, for example, can hear frequencies as high as

By permission of Johnny Hart
and Field Enterprises, Inc.

120,000 vibrations per second. Are the wavelengths of these sounds very long or very short?

The weight of a cube of ice is a function of the length of its edge. The formula is

$$w = 0.033e^3,$$

where w is the weight in pounds and e is the length of the edge in inches. A table for this function is

e	0	2	4	6	8	10	12
w		0.3	2	7	17		57

4. What are the two missing numbers in the table?

5. How much does one cubic foot of ice weigh? (The answer is in the table.)

6. Draw a graph for the function, $w = 0.033e^3$, using the table above and letting the x-axis represent edge length

and the y-axis represent weight. Let 1 unit along the y-axis represent 5 pounds.

7. Use your graph to estimate the weight of a cube of ice that measures 7 inches on an edge.

Courtesy of Wham-O Manufacturing Co., Inc.

Set III

The Wham-O Company introduced the "Super Ball" to the world in 1965. It is made of a highly compressed synthetic rubber and when dropped on a hard floor from a height of several feet, it will keep bouncing for more than a minute. Each bounce is about 90 percent as high as the previous one.

The height of each bounce is a function of the bounce number and a table for this function is:

n (bounce number)	0	1	2	3	4	5	6	7
h (height of bounce in feet)	5	4.5	4.1	3.6	3.3	3.0	2.7	2.4

n	8	9	10	11	12	13	14	15
h	2.2	2.0	1.8	1.6	1.4	1.3	1.1	1.0

The table shows that the ball was dropped from a height of 5 feet, came up 4.5 feet on the first bounce, 4.1 feet on the second bounce, and so on.

1. Draw a graph of this "bouncing ball" function, showing the bounce numbers along the x-axis and the heights along the y-axis. Let 2 units on the y-axis represent 1 foot.
2. Do you think that it makes any sense to draw a curve through the points on your graph?
3. On the basis of your graph, predict what happens to the Super Ball after the 15th bounce.

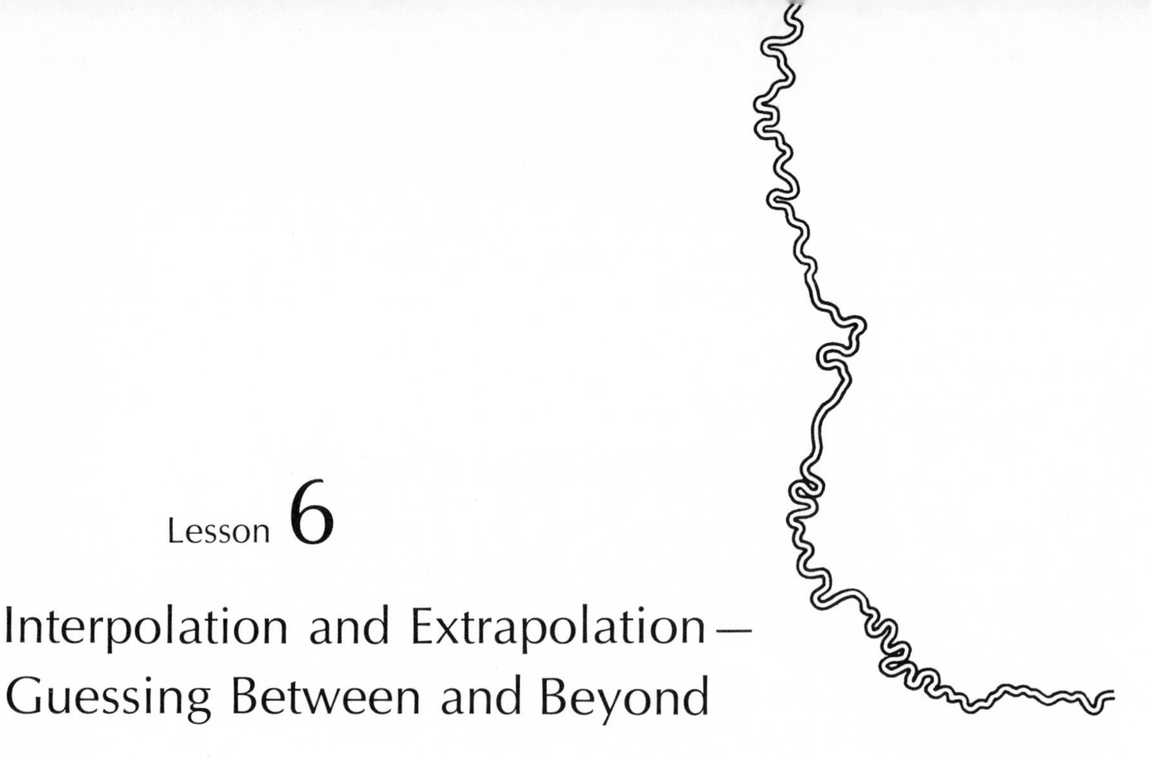

Lesson 6

Interpolation and Extrapolation—Guessing Between and Beyond

MARK Twain once remarked that in eternity he planned to spend eight million years on mathematics. In his book *Life on the Mississippi,* Twain made a strange prediction about the future of the Mississippi River. The river is extremely crooked with many curves, as you can see from the map showing part of it. From time to time "cut-offs" take place in which the river changes its course from a wide bend to a more direct path. As a result, the length of the Mississippi is becoming shorter and shorter. Twain gave some figures: "The Mississippi between Cairo and New Orleans was 1,215 miles long 176 years ago. It was 1,180 after the cut-off of 1722 . . . its length is only 973 miles at present [1875]."

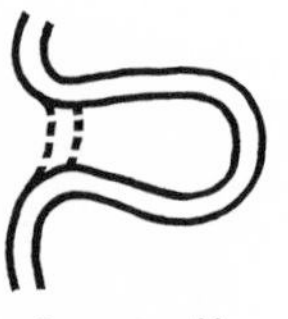

A cut-off

Since the length is a function of time, we can make a table from these numbers. Rounding them off a bit, we get:

time (the year)	1700	1720	1875
length (in miles)	1,215	1,180	975

We have no formula for this function, but we can still plot the points on a graph (top of facing page), and when we do, they seem to lie along a straight line. If we extend the line until it intersects the time axis, it looks as if sometime about the year 2600 there won't be much left of the Mississippi.

Twain jokingly made this prediction: "Any calm person, who is not blind or idiotic, can see . . . that 742 years from now

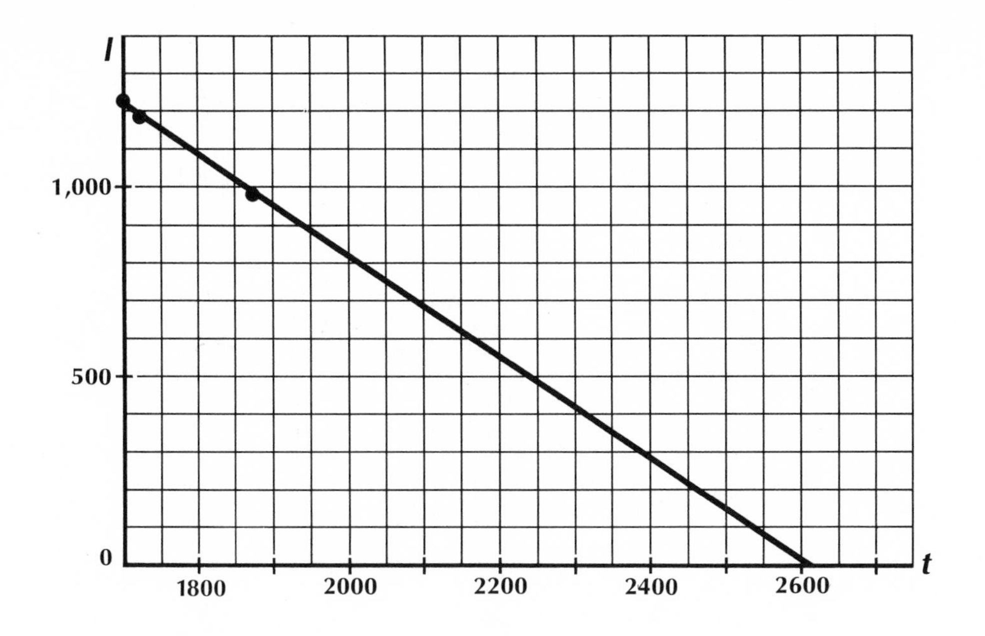

the Lower Mississippi will be only a mile and three-quarters long. . . ."

What's wrong with this reasoning? The trouble is that although the three points may *seem* to lie along a straight line, the graph must actually *curve* later on. The river cannot become any shorter after all of the bends are gone. When we draw the line past the third point, we are using a method called *extrapolation.*

► To **extrapolate** is to guess other values of a variable *beyond* the ones we know.

If the function continues to behave in the same way, then our guess may be very close; otherwise, you can see what may happen.

EXERCISES

Set I

Perhaps you have learned how to find the square root* of a number by arithmetic. The methods used are rather tedious and

* The *square root* of a given number is *that number whose square* is the given number.

"It hurts when I do square roots."

take a lot of time. We will draw a graph from which we can get approximate square roots in an easy way.

Some numbers have *exact* square roots. For instance, the square of 5 is 25, so the square root of 25 is 5 (written: $\sqrt{25} = 5$). Most numbers, however, have square roots that can be expressed only *approximately;* the square root of 10 is about 3.2, and this is the best we can do if we want to write the root to only one decimal place.

Making a table of squares of some whole numbers, we get:

x	0	1	2	3	4	5
y	0	1	4	9	16	25,

where x represents any number and y represents its square. You know that the formula for this function is

$$y = x^2.$$

We have seen the graph of this function before (in Lesson 4).

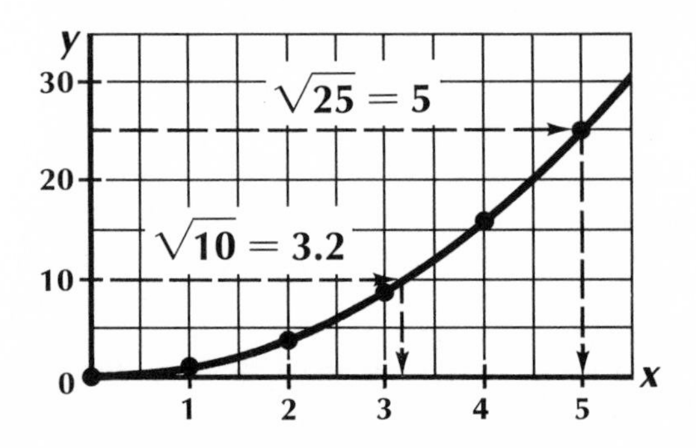

To find the square root of a number from this graph, we find the number on the y-axis, read straight across to the curve and then directly down to the x-axis. When we look for the square root of 25, we come to one of the original points we had plotted; looking for the square root of 10 leads us to part of the curve *between* two plotted points, and the root is between 3 and 4. When we estimate that it is about 3.2, we are *interpolating.*

► To **interpolate** is to guess other values of a variable *between* values we know.

1. What is the difference between interpolation and extrapolation?
2. Copy the table of squares and continue it through the square of 10.
3. Copy the graph of the function $y = x^2$ and extend it to include the rest of the points in your table. Draw as accurate and as smooth a curve as you can through the points.
4. Use your graph to estimate the square roots of 20, 30, 40, . . ., 90 in the same way as has been illustrated for the square root of 10. Estimate each root to one decimal place and write each root on the graph.
5. Is it possible, using the graph you have drawn, to estimate the square roots of other numbers, such as 15 or 37?
6. Do you think square roots can be accurately read from your graph to more than one decimal place?

Set II

We hear a lot these days about the "population explosion." The earth's population is a function of time; here is a table of numbers for this function:

time (the year)	1650	1700	1750	1800	1850	1900	1950
population (in billions)	0.5	0.6	0.7	0.9	1.1	1.6	2.5.

1. Draw a graph of this function, labeling your axes as shown on the following page.
2. Use your graph to estimate the earth's population in 1925.

Courtesy of Irv Phillips; by permission of Publishers-Hall Syndicate.

3. Did you use *interpolation* or *extrapolation* in making your estimate?

4. Has the earth's population passed 3 billion yet?

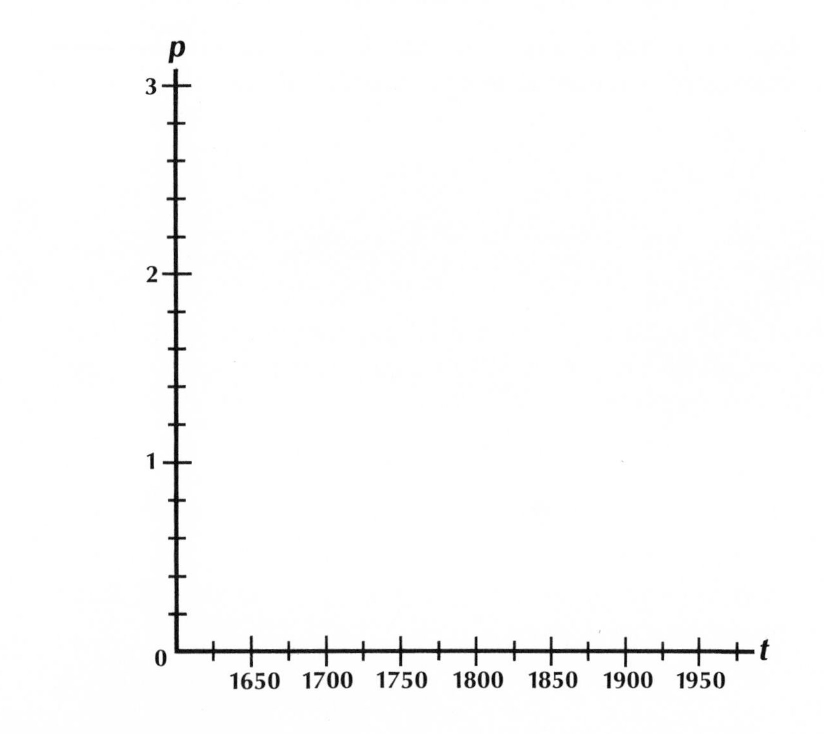

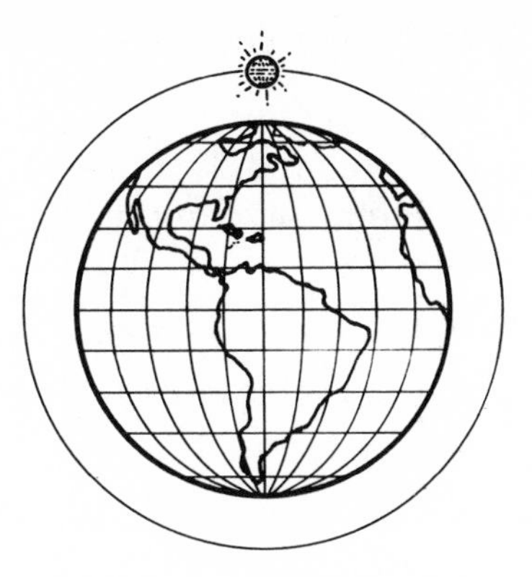

Set III

Telstar, the first privately owned communications satellite, was launched in 1962. The time that it takes a satellite at a certain altitude to circle the earth is a function of its speed. The formula for an altitude of 100 miles is

$$t = \frac{1{,}540}{s},$$

where t is the time in minutes and s is the speed of the satellite in thousands of miles per hour. A table for this function is:

s	18	19	20	21	22
t	86	81	77	73	70.

1. Draw a graph of this function. A convenient way to label your axes is shown below. Notice the wavy lines near the point where they meet. They show that we are skipping ahead in the numbering in order to save space.

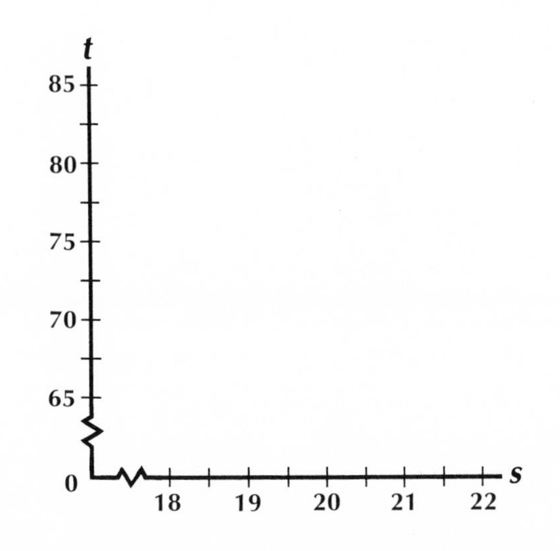

2. Do you think the graph you have drawn is a straight line or a curve?

3. Estimate by interpolating from your graph the time it would take a satellite traveling at 20,500 miles per hour to circle the earth.

4. Try to estimate by extrapolating from your graph the speed a satellite in the same orbit must have to circle the earth in one hour.

Chapter 3 / Summary and Review

IN this chapter we have become acquainted with:

The idea of a function *(Lesson 1)* A pairing of two number sequences, in which any term of the first sequence determines exactly one term of the second is a function.

A function may be represented by:

a table of numbers,
a formula,
a graph.

The coordinate graph *(Lesson 2)* The coordinate graph was invented by the French mathematician and philosopher, René Descartes.

Each point on a coordinate graph is located by a pair of numbers, called its coordinates, which are the distances of the point from the x- and y-axes.

Functions with linear graphs *(Lesson 3)* To draw the graph of a function for which you know the formula, it is first necessary to write a table. The pairs of numbers in the table are coordinates of points of the graph.

Functions with curved graphs *(Lessons 4 & 5)* In drawing a curved line graph, it is necessary to include points with negative

x-coordinates to get a complete picture.

Here are the formulas and graphs of some functions we have studied.

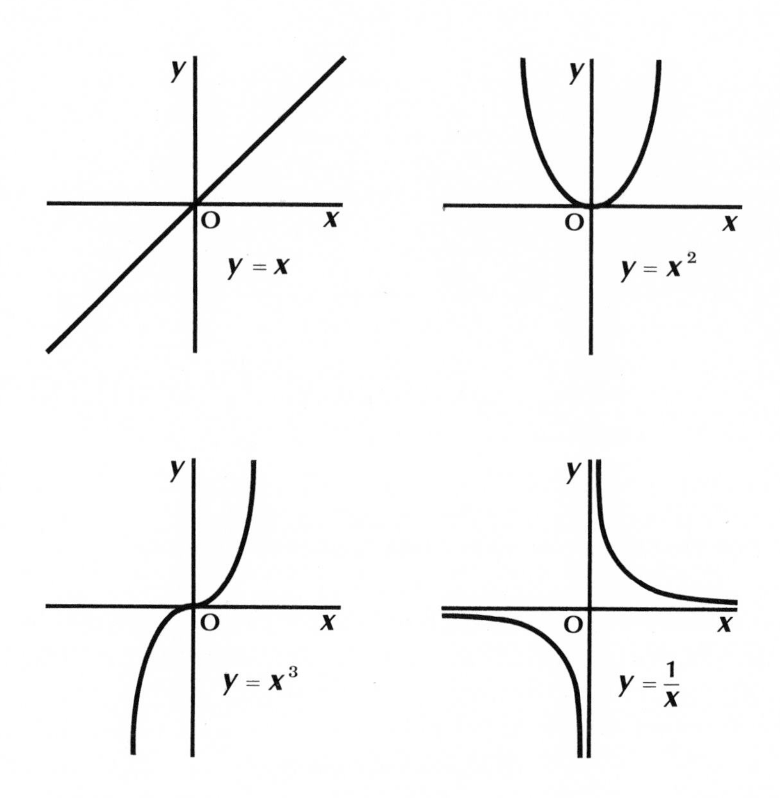

Interpolation and extrapolation *(Lesson 6)* Interpolation is guessing another value of a variable *between* values we know; extrapolation is guessing another value *beyond* ones we know.

EXERCISES

Set I

Write a formula for the function represented by each of these tables.

1.

x	1	2	3	4	5
y	3	6	9	12	15

2.

x	0	1	2	3	4
y	5	6	7	8	9

3.

x	2	3	4	5	6
y	4	9	16	25	36

4.

x	0	10	20	30	40
y	7	27	47	67	87

Use the formulas for the following functions to complete the tables.

5. $y = x + 6$

x	−4	−3	−2	−1	0	1	2	3	4
y	2								

6. $y = x^2 + 6$

x	−3	−2	−1	0	1	2	3
y	15			6			

7. Draw and label a pair of axes like the pair shown here. Use your tables to draw the graphs of $y = x + 6$ and $y = x^2 + 6$ on your pair of axes. Write each formula along its line or curve.

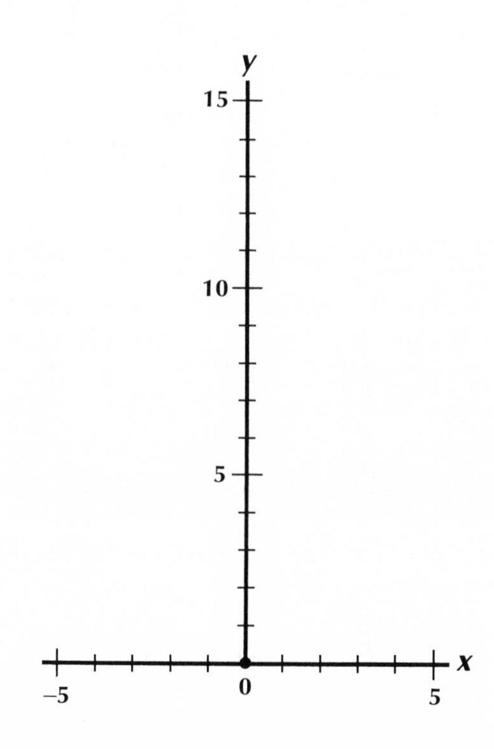

By permission of Johnny Hart and Field Enterprises, Inc.

The lever is a simple machine, which has been known for thousands of years. By using a lever, it is possible to move a heavy weight with a small force.

Suppose a weight of 300 pounds is located 2 feet from a lever's pivot point. The force necessary to move the weight is a function of the distance it is applied from the other side of the pivot. In this case, the formula for the function is

$$f = \frac{600}{d},$$

where f is the force in pounds and d is the distance in feet from the pivot.

8. Complete the following table for the function $f = 600/d$.

d	15	12	10	8	6	4
f	40	▒	▒	▒	▒	▒

9. Draw a graph, letting the x-axis represent the distance and the y-axis represent the force. Label the axes as shown below.

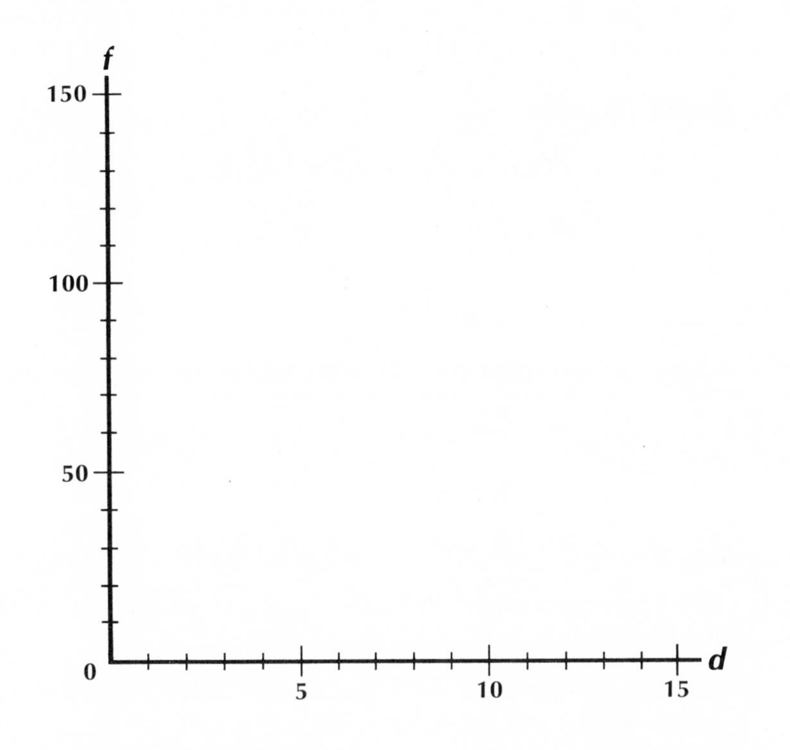

10. Use your graph to estimate the force at a distance of 7 feet from the pivot that would be needed to move the weight.

11. In estimating, did you *interpolate* or *extrapolate?*

Set II

The following puzzle was given on a radio quiz program*: "You and I have 35 apples together. You have $\frac{2}{5}$ of what I have. How many do we each have?"

---

*"Testing One Two Three," KNX radio, Los Angeles.

The first person calling in who was able to figure it out would win five dollars. yet no one of the three people given the puzzle could give the correct answer.

One way to solve it is with a graph. If we let x represent the number of apples I have and y represent the number you have, we can write the first statement,

You and I have 35 apples together,

as:

$$x + y = 35 \quad \text{or} \quad y = 35 - x.$$

The second statement,

You have $\frac{2}{5}$ of what I have,

would be:

$$y = \tfrac{2}{5}x.$$

1. Copy and complete the following tables for these formulas.

$y = 35 - x$

x	0	5	10	15
y	35	30		

$y = \frac{2}{5}x$

x	0	5	10	15
y	0	2		

2. Plot the points of the first table on a pair of axes on which 1 unit on each axis represents 2 apples.

3. Draw a line through the 4 points and extend it across the graph. Label the line:

$$y = 35 - x.$$

4. Does the line meet the x-axis in the right place? Explain your answer.

5. Plot the points of the second table on the *same* pair of axes.

6. Draw another line through these 4 points and extend it across the graph. Label the line:

$$y = \tfrac{2}{5}x.$$

7. What are the coordinates of the point where the 2 lines intersect?

8. What is the answer to the puzzle?

Set III

If several points are chosen on a circle and connected with straight lines, the number of lines that can be drawn is a function of the number of points. For instance, if 2 points are chosen, then 1 line can be drawn through them. If there are 3 points, 3 lines can be drawn, and 6 lines can be drawn through 4 points. The circles below show the results of joining from 2 to 8 points with every possible line.

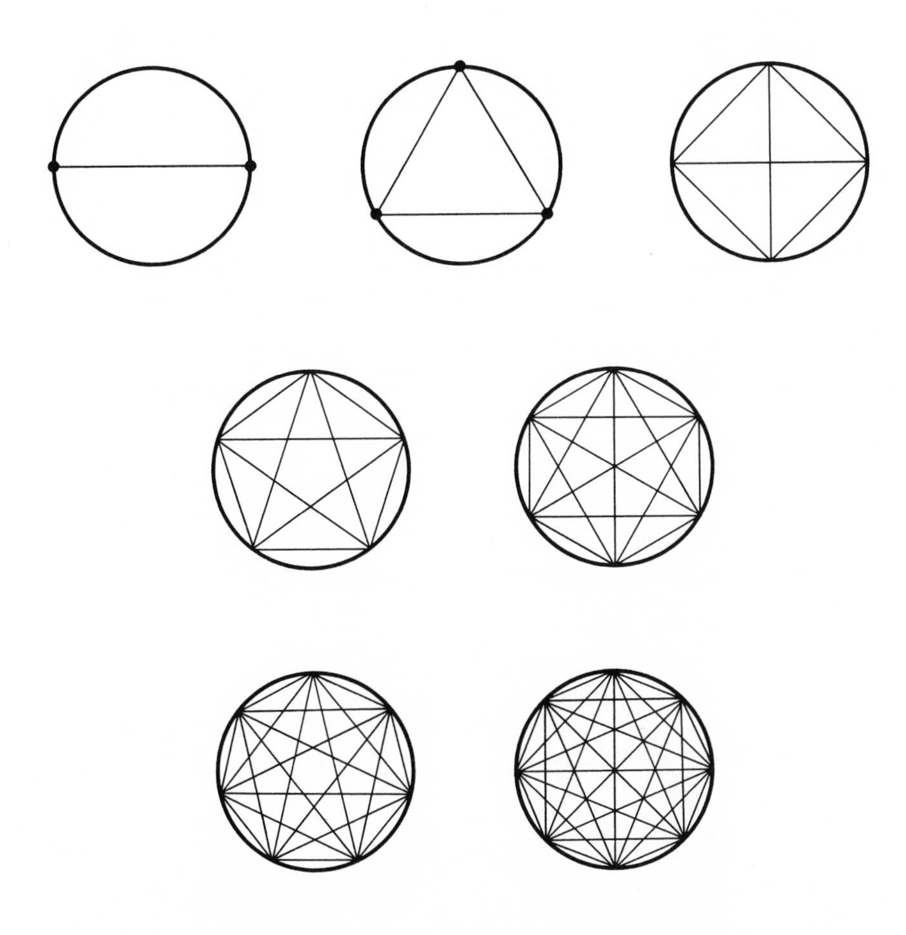

Here is a table showing the number of lines in each:

number of points, p	2	3	4	5	6	7	8
number of lines, l	1	3	6	10	15	21	28.

1. Draw a graph of this function, labeling your axes as shown on the following page.

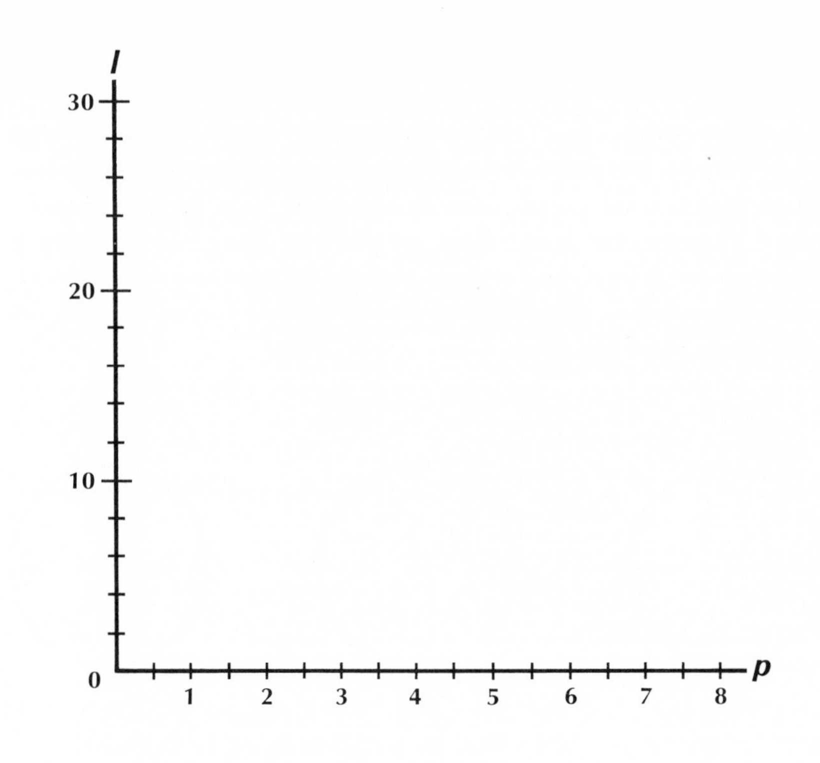

2. What do you think the value of l would be when $p = 1$? Does your answer make any sense?

3. Use your graph to decide the value of l when $p = 6.5$. Does your answer mean anything?

The circle below has lines drawn on it from each of 21 points.

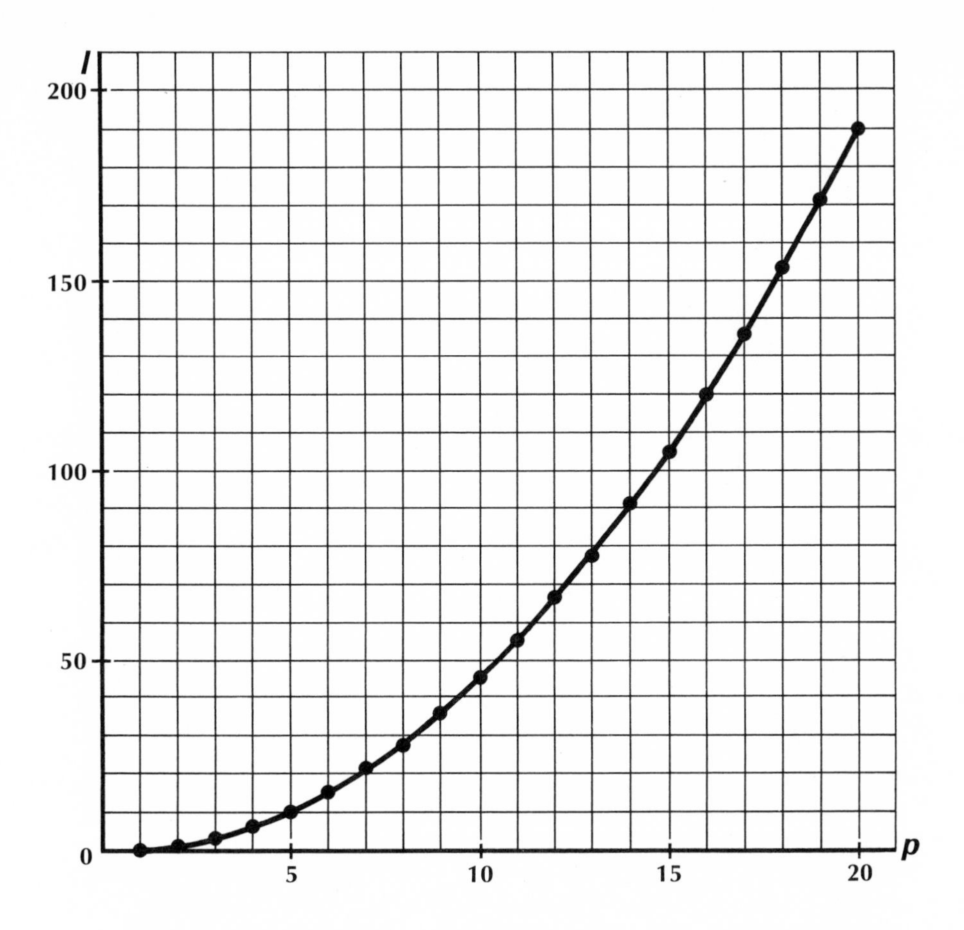

4. Here is a graph of the same function as given on page 135 drawn on a different scale and including circles having up to 20 points. Use it to estimate the number of lines in the circle with 21 points by extrapolation.

Chapter 4

LARGE NUMBERS AND LOGARITHMS

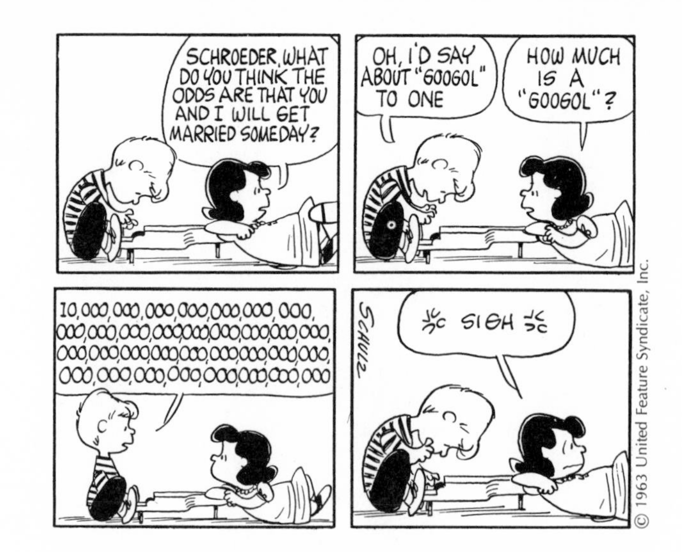

Lesson 1

Numbers Beyond Our Imagination

WHAT is the largest number you can think of? A billion? A trillion? If you wrote down a one and started adding zeros, it wouldn't take you very long to write a number so big that it would be beyond the names for large numbers that even a mathematician knows. Schroeder has thought of a googol, which, believe it or not, is the accepted mathematical name for the number 1 followed by a hundred zeros!* Although it has a name, this number is far too big for anyone to be able to understand it. In fact, it is difficult for us to appreciate the size of a number even as "small" as a million, since we have had no personal experience with such a number.

Do you know how large a million is? It is easy to write 1,000,000, but how big is that? If you counted to one million and were able to name a number every second without stopping, it would take you nearly 12 days! And how long is a million days? A million days ago was in the 8th century B.C. How far is a million inches? Almost 16 miles.

If our idea of the size of a million is fuzzy, our notions of larger numbers must be even fuzzier. To take another example, how large is a billion? Oddly enough, it depends upon where you

* The name "googol" was invented by a young nephew of the American mathematician Edward Kasner when the boy was asked to make up a name for a very large number.

live! In the United States a billion is 1,000,000,000, while in England it is 1,000,000,000,000. Perhaps the reason for this difference is that until recently in history there was no need for such a large number and so everyone has not yet agreed on how to name it. Using the smaller version, a billion seconds is still a long time; a billion seconds from now is in the beginning of the 21st century.

Is there much difference in the size of two numbers such as

$$100000000 \quad \text{and} \quad 1000000000?$$

At a glance they look about the same, yet when we add a zero to the first number to give the second, we have multiplied it by ten. Don't be fooled into thinking that since zero equals nothing, adding a zero to a number doesn't make much difference.

Each number in the sequence

$$1 \quad 10 \quad 100 \quad 1{,}000 \quad 10{,}000 \quad 100{,}000 \quad 1{,}000{,}000 \ \ldots$$

is ten times larger than the number before. Another way to write this sequence is

$$1 \quad 10 \quad 10^2 \quad 10^3 \quad 10^4 \quad 10^5 \quad 10^6 \ \ldots,$$

and this suggests that using exponents is a compact way of writing large numbers. Notice that the *exponent* of each 10 is also the *number of zeros* that follow the 1 if the number is written the long way. This pattern suggests that it makes sense to write

$$10 \text{ as } 10^1$$

and even

$$1 \text{ as } 10^0 \text{ (1 followed by 0 zeros)}$$

The beginning of the sequence can then be written as

$$10^0 \quad 10^1 \quad 10^2 \quad 10^3 \ \ldots .$$

Here is a table listing the names and exponential forms of some large numbers.

10^2	hundred	10^{18}	quintillion
10^3	thousand	10^{21}	sextillion
10^6	million	10^{24}	septillion
10^9	billion	10^{27}	octillion
10^{12}	trillion	10^{30}	nonillion
10^{15}	quadrillion	10^{33}	decillion

EXERCISES

Set I

1. How many zeros follow the 1 if the number one quadrillion is written the long way?
2. Write in exponential form:
   a) ten million,
   b) one hundred septillion,
   c) one thousand decillion.
3. Write a googol, using the exponential form.
4. What does 10^0 mean?
5. Guess, without counting anything, how many dots are in this figure. Write your guess as a power of 10.

6. The human brain contains about 10^{10} cells, called neurons, and is far more complex than any computer that has ever

been built. Write the number of cells the long way and then name the number in words. If this number was represented by a line of dots like this,

. . . .,

it would stretch more than half way around the earth!

7. A tourist from England visited Yankee Stadium to see a baseball game. He didn't understand the game at all and left when the scoreboard read:

   1 0 0 0 0 0 0 0 0
   1 0 0 0 0 0 0 0 0

   When a boy outside the gate asked him the score, he answered: "It's up in the millions." How did the man get so confused?

8. Our solar system is part of the Milky Way galaxy, which is estimated to contain more than one hundred billion stars. Write this number in two different ways.

Photograph from the Mount Wilson and Palomar Observatories

Star clouds in the Sagittarius region of our galaxy, the Milky Way.

9. Suppose only one star out of every million was the sun of a planet with intelligent life on it. Approximately how many planets with intelligent life would there be in the Milky Way?

10. Water molecules are so small that if you could pour a *million* of them into a quart bottle every second and never stop until the bottle was full, it would take more than 10^8 years. If you poured at the rate of a *billion* per second instead, how long would it take? (Since you are pouring at a faster rate, of course it would take less time.)

11. How long at the rate of a *trillion* per second?

You know that 10^3 means "1 followed by 3 zeros," or 1,000. Don't assume, however, that this is true if the number is written as a power of a number other than 10. For example, to find out what 100^3 means, you should think: $100^3 = 100 \times 100 \times 100 = 1{,}000{,}000$.

12. Which of these numbers is largest?

$$100^4 \qquad 1{,}000^3 \qquad 10{,}000^2$$

Write each number the long way and see.

Notice that $100^4 = (10^2)^4$ and that $100^4 = 10^8$, since you have just written 100^4 as 100,000,000. So $100^4 = (10^2)^4 = 10^{2\times4} = 10^8$.

13. Now use this method to change $1{,}000^3$ and $10{,}000^2$ to powers of 10.

14. Can you express a googol squared as a power of 10?

15. How about a googol cubed?

Set II

The Greek mathematician Archimedes, who lived in the third century B.C., wrote a book called *The Sand Reckoner* in which he devised a method for forming large numbers.

1. The Greek name for 10,000 was "myriad." Write this number, using an exponent.

2. Archimedes began by thinking of a "myriad of myriads." What did he mean by this?

3. What would we call this number?

4. Write it in exponential form.

5. Archimedes called this number an "octade." Can you tell why? (Hint: What do an octet and an octopus have in common?)

6. He named the number 10^{16} "the second octade." What is our name for this number?

Notice that $10^{16} = (10^8)^2$, so that "the second octade" could also be called "an octade squared."

7. What number do you suppose Archimedes meant by "the third octade"? Write it as a power of 10 and give our name for this number.

8. An extraordinarily large number thought of by Archimedes was "an octade of octades," which we would write as

$$10^{800,000,000}$$

If this number were to be written out in full, what would you have to do?

Notice that it may be quite easy to *write* a large number such as a million in full: 1,000,000; it is *not* easy, however, to count from one to a million.

9. Do you think that a computer could *print out* Archimedes' "octade of octades" in the long form?

10. Do you think it could *count* up to this number?

Set III

A telephone cable across the Atlantic ocean connects Newfoundland with Scotland. To keep the sound loud enough to hear, there are 51 amplifiers spaced fairly evenly along its length. Each one of these amplifiers increases the signal strength by about a million times to make up for fading of the signal along the cable. The first amplifier on the way from Newfoundland to Scotland strengthens the signal 10^6 times. After it passes through the second amplifier, it has been boosted $10^6 \times 10^6$ times.

1. How many times is that: 10^{12} or 10^{36}? Write the numbers out and see.

2. Upon passing through the third amplifier, the signal has been strengthened $10^6 \times 10^6 \times 10^6$ or $(10^6)^3$ times. Altogether how many times has it been boosted by the time it passes through the 51st amplifier just before reaching Scotland?

By permission of Johnny Hart and Field Enterprises, Inc.

Lesson 2

The Mathematics of Large Numbers

SCIENTISTS in many fields need to write large numbers and make calculations with them. If these numbers are written the "long way," they are awkward to work with and it is easy to make a mistake in putting down the proper number of zeros. Here are some examples. The astronomer uses a unit of distance called a light-year, which is about 5,880,000,000,000 miles. The physicist has found that x-rays have frequencies of more than 10,000,000,000,000,000 vibrations per second. The chemist can figure out that an ounce of gold contains approximately 8,650,000,000,000,000,000,000 atoms.

You have already had some experience with writing numbers in a compact way by using exponents. Let's see how these numbers from science could be written in a more convenient form.

Peter says that the sun is "93 million miles high"; we can write this number as 93,000,000, or, since

$$93{,}000{,}000 = 93 \times 1{,}000{,}000$$

as

$$93 \times 10^6.$$

It is also true that

$$93{,}000{,}000 = 9.3 \times 10{,}000{,}000,$$

so the sun's "long fall" can be written as

$$9.3 \times 10^7$$

miles. When we write it *this* way, the number is said to be in *scientific notation.*

► When a number is written as some number, which is at least 1 but less than 10, multiplied by 10 raised to some power, it is expressed in **scientific notation.**

Compare 93,000,000 and 9.3×10^7. In the scientific form, the decimal point has been moved from the end of the number to just after its first digit. The point has been moved 7 decimal places and the 7 becomes the exponent of the 10.

$$9_{\uparrow}3,000,000 = 9.3 \times 10^7$$

7 6 5 4 3 2 1

Following the same procedure, the astronomer's "light-year" number, 5,880,000,000,000, can be written as 5.88×10^{12}; the physicist's "x-ray" number, 10,000,000,000,000,000, can be written as 1×10^{16}; and the chemist's "gold atom" number, 8,650,000,000,000,000,000,000, can be written as 8.65×10^{21}. It is easy to see that the larger the number, the greater the advantage of writing it in scientific notation.

The biologist knows that a single red cell of human blood contains 2.7×10^8 hemoglobin molecules. How many is that? We can do what we did before, but in reverse, to find out.

$$2.7 \times 10^8 = 2\,7\,0\,0\,0\,0\,0\,0\,0_{\uparrow}$$

1 2 3 4 5 6 7 8

Writing in a couple of commas, we have 270,000,000, or two hundred and seventy million hemoglobin molecules in one red cell!

EXERCISES

Set I

1. How do you think the number of birds in the picture on the facing page compares to the number of students who go to your high school? Make an estimate of the number and write it in scientific notation.

2. The earth picks up approximately 2.6×10^7 pounds of dust from the sky each day. Write this number in the long form and then name the amount in words.

Courtesy of The American Museum of Natural History.

Peruvian booby birds on South Guanape Island, Peru.

3. It would take 3,000,000,000,000,000,000,000,000,000 candles to give as much light as the sun. Write this number in scientific notation.

4. The number of grains of sand in Malibu Beach, California, is about 1×10^{14}. Write this number in the long form and then name the amount in words.

5. The weight of the water in all of the oceans of the world is 1,580,000,000,000,000,000 tons. Write this number in scientific notation.

6. The number of different hands which it is possible for you to be dealt in a game of bridge is about 6.35×10^{11}. Write this number in the long form and then name it in words.

7. The *total* number of grains of wheat for which the inventor of chess is said to have asked the King of Persia* was

---

* See the story on page 54.

almost 18,450,000,000,000,000,000. Write this number in scientific notation.

8. In 1626, Peter Minuit paid the Indians of New York \$24 for Manhattan Island. If today he completely covered Manhattan with dollar bills, it would take more than 5.94×10^9 of them and this would not be nearly enough to pay for the island (including its property)! How much money is this? Name the amount in words.

9. Proxima Centauri, the nearest star beyond the sun, is 25 trillion miles away. One way of writing this distance is 25×10^{12} miles, but this number is not in scientific notation. Why not? Rewrite it in scientific notation. (Hint: $25 = 2.5 \times 10$; $10 \times 10^{12} = ?$)

10. Inflation of the value of money is a serious economic problem. In 1946, inflation of the currency was so bad in Hungary that the gold pengo was worth 130 quintillion paper pengos. Write this number in scientific notation. (The pengo was replaced that year by another unit of money.)

A bill worth 100 quintillion pengos.

Courtesy of the Chase Manhattan Bank Money Museum.

Set II

One way to multiply two numbers written in scientific notation is to change them to their long forms before multiplying them. For example,

$$(2 \times 10^2) \times (3 \times 10^3) =$$
$$200 \quad \times \quad 3{,}000 \quad = 600{,}000.$$

Notice that the result in scientific notation is 6×10^5. We can also get that answer in this way:

$$(2 \times 10^2) \times (3 \times 10^3) = 6 \times 10^5.$$

1. When powers of 10 are multiplied, are the exponents *multiplied* or *added?*

2. Multiply the two numbers (4×10^2) and (2×10^4) by first changing them to their long forms. Then change the result to scientific notation. Does it agree with your answer to the previous question about exponents?

3. Copy and complete the following statement: "To multiply two numbers in scientific notation, ______ the numbers in front and ______ the exponents."

Notice that in this problem,

$$(5 \times 10^3) \times (6 \times 10^4) = 30 \times 10^7,$$

the result is not in scientific notation, but

$$30 \times 10^7 = 3 \times 10 \times 10^7 = 3 \times 10^8.$$

Write the results in each of the following problems in scientific notation.

4. $(8 \times 10^2) \times (5 \times 10^6)$

5. $(3 \times 10^{11}) \times (7 \times 10^5)$

6. $(9 \times 10^3) \times (1.5 \times 10^{10})$

7. The earth travels about 5.8×10^8 miles in its trip around the sun each year. What distance does it travel around the sun in 1,000 years?

8. Our solar system is about 3×10^4 light-years from the center of the Milky Way galaxy. Assuming a light-year to be approximately 6×10^{12} miles, how far are we in miles from the center of the Milky Way?

9. During one summer, it is possible for a couple of house flies to become parents and ancestors of 1.9×10^{20} flies. Suppose the Swindle Swatter Company decides to improve its business prospects by breeding 5×10^9 pairs of house flies in strategic locations all over the country. How many flies could theoretically be raised during the summer?

10. Every minute more than 8.4×10^{11} drops of water flow over Niagara Falls. Each drop contains 1.7×10^{21} molecules. How many molecules of water pass over Niagara Falls in one minute?

Set III

In one of their ads, the Volkswagen Company claims that "*exactly* 1,612,462 beans" can be put into a Volkswagen station wagon.

1. Write this number in scientific notation.
2. Is this way of writing the number more compact than writing it in the long form?

Suppose we round off the number of beans and say that a V.W. wagon will hold *about* 1,600,000 beans.

3. How does this number look in scientific notation?
4. Is this way of writing the number any shorter than writing it in the long form?
5. When a scientist deals with large numbers, do you think they are *exact* or *approximate?*
6. For which kind of numbers is scientific notation more appropriate: exact numbers or approximate numbers?

Lesson 3

An Introduction to Logarithms

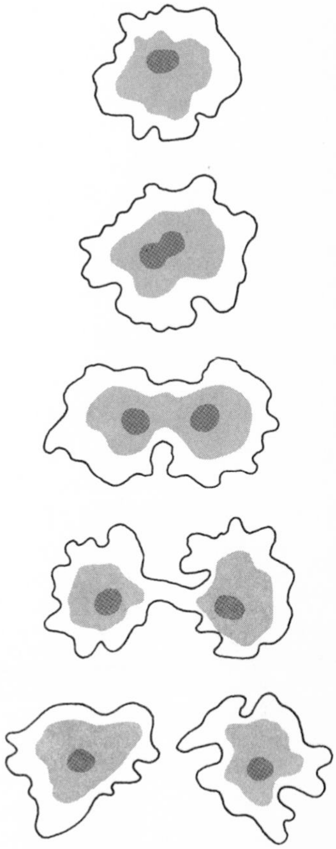

AN amoeba can do an unusual mathematical trick: it multiplies by dividing! After it has grown to a certain size, the amoeba's single cell divides in half and there are two amoebas where there was only one before. In about a day the two amoebas have grown to the point where they are ready to divide and form four; the day after that, there are eight amoebas, and so forth. How many amoebas will there be at the end of a week?

The number of amoebas is a function of the time that has passed. A table for this function looks like this:

time (in days)	0	1	2	3	4	5	6	7	8	. . .
number of amoebas	1	2	4	8	16	32	64	128	256	. . . .

The answer is in the table: at the end of 7 days there will be 128 amoebas.

You will recognize the second line of numbers in this table as the binary sequence, a geometric sequence in which each term is twice the one it follows. The first line is an arithmetic sequence in which each term is one more than the one before.

There is something remarkable about this function that was discovered by a Scottish mathematician named John Napier in the early 17th century. Suppose we choose two numbers from the second sequence, say 4 and 32. If we multiply these numbers,

$$4 \times 32 = 128,$$

we get another number of the sequence. Now look at the numbers in the first sequence that correspond to 4, 32, and 128. They are 2, 5, and 7, and here is the remarkable part:

$$2 + 5 = 7.$$

We have discovered that a *multiplication* in the second sequence corresponds to an *addition* in the first sequence. Let's try this again, but in reverse:

$$4 + 1 = 5$$
$$16 \times 2 = 32.$$

This suggests that if we don't want to *multiply,* we can *add* instead by using the table.

For instance, what is 64 × 4? Finding these numbers in the second sequence, we look above them and see 6 and 2; add 6 and 2 to get 8, and look below 8 to see 256. So, without doing any multiplying, we have figured out that 64 × 4 = 256. We simply added 6 and 2!

$$6 + 2 = 8$$
$$64 \times 4 = 256$$

The numbers of the first sequence are called the **logarithms** of the corresponding numbers of the second sequence:

logarithms	0	1	2	3	4	5	6	7	8	. . .
numbers	1	2	4	8	16	32	64	128	256	. . . .

A formula for this function is

$$n = 2^x,$$

where n represents a number and x represents its logarithm. Notice that in the formula, the logarithm, x, is the *exponent* of the 2. For example, $32 = 2^5$; the logarithm of the number 32 is 5, the exponent of the 2. Check this in the table above.

EXERCISES

Set I

1. Here is our logarithm table arranged in two columns with the terms of the binary sequence written in the first column and their logarithms written in the second.

Number	Logarithm
1	0
2	1
4	2
8	3
16	4
32	5
64	6
128	7
256	8
512	9
1,024	10

Copy the table and continue it for 20 more lines. The last line should read:

Number	Logarithm
1,073,741,824	30

Use your table to find the answers to the following problems without doing any multiplying.

2. 64×128

3. $512 \times 2{,}048$

4. $8{,}192 \times 131{,}072$

5. $16 \times 64 \times 32{,}768$

6. $(128)^2$

7. $(4{,}096)^2$

By this point, you are probably ready to agree that using logarithms can make multiplication problems much easier.

8. Look again at problems 6 and 7 to help in answering this question. To square a number, by what number should you multiply its logarithm to get the logarithm of the answer?

9. Can you find what number is the *square root* of 65,536 by using your table?

10. Use the table to find the answer to this problem: $(32)^3$.

11. To cube a number, by what number should you multiply its logarithm to get the logarithm of the answer?

12. Now use the table to find the answer to this problem: $(64)^4$.

13. Find the answer to this problem: $(16)^5$.

14. Copy and complete the following rule: "To raise a number to any power, multiply its logarithm by ________ to get the logarithm of the answer."

Set II

Why does the logarithm table operate the way it does? It doesn't take much detective work to find out. First, we will express each number in the first column of the table as a power of 2; it is convenient to write 1 as 2^0 as you will see.

Number	Logarithm
$1 = 2^0$	0
$2 = 2^1$	1
$4 = 2^2$	2
$8 = 2^3$	3
$16 = 2^4$	4
$32 = 2^5$	5
$64 = 2^6$	6

This table makes it easy to see that the *logarithm* numbers are the *exponents* of the 2's.

1. What would be the logarithm of 2^{100} if the table were continued that far?

Let's see how the table works by multiplying 4 and 8. The logarithms of these numbers are 2 and 3; $2 + 3 = 5$, and 5 is the logarithm of 32.

2. You know that $4 = 2^2$, which means 2×2. What does this mean: $8 = 2^3$?

3. Copy and complete the following:

$$4 \times 8 = 2^2 \times 2^3 = 2 \times 2 \times \_\_\_\_\_\_\_\_ = 2^{\_\_\_}$$

4. So $2^2 \times 2^3 = 2^{2\,\_\_\,3} = 2^{\_\_\_}$, and this is how multiplication is replaced by addition.

Copy and complete the following:

5. $2^1 \times 2^4 = 2^{\square} = 2^{\square}$

6. $2^0 \times 2^6 = 2^{\square} = 2^{\square}$

7. $2^1 \times 2^5 \times 2^4 = 2^{\square} = 2^{\square}$

8. $(2^5)^2 = 2^5 \times \square = 2^{\square}$

9. When a number with an exponent is squared, what happens to the exponent?

10. $(2^4)^3 = 2^4 \times \square \times \square = 2^{\square}$

11. When a number with an exponent is cubed, what happens to the exponent?

12. What do you think happens to the exponent of a number when that number is raised to any power?

Set III

Let's get back to the amoebas with which we started. Your table in Set I shows the number of amoebas at the end of every day for a month, after starting with one at the beginning of the first day. At the end of the first week, there are 128 amoebas. If we round this number to two figures, 130, and write it in scientific notation, we can say that at the end of the first week, there are about 1.3×10^2 amoebas.

Round your answers to each of the following to two figures and write them in scientific notation.

1. How many amoebas are there at the end of the second week?
2. How many are there at the end of the third week?
3. How many at the end of the fourth week?
4. There is a pattern in the exponents of the numbers of amoebas at the end of each of the four weeks. What is it?

Lesson 4

More About Logarithms

IT is hard to imagine how man could have ever made his first trip to the moon without the help of computers. The problems of space flight are so complex and require such rapid answers that they are beyond the power of human calculators with only pencil and paper to solve.

Before the invention of the computer, man had something else to help make doing arithmetic easier: logarithms. You have already seen how logarithms can be used to simplify multiplication problems. Here is the beginning of the table of logarithms you have been using:

numbers	1	2	4	8	16	32	64	128	256 . . .
logarithms	0	1	2	3	4	5	6	7	8 . . . .

This table is not as useful as it might be, since it is based on the binary sequence, and our number system is not built around the number *two,* but the number *ten.* A logarithm table based on the number ten is much more convenient to have. It starts out like this:

numbers	1	10	100	1,000	10,000	100,000	1,000,000 . . .
logarithms	0	1	2	3	4	5	6 . . . .

Again the numbers of a geometric sequence are paired with numbers of an arithmetic sequence. This is the key to how this and every logarithm table works. The terms of a geometric se-

"You've got to hand it to those computers."

quence are found by *multiplication* while the terms of an arithmetic sequence are found by *addition,* and you know that *multiplication* of two numbers corresponds to *addition* of their logarithms.

A formula for this table is

$$n = 10^x,$$

where n represents a number and x represents its logarithm. For example, $10{,}000 = 10^4$, so the logarithm of the number 10,000 is 4.

Let's try a problem with the new table. To multiply the numbers 100 and 1,000, we add their logarithms: $2 + 3 = 5$. The number 5 is the logarithm of 100,000 so $100 \times 1{,}000 = 100{,}000$. All this work seems rather silly, however, since 100 and 1,000 are so easy to multiply in the first place.

Here's another problem that seems to be even simpler: multiply the numbers 2 and 3 by adding their logarithms. Another look at our table shows that we're in trouble. The table shows the logarithms of the numbers 1 and 10, but leaves out the logarithms of all the numbers in between, including 2 and 3. (Notice that we can't even do this problem with our original binary logarithm table since, although the logarithm of 2 is included, the logarithm of 3 is not.) This is ridiculous. Our logarithm table isn't much good if we can't even solve a problem as simple as 2×3!

Let's rewrite the first part of the table, leaving room for the logarithms of the numbers 2 through 9:

numbers	1	2	3	4	5	6	7	8	9	10
logarithms	0									1.

The missing logarithms are apparently all numbers between 0 and 1. Mathematicians have a method for finding out what these numbers are that we will not go into here. When the missing logarithms are computed and filled into the table, rewritten in two columns, it looks like this:

Number	Logarithm
1	0.000
2	0.301
3	0.477
4	0.602
5	0.699
6	0.778
7	0.845
8	0.903
9	0.954
10	1.000

Now we are ready to multiply 2 and 3. The logarithms of 2 and 3 are 0.301 and 0.477:

$$0.301 + 0.477 = 0.778.$$

The answer is the number whose logarithm is 0.778.

Perhaps at this point, using logarithms seems to you like more trouble than it's worth, but if the numbers get very complicated, the method really pays off. With logarithms, even a problem such as multiplying 2.856 and 3.974 is no more difficult than 2 times 3! The man who figured out one of the first tables of logarithms said: "Logarithmes are numbers invented for the more easie working of questions in arithmetike and geometrie . . . by them all troublesome multiplications . . . are avoided and performed only by addition. . . . In a word, all questions not only in arithmetike and geometrie but in astronomie also are thereby most plainely and easily answered."*

*John Briggs, *Logarithmall Arithmeticke,* 1631.

EXERCISES

SET I

Although we now know the logarithms of the numbers from 1 through 10, we still don't know the logarithms of any numbers larger than 10, except for those numbers that are powers of 10: 100, 1,000, and so on. Many other logarithms can be found by simply adding the ones we already have. For example, the logarithm of 12 is equal to the sum of the logarithms of 3 and 4, since $12 = 3 \times 4$.

1. Find the logarithm of 12 by adding the logarithms of 3 and 4.
2. Since it is also true that $12 = 2 \times 6$, the logarithm of 12 can be found by adding the logarithms of 2 and 6. Check your answer to problem 1 by doing this.
3. Find the logarithm of 36 by adding the logarithms of 4 and 9.
4. Check your answer to problem 3 by adding the logarithms of 3 and 12.

Copy each of the following equations, and fill in the missing number. Then find the logarithm of the number on the left of the equal sign. To save space, we will abbreviate the "logarithm of n" as "log n."

5. $\log 20 = \log 4 + \log$ ▯
6. $\log 20 = \log 10 + \log$ ▯
7. $\log 21 = \log$ ▯ $+ \log 7$
8. $\log$ ▯ $= \log 8 + \log 9$
9. $\log$ ▯ $= \log 3 + \log 5 + \log 6$
10. $\log 90 = \log 9 + \log$ ▯

To find the logarithm of the square of a number, you can double the logarithm of the number. To see why, look at the example below.

$$\log 25 = \log 5 + \log 5 = 2 \times \log 5$$
$$\text{So } \log 25 = 2 \times 0.699 = 1.398.$$

Find the missing numbers in each of the following.

11. $\log 9 = \log ▒ + \log ▒ = 2 \times \log ▒$

12. $\log 5^3 = \log ▒ + \log ▒ + \log ▒ = 3 \times \log ▒$

Use any method to find each of the following logarithms.

13. log 49

14. log 27

15. $\log 4^5$

16. log 1,000,000

SET II

The advantage of using "decimal" logarithms is that they are closely related to the way we write numbers in scientific notation. To see what this relationship is, we will write some numbers in scientific notation and then compare their logarithms.

1. The number 20 written in scientific notation is 2×10^1. Write the numbers 200, 2,000, and 20,000 in a column in scientific notation.

2. What do the three numbers you have written have in common?

3. You have already found the logarithm of 20 by the following method:

   $$\log 20 = \log 10 + \log 2 = 1 + 0.301 = 1.301.$$

   Notice that log 200 = log 100 + log 2. Find log 200.

4. Find log 2,000 and log 20,000.

5. What do the three logarithms you have written have in common?

6. Write the three logarithms in a column to the right of the column in which you wrote the corresponding numbers in scientific notation. The *exponent of the 10* in a number written in scientific notation also appears in the logarithm of the number. Where?

7. What can you say about the logarithm of the number 2×10^7?

8. Copy the table below, and fill in the missing numbers.

$$70 = 7 \times 10^1 \text{ and } \log 70 = 1.845$$
$$700 = 7 \times 10^{\square} \text{ and } \log 700 = \square.845$$
$$7{,}000 = \square \text{ and } \log 7{,}000 = \square$$

9. What part of a number's logarithm is determined by the *number in front* in the scientific notation form of the number?
10. Can you write the logarithm of 3×10^5 without doing any arithmetic?

Now let's try this in reverse. What number has the logarithm 3.602? First, we will try to find the way it is written in scientific notation. Remember that the *exponent of the 10* in a number written in scientific notation appears in its logarithm.

11. What is the exponent of the 10 in the number whose logarithm is 3.602?

Also remember that the fraction part of a number's logarithm is determined by the *number in front* in the scientific notation form of the number.

12. The number having a logarithm of .602 can be found by looking in the table. What is it?

    So the number that has 3.602 for its logarithm is 4×10^3, or 4,000.

Find the missing numbers in each of the following:

13. $\log (6 \times 10^{\square}) = 5.778$
14. $\log (\square \times 10^2) = 2.903$
15. $\log (\square \times 10^{\square}) = 8.699$
16. Can you figure out the name of the number that has the logarithm 100.000? (Hint: See page 140.)

Set III

Suppose that some day in the future we manage to communicate with intelligent beings who live on some other world in outer space. Although their number system might very likely be different from our own, it is possible that they might also have developed the idea of logarithms. Suppose that after translating their numbers into our own and learning some of their logarithms, we have the following table:

number	1	2	3	4	5	6	7	8	9	10
logarithm	0	0.39	0.61		0.90		1.09			.

1. Assuming that their logarithms behave just as ours do, can you figure out the missing logarithms in this table?
2. Our number system is based on the number ten. On what number do you think their system is based?
3. What number would have 2 for its logarithm in their system?

Lesson 5

Computing with Logarithms

DO you remember what a chore it was to learn the "times-table"? People have not always memorized the multiplication table; in fact, as late as the Renaissance it was referred to in the same way that a modern mathematician might look up numbers in a table of logarithms. The table he uses, to be of any real use, is necessarily much longer than the short table you have been using. In fact, some logarithm tables fill entire books!

On the next page is a table that lists the logarithms of numbers from 1.01 to 10.0. It is not necessary to go past 10, because of the relationship between numbers written in scientific notation and the decimal logarithms of the numbers. For example, the logarithm of 7 is 0.845, and you have shown that

$$70 = \boxed{7} \times 10^{①} \text{ and } \log 70 = ①.\boxed{845}$$
$$700 = \boxed{7} \times 10^{②} \text{ and } \log 700 = ②.\boxed{845}$$
$$7{,}000 = \boxed{7} \times 10^{③} \text{ and } \log 7{,}000 = ③.\boxed{845}.$$

► The *whole number* part of the logarithm *is the exponent of the 10* when the number is written in scientific notation. The *fraction* part of the logarithm is looked up in the table and *is determined by the number in front* in the scientific form of the number.

Since, when a number is written in scientific notation, the number in front is always less than 10, the logarithm table does

TABLE OF LOGARITHMS

No.	Log.	No.	Log.	No.	Log.	No.	Log.	No.	Log.	No.	Log.	No.	Log.	No.	Log.	No.	Log.
1.01	.004	1.21	.083	1.41	.149	1.61	.207	1.81	.258	2.1	.322	4.1	.613	6.1	.785	8.1	.908
1.02	.009	1.22	.086	1.42	.152	1.62	.210	1.82	.260	2.2	.342	4.2	.623	6.2	.792	8.2	.914
1.03	.013	1.23	.090	1.43	.155	1.63	.212	1.83	.262	2.3	.362	4.3	.633	6.3	.799	8.3	.919
1.04	.017	1.24	.093	1.44	.158	1.64	.215	1.84	.265	2.4	.380	4.4	.643	6.4	.806	8.4	.924
1.05	.021	1.25	.097	1.45	.161	1.65	.217	1.85	.267	2.5	.398	4.5	.653	6.5	.813	8.5	.929
1.06	.025	1.26	.100	1.46	.164	1.66	.220	1.86	.270	2.6	.415	4.6	.663	6.6	.820	8.6	.934
1.07	.029	1.27	.104	1.47	.167	1.67	.223	1.87	.272	2.7	.431	4.7	.672	6.7	.826	8.7	.940
1.08	.033	1.28	.107	1.48	.170	1.68	.225	1.88	.274	2.8	.447	4.8	.681	6.8	.833	8.8	.944
1.09	.037	1.29	.111	1.49	.173	1.69	.228	1.89	.276	2.9	.462	4.9	.690	6.9	.839	8.9	.949
1.10	.041	1.30	.114	1.50	.176	1.70	.230	1.90	.279	3.0	.477	5.0	.699	7.0	.845	9.0	.954
1.11	.045	1.31	.117	1.51	.179	1.71	.233	1.91	.281	3.1	.491	5.1	.708	7.1	.851	9.1	.959
1.12	.049	1.32	.121	1.52	.182	1.72	.236	1.92	.283	3.2	.505	5.2	.716	7.2	.857	9.2	.964
1.13	.053	1.33	.124	1.53	.185	1.73	.238	1.93	.286	3.3	.519	5.3	.724	7.3	.863	9.3	.968
1.14	.057	1.34	.127	1.54	.188	1.74	.241	1.94	.288	3.4	.531	5.4	.732	7.4	.869	9.4	.973
1.15	.061	1.35	.130	1.55	.190	1.75	.243	1.95	.290	3.5	.544	5.5	.740	7.5	.875	9.5	.978
1.16	.064	1.36	.134	1.56	.193	1.76	.246	1.96	.292	3.6	.556	5.6	.748	7.6	.881	9.6	.982
1.17	.068	1.37	.137	1.57	.196	1.77	.248	1.97	.294	3.7	.568	5.7	.756	7.7	.886	9.7	.987
1.18	.072	1.38	.140	1.58	.199	1.78	.250	1.98	.297	3.8	.580	5.8	.763	7.8	.892	9.8	.991
1.19	.076	1.39	.143	1.59	.201	1.79	.253	1.99	.299	3.9	.591	5.9	.771	7.9	.898	9.9	.996
1.20	.079	1.40	.146	1.60	.204	1.80	.255	2.0	.301	4.0	.602	6.0	.778	8.0	.903	10.0	1.000

not need to go beyond 10. Here is an example of how to use it to find the logarithm of a number larger than 10. Most people do not learn the "times-table" past $12 \times 12 = 144$. What is the logarithm of 144? First, write it in scientific notation: 1.44×10^2. Second, write the whole number part of the logarithm: it is 2. Third, since 1.44 is less than 10, we can look up the fraction part of the logarithm beside 1.44 in the table: it is .158. So log 144 = 2.158.

Now let's try out the table of logarithms by solving the problem Charlie Brown has given his sister: "How much is 3 times 4?" Since 3 and 4 are both less than 10, we can find their complete logarithms in the table. The logarithm of 3 is .477 and the logarithm of 4 is .602. To multiply the numbers, we add their logarithms:

$$\begin{array}{rr} \log 3 = & .477 \\ + \log 4 = & .602 \\ \hline & 1.079 \end{array}$$

The result is the logarithm of the answer. Looking for the fraction part, .079, in the table, we find 1.2 beside it. The whole number part is the exponent of the 10.

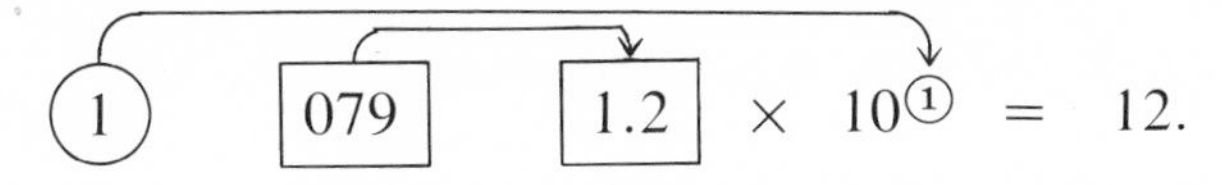

We have chosen this example because it is simple, and so the method seems as roundabout as Sally's series of wild guesses. With very complicated calculations, it is much less "roundabout" than working them out the long way. In fact, logarithms can be used to solve with ease problems that would be impossible to handle without them.

EXERCISES

Set I

Write the numbers in each of the following statements in scientific notation and then refer to the table to write their logarithms.

1. Upper Yosemite Falls is 1,430 feet high.
2. One cubic foot of walnut wood weighs about 40 pounds.
3. The STRETCH computer of the United States Weather Bureau can do approximately 700,000 additions in one second.
4. The temperature in the center of the sun is 25,000,000 degrees Fahrenheit.

Find the answers to the following problems by using logarithms.

5. 6.5×1.29

   (Arrange your work like this:

   log 6.5 = ▒▒▒▒
   + log 1.29 = ▒▒▒▒
   log of answer = ▒▒▒▒
   answer = ▒▒▒▒

6. $1.48 \times 1.87 \times 2.6$
7. $(1.34 \times 10^5) \times (4.1 \times 10^2)$
8. $9.1 \times 8{,}800$

Nearly all logarithms are not exact, but only approximate. Suppose you get a number such as 2.077 for the logarithm of an answer. When we look in the table for the fraction .077, the closest one we find is .076 and so we choose it. The number whose logarithm is 2.077 is 1.19×10^2 or 119.

Find answers to these problems.

9. 1.15×5.4

10. $(1.76)^2$

11. $(7.1 \times 10^3) \times (3.4 \times 10^8)$

By permission of Johnny Hart and Field Enterprises, Inc.

Set II

The "big idea" is Einstein's famous formula relating the energy contained in something to its mass (the amount of material it contains). The mass, represented by *m,* is multiplied by the square of the speed of light, represented by *c*. The speed of light is 186,000 miles per second.

1. Write this number in scientific notation.
2. Write its logarithm.
3. To square a number, what do you do with its logarithm?
4. Write the logarithm of the square of the speed of light.
5. What is the approximate square of the speed of light? Write it in scientific notation.
6. Write the square of the speed of light in the long form and name it in words.

The first X-15 rocket plane was built by North American Aviation in 1960. It has flown at a speed of more than 1.14 miles per second.

7. Find the logarithm of 1.14.
8. How many seconds are in one hour? Change your answer to scientific notation.
9. Find the logarithm of this number.
10. Use these logarithms to find the speed of the rocket plane in miles per hour.

Set III

You will remember that back in 1626 the Indians of New York were paid $24 for Manhattan Island. Suppose they had invested the $24 in the Big Heap Moola Savings and Loan Company at 5% interest compounded annually.

At the end of the first year their money would be worth

$$24 \times 1.05 \text{ dollars.}$$

At the end of the second year it would be worth

$$24 \times (1.05)^2 \text{ dollars;}$$

after the third year,

$$24 \times (1.05)^3 \text{ dollars,}$$

and so on. If the Indians had left all of the money in their account until 1970, how much money would they have? To answer this, we can reason as follows.

If this were the end of the 344th year in the account, the amount would apparently be

$$24 \times (1.05)^{344} \text{ dollars.}$$

In order to find this amount without using logarithms, we would have to multiply 24 by 1.05 over and over again, 344 times altogether! With logarithms, it is a different story. The logarithm of the answer can be found by adding the logarithm of 24 and 344 times the logarithm of 1.05.

1. What is log 1.05?
2. Multiply log 1.05 by 344.

3. What is log 24?
4. Add the results to problems 2 and 3 and you will have the logarithm of the answer.
5. Write the answer in scientific notation.
6. Change the answer to the long form, and name in words the amount of money the Indians would have.

   This still wouldn't be enough money to buy back Manhattan, but you can see that, so far as the Indians were concerned, their Savings and Loan Company has an appropriate name!

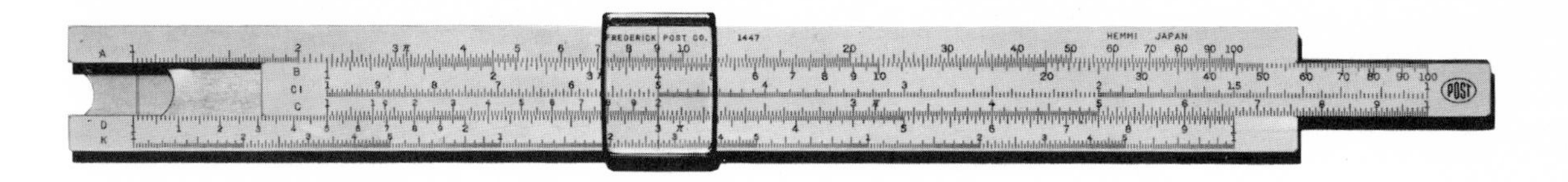

Lesson 6

The Slide Rule and Other Logarithmic Scales

ASK an engineer to figure out any of the problems we have been solving with logarithms and he will be able to give you the answer in seconds, without looking in a table or writing a single number! How does he do it? With a small computer that works without electricity and has only two moving parts. It was invented more than three centuries ago and has been used by scientists and mathematicians ever since. The engineer uses a slide rule, an instrument accurate enough to give a useful answer to almost any practical problem.

How is a slide rule made and how does it work? We can get some clues by looking at the way the numbers are arranged on its scales. One scale on every slide rule looks something like this:

1 2 3 4 5 6 7 8 9 10 20 30 40 50 60 70 80 90 100

The number 1 appears at the left end of the scale and the number 100 at the right end. Notice the spacing of the numbers in-between; it is certainly not like the spacing of numbers on a ruler. The numbers are squeezed closer and closer together as we look from left to right. There is something special about their arrangement and we can notice it more easily by looking at certain sequences of numbers on the scale.

We have noticed that the numbers in the arithmetic sequence,

1 2 3 4 5 6 7 . . .,

are not evenly spaced. Let's look at the binary sequence:

1 2 4 8 16 32 64 . . . .

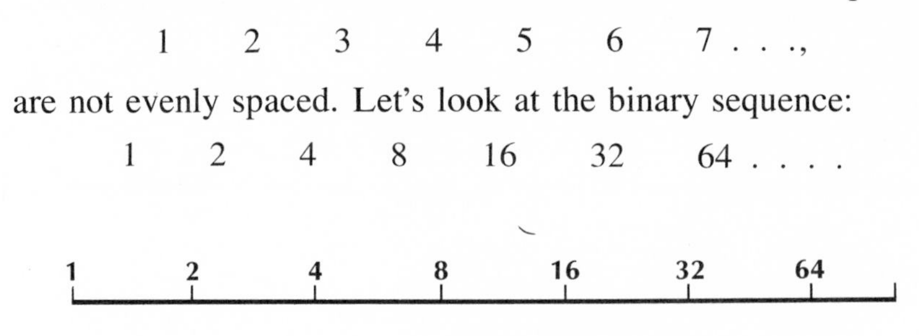

That's interesting! These numbers *are* evenly spaced. What about the numbers in the decimal sequence,

1 10 100 1,000 10,000 . . .?

1 10 100

The three that appear on the scale are also evenly spaced.

Going back to the binary sequence, we rewrite each term as a power of 2 and number below the scale like an ordinary ruler, that is, with 0, 1, 2, 3, and so on.

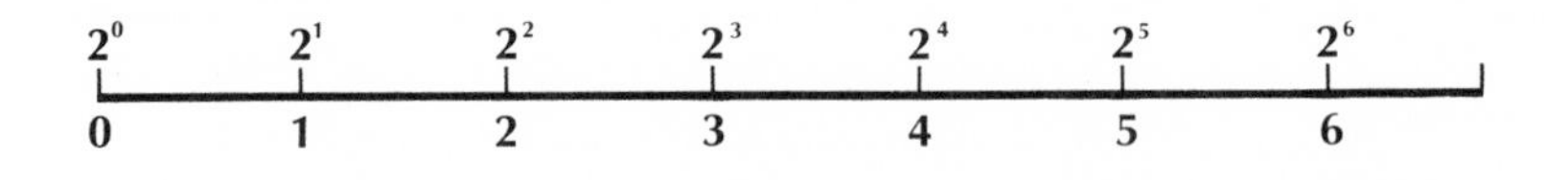

Now, look at the exponents and you will see that the numbers are arranged so that it is the *exponents* that are *evenly spaced.*

If we rewrite each term in the decimal sequence as a power of 10, the numbers are again arranged so that the exponents are evenly spaced.

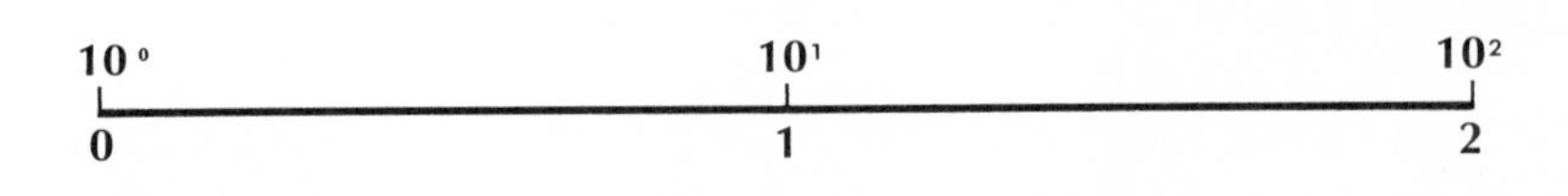

Since *logarithms are exponents* (we got them from these geometric sequences in the first place), the numbers on a slide rule are arranged so that their *logarithms* are evenly spaced. You will see better just how the slide rule works when you make one.

EXERCISES

Set I

We have noticed that these numbers in the binary sequence,

1 2 4 8 16 32 64,

are evenly spaced on a slide rule.

1. Copy and complete the following table by looking up the decimal logarithms of these numbers in the table on page 166.

Number	1	2	4	8	16	32	64
Logarithm	0	0.301	▮▮▮	▮▮▮	1.204	▮▮▮	▮▮▮

2. What do you notice about the sequence of logarithms on the second line of the table?

3. Here is a table of logarithms that have been rounded to two decimal places each:

Number	1	1.5	2	3	4	5	6	7	8	9	10
Logarithm	0	.18	.30	.48	.60	.70	.78	.85	.90	.95	1.00.

Draw a graph of this function on graph paper that has 10 squares to the inch. Let the x-axis represent the numbers and 5 small units on it stand for 1. Let the y-axis represent the logarithms and 5 small units on it stand for .1. Mark each point with a heavy dot and join the points with a smooth curve. The lower left-hand corner of your completed graph should look like this.

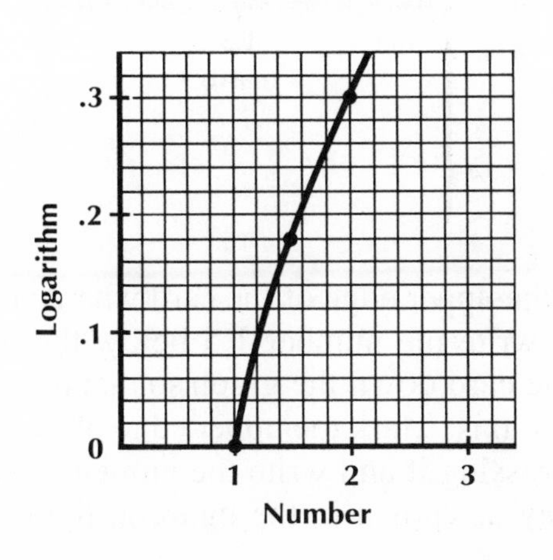

4. As you look at the curve from left to right, does its steepness *increase* or *decrease?* It is called a logarithmic curve.

Set II

Experiment. A Slide Rule

Take a 4 × 6 card, draw an accurate line dividing it in half like this, and cut it into two pieces along the line. Write the number 1 in the two corners indicated.

Now turn the graph of the logarithmic curve you drew in Set I sideways, so that the logarithm axis is at the top and the number axis is at the left. Take the lower piece of your card and place it on the graph so that the left edge falls along the number axis and the upper left-hand corner is at 1.5, as shown in the figure below.

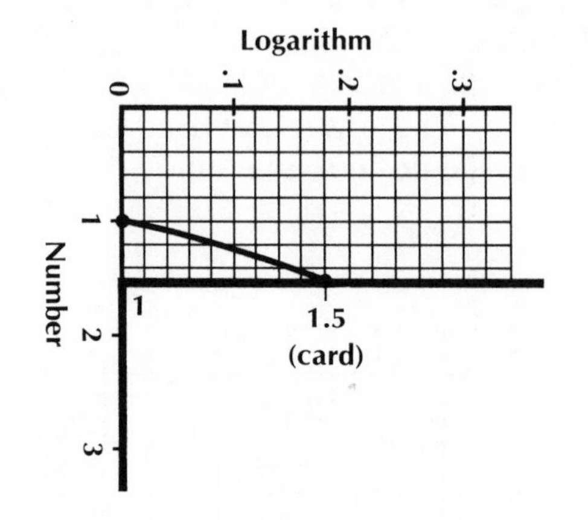

Put a mark on the upper edge of the card where you see the curve crossing it and write the number 1.5 below the mark.

Now slide the card down the graph so that the corner comes to 2 on the number axis. Mark the upper edge of the card where you see the curve crossing it and write the number 2 below the mark, in the same way as shown in the figure at the top of page 175.

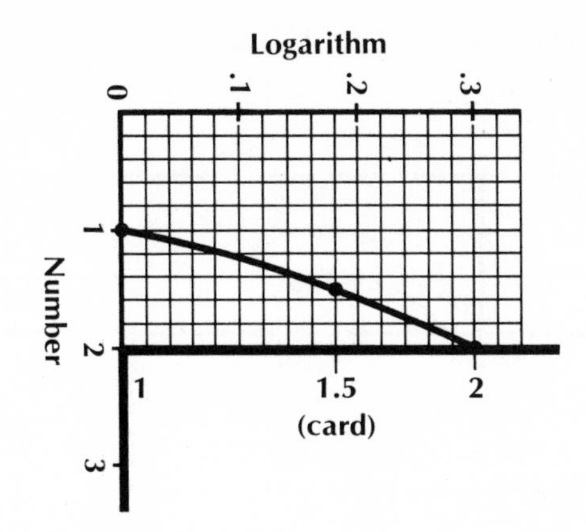

Slide the card down to 3, 4, 5, 6, 7, 8, 9, and 10 on the number axis and do the same thing: put a mark on the edge of the card at the point where the curve crosses it and write the corresponding number below the mark.

Your finished card should look like this:

1 1.5 2 3 4 5 6 7 8 9 10

Now put it below the other half of the card and copy the scale and the numbers from the finished card onto the bottom of the other card, as shown in the figure below. Your slide rule is now ready to use.

1 1.5 2 3 4 5 6 7 8 9 10

1 1.5 2 3 4 5 6 7 8 9 10

Of course commercially produced slide rules have many more scales, and each scale has many more marks, than the one you have made. Look again at the photograph of the relatively simple slide rule at the top of page 171. The scales you have made correspond to those labeled C and D on it.

Here is how your slide rule can be used to multiply 2×3.

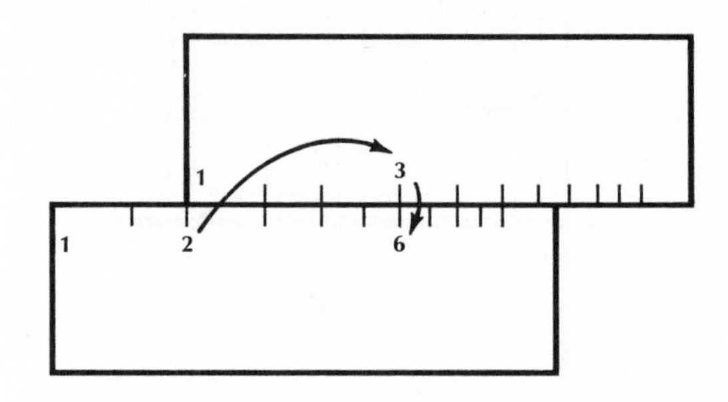

If you are interested in learning more about using a slide rule, a good booklet to read is: *How Do You Use a Slide Rule,* by Arthur A. Merrill, Dover, 1961.

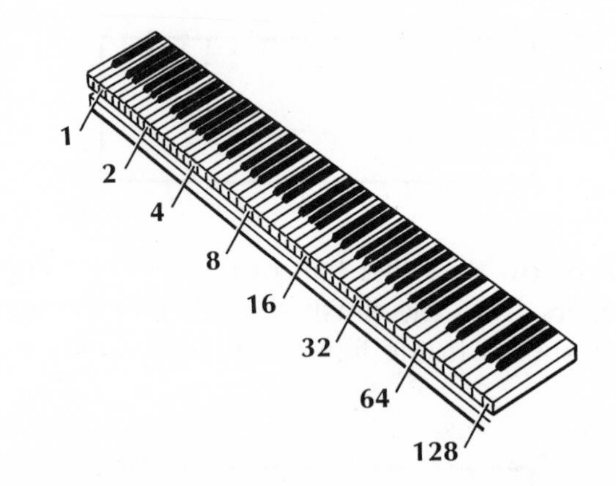

Set III

Here is a drawing of the keyboard of a piano. The keys named C have been marked with lines and numbers. The numbers show the relationship between the frequencies of the notes produced by these keys. Starting from the left end of the keyboard, each successive C note has twice the frequency of the C note before it.

1. What kind of a sequence do these frequency numbers form?
2. How is a piano keyboard like a slide rule?
3. What do you notice about the arrangement of the frets (ridges to guide the fingers) in this photograph of a guitar?
4. On what kind of a scale (mathematical, not musical!) do you think the frets of a guitar are arranged?

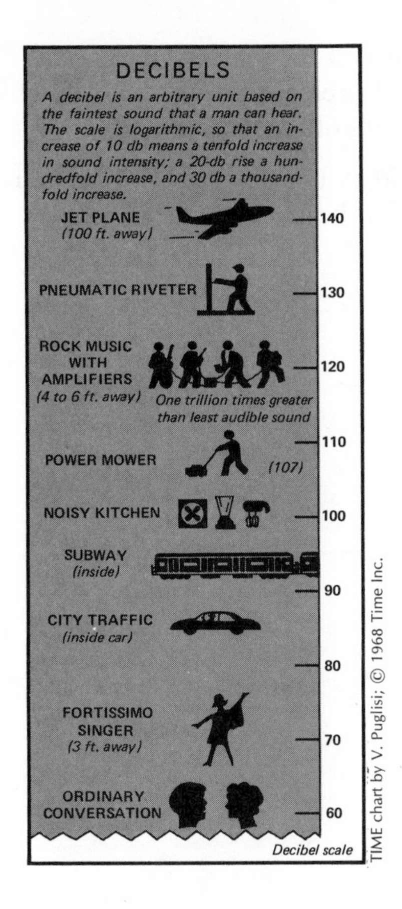

A scale for measuring the loudness of sounds is the decibel scale. It is explained in this chart from *Time* magazine.

The softest sound that can be heard is said to be at the "threshold of hearing" and is rated as 0 decibels.

5. Copy and complete the following table, in which 0 decibels has been assigned a relative loudness of 1 unit.

Relative loudness	1						
Decibels	0	10	20	30	40	50	60

6. Leaves rustling in a breeze have a loudness of 20 decibels. How many times louder is ordinary conversation than the rustle of leaves?

7. The chart says that an amplified rock group is a trillion times louder than the least audible sound. Write a trillion in scientific notation.

8. How does the exponent of the number you have written compare with the loudness of rock music in decibels?

9. The loudness of the water at the foot of Niagara Falls is a billion times greater than the least audible sound. Write a billion in scientific notation.

10. What would you expect the loudness in decibels of Niagara Falls to be?

Chapter 4 / Summary and Review

IN this chapter we have become acquainted with:

The names of some large numbers and the use of exponents *(Lesson 1)* In writing a large number as a power of 10, the exponent is the number of zeros that follow the 1 when the number is written the long way:

$$10^1 = 10;\ 10^0 = 1.$$

Scientific notation *(Lesson 2)* A large number in scientific notation is written as some number, which is at least 1 but less than 10 (shown here as a), multiplied by 10 raised to some power (shown here as b):

$$a \times 10^b.$$

Logarithms *(Lessons 3, 4, & 5)* Logarithms are exponents of numbers that are in a geometric sequence.

To find the logarithm of a number, first write the number in scientific notation: $a \times 10^b$. The whole number part of the logarithm is the power of 10, b, and the fraction part of the logarithm is found beside the number in front, a, in a logarithm table.

To multiply two numbers, add their logarithms to get the logarithm of the answer.

To raise a number to a power, multiply its logarithm by that power to get the logarithm of the answer.

The slide rule and other logarithmic scales *(Lesson 6)* The numbers on a slide rule are arranged on a logarithmic scale (their logarithms are evenly spaced).

EXERCISES

Set I

1. At the time of Apollo 11's first landing on the moon, the director of NASA's Lunar Receiving Laboratory said: "The chance of bringing anything harmful back from the moon is probably one in a hundred billion." Write one hundred billion the long way and also in exponential form.
2. Did you know that Hershey chocolate bars are never advertised? Write the number in the cartoon in scientific notation.

"He just learned that the Hershey Corporation grossed one hundred and eighty million dollars last year without spending one red cent on advertising."

3. If a silver coin were made whose diameter and thickness were each 1,000 times that of a dime, its volume would be $1{,}000^3$ times as great. What would be the value of this coin in dollars, assuming it is worth $1{,}000^3$ dimes?
4. Name the number in words and write it in scientific notation.
5. The gigantic Houston Astrodome encloses 41 million cubic feet of space. Write this number in scientific notation.
6. Write the logarithm of this number.
7. One cubic foot contains approximately 1,730 cubic inches. Write this number in scientific notation.
8. Find its logarithm.
9. Use logarithms to find the approximate number of cubic inches of space in the Astrodome. Write it in scientific notation and name the number.

Set II

Here is how another sequence of numbers appears on a scale of a slide rule.

1 3 9 27 81

1. What kind of number sequence is this?
2. Copy and complete the following table by looking up the logarithms of these numbers.

Number	1	3	9	27	81
Logarithm	0			1.431	

3. What do you notice about the sequence of logarithms on the second line of the table?

As you can see from the newspaper article on page 182, the scale used to measure the intensity of an earthquake is logarithmic.

4. Copy and complete this table in which an earthquake rated

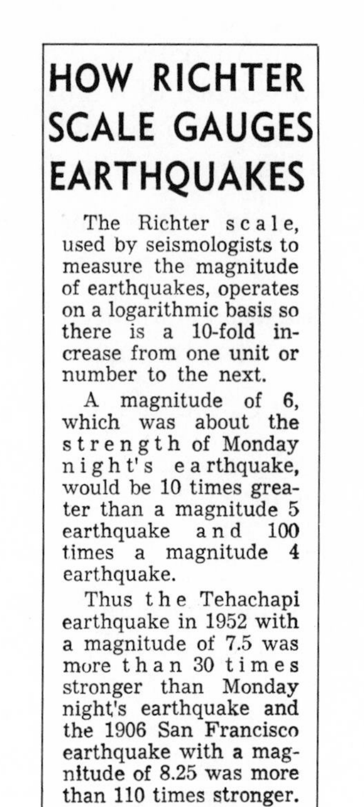

HOW RICHTER SCALE GAUGES EARTHQUAKES

The Richter scale, used by seismologists to measure the magnitude of earthquakes, operates on a logarithmic basis so there is a 10-fold increase from one unit or number to the next.

A magnitude of 6, which was about the strength of Monday night's earthquake, would be 10 times greater than a magnitude 5 earthquake and 100 times a magnitude 4 earthquake.

Thus the Tehachapi earthquake in 1952 with a magnitude of 7.5 was more than 30 times stronger than Monday night's earthquake and the 1906 San Francisco earthquake with a magnitude of 8.25 was more than 110 times stronger.

Courtesy of
Los Angeles Times.

6 on the Richter scale has been assigned a relative intensity of 1 unit.

Relative intensity		.1	1				
Richter scale	4	5	6	7	8	9	10

5. Someone who didn't understand the Richter scale might think that an earthquake that measures 8 would be twice as strong as one that measures 4. How do the two actually compare in intensity?

Atoms of a radioactive element slowly break apart into atoms of other elements. This table shows how much of a certain kind of uranium is left at the end of different times, if we start with 64 pounds of it.

Time (in years)	0	70	140	210	280	350	420	. . .
Amount (in pounds)	64	32	16	8	4	2	1	. . .

6. The "half-life" of a radioactive element is the amount of time it takes half of the atoms of the element to disintegrate. What is the half-life of this kind of uranium?

7. In what kind of sequence are the numbers on the second line of this table?
8. Do you think that the amount of uranium left is a logarithmic function of the time that has passed by?

Set III

A great English astronomer and physicist, Sir Arthur Eddington (1882-1944), once claimed that the number of electrons in the universe is

$$136 \times 2^{256}.$$

How large a number is that? To find out, we can use logarithms.

1. What is log 2?
2. Multiply log 2 by 256.
3. What is log 136?
4. Add the results of problems 2 and 3 to get the logarithm of the (approximate) answer.
5. Write the answer in scientific notation.
6. If this number were written out in full, how many digits would it contain?

Of course, Eddington didn't count the number of electrons in the universe, but he had a theory upon which he based his claim that this was actually the correct number!

Chapter 5

REGULAR POLYGONS

Lesson 1

Regular Polygons and Symmetry

THOR, the inventor of the wheel, has come up with a new design. His wheel is in the shape of an equilateral triangle, the simplest member of a family of geometric figures called the regular polygons.

► A **regular polygon** is a figure all of whose *sides are the same length* and all of whose *angles are equal.*

Thor thinks his new design is better than a wheel in the shape of a square, the regular polygon having four sides.

Equilateral triangle

The names of the rest of the regular polygons are based on their numbers of sides. These figures were studied in detail by the early Greek mathematicians and the names they gave them have been used ever since.

You can see how the Greeks got these names by looking at their names for the numbers:

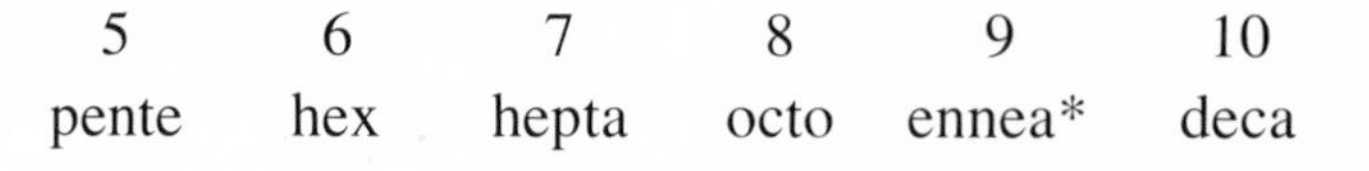

5	6	7	8	9	10
pente	hex	hepta	octo	ennea*	deca

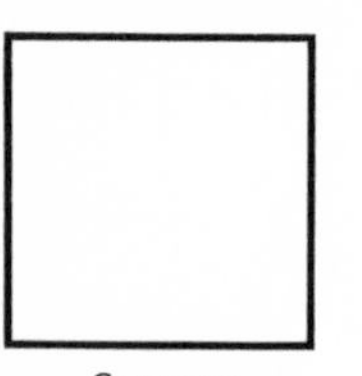

Square

The family of regular polygons does not end with the regular decagon, but goes on indefinitely. As the number of sides increases, it becomes more and more difficult to tell the polygon

*The Greek name for the regular polygon having nine sides was "enneagon." We now use the word "nonagon," which comes from the Latin word "nonus" meaning "nine."

By permission of Johnny Hart and Field Enterprises, Inc.

from a circle. Is the last figure a regular polygon with 100 sides or is it a circle? We might think of a circle as a regular polygon with "an infinite number" of sides. Needless to say, Thor's ingenuity is leading him in the wrong direction. If he could *increase* the number of sides of his wheel enough times, the bumps could be eliminated entirely.

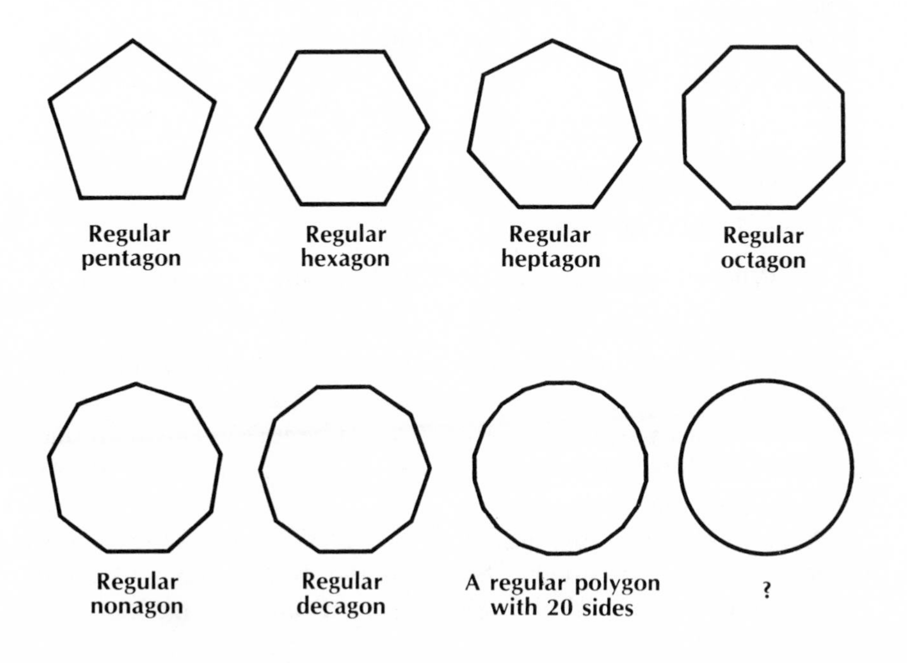

EXERCISES

Set I

Regular polygons are *symmetrical.* A flat figure can have two kinds of symmetry: line symmetry and point symmetry.

► A figure has **line symmetry** if a line can be drawn through it so that each point on one side of the line has a matching point on the opposite side at the same distance from the line. This means that if the figure is folded along the line of symmetry, the two halves will coincide (fit exactly together).

This figure in the shape of a kite has line symmetry. Several pairs of matching points are indicated. If the figure is folded along the line of symmetry, these points will coincide.

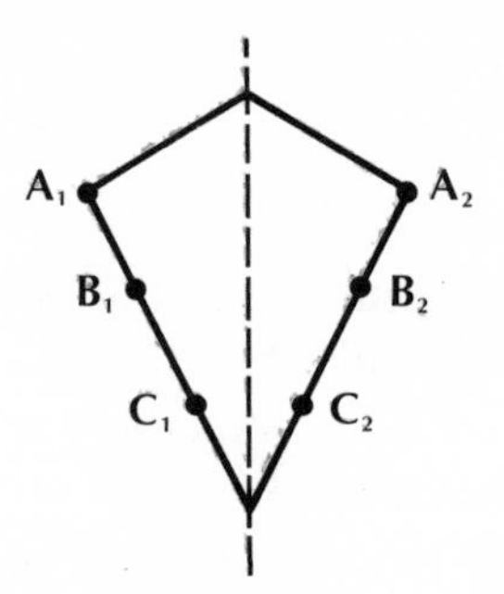

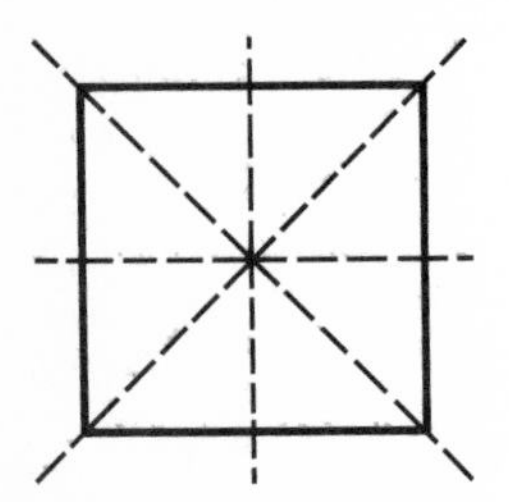

A figure can have more than one line of symmetry. For example, a square has four lines of symmetry.

Make a sketch of each of the following figures and draw in all lines of symmetry.

1. An equilateral triangle.
2. A regular pentagon.
3. A regular hexagon.
4. Without drawing a figure to find out, can you tell how many lines of symmetry a regular decagon has?
5. How many lines of symmetry do you think a circle has?

► A figure has **point symmetry** if it can be turned less than one revolution (360°) about a point so that it coincides with its original position.

A parallelogram (the figure shown at the top of the facing page) has point symmetry. Some of the points that will trade positions if it is turned 180° about point P are indicated.

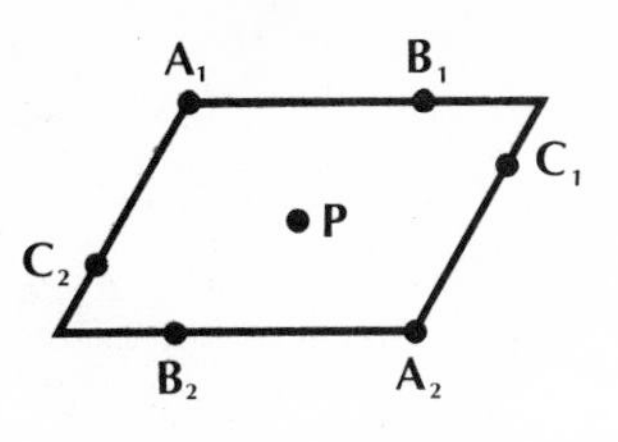

A figure with point symmetry can often be turned through more than one angle so that it matches its original position. For example, an equilateral triangle can be turned clockwise through two different angles: either 120° or 240°.*

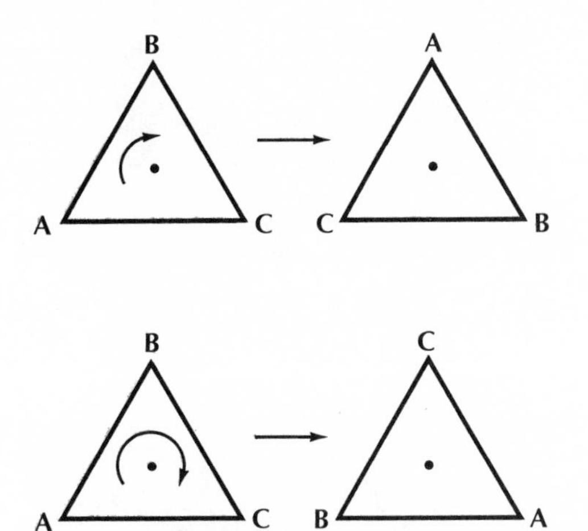

6. Make a series of lettered drawings like those for the equilateral triangle to show the different angles less than 360° through which a square can be turned clockwise so that it coincides with its original position.

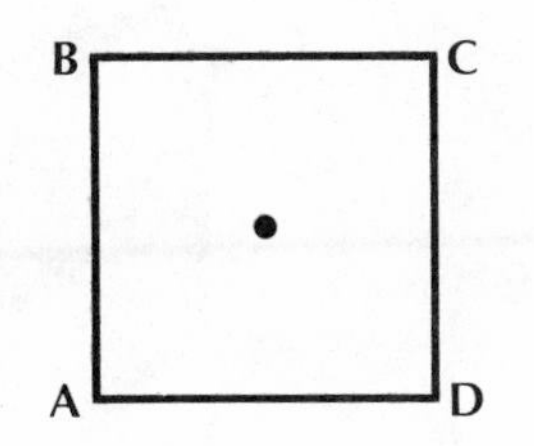

*Angles and their measures are discussed on pages 494–496.

Crest

Symplexity

Courtesy of California Computer Products, Inc.

7. Without drawing any figures, can you tell through how many different angles less than 360° a regular pentagon can be turned to match its original position?

8. Through how many angles do you think a circle can be turned?

Two examples of computer art are shown on the facing page. They were drawn on a CalComp plotter, called by its inventors "the machine that teaches the computer to draw." The first figure, titled *Crest,* took 5 minutes to produce, and the second, titled *Symplexity,* took 1 hour. As you can see, both are highly symmetric.

9. How many lines of symmetry does each figure have?

Both figures also possess point symmetry. The smallest angle through which *Crest* can be turned so that it coincides with its original position is one third of a revolution: 360°/3 = 120°.

10. Can you figure out the smallest angle through which *Symplexity* can be turned to coincide with its original position?

Set II

Experiment. A Kaleidoscope for Producing Regular Polygons

Since every regular polygon has lines of symmetry, a kaleidoscope can be used to form them. Our kaleidoscope is made with two mirrors hinged together with tape, like this.

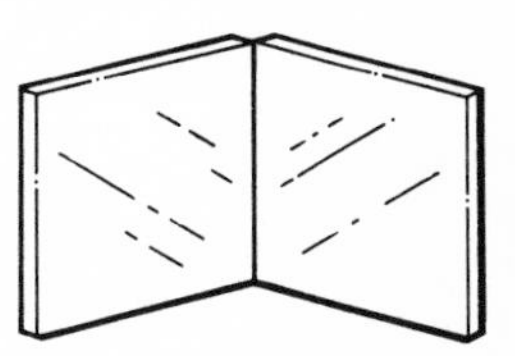

Draw a horizontal line segment about 4 inches long on your paper, and place the mirrors so that they are standing on the segment as shown in the diagram below.

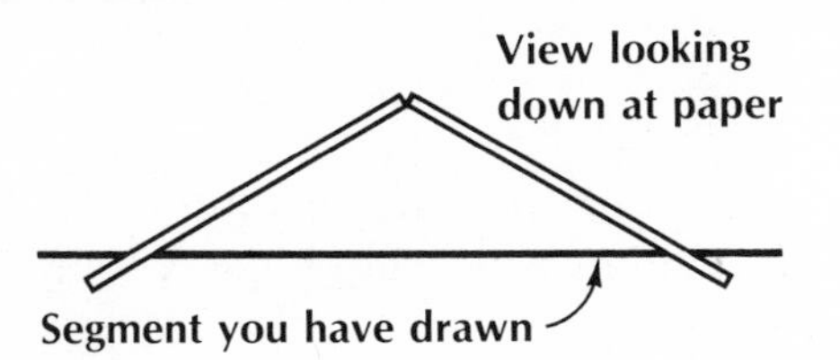

Now adjust the mirrors until you see an equilateral triangle. One third of the triangle will be in front of the mirrors and the rest will appear "behind" the mirrors.

Slowly turn the mirrors toward each other until you see a square and then continue until you have seen a regular pentagon, hexagon, heptagon, and octagon. By the time you get to the octagon, the mirrors will be so close together that it would be difficult to go much further.

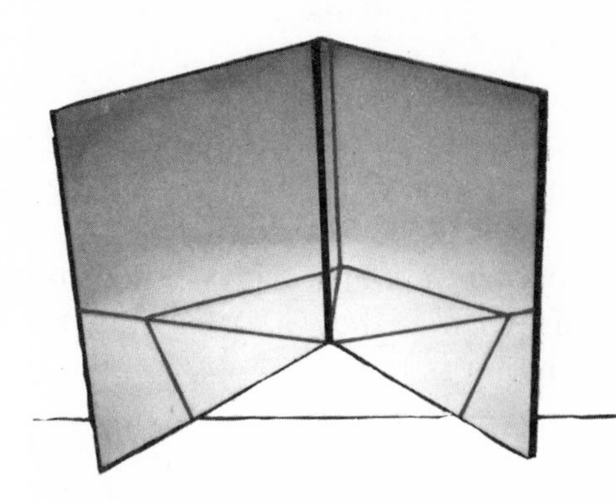

The kaleidoscope adjusted to produce a regular pentagon.

1. What do you think the figure would look like if you could see it when the mirrors were close enough together to be almost touching?

2. Move the mirrors apart to form an equilateral triangle again. Then try to form a figure with less than 3 sides. What happens?

Carefully adjust the mirrors so that you can see the equilateral triangle again. Move them back from the line to make the triangle as large as possible. Hold the mirrors firmly in place with one hand and trace the angle they form with your pencil. This angle is called a *central angle* of the triangle. Take the mirrors away and label the result, "central angle of equilateral triangle." It should look something like this:

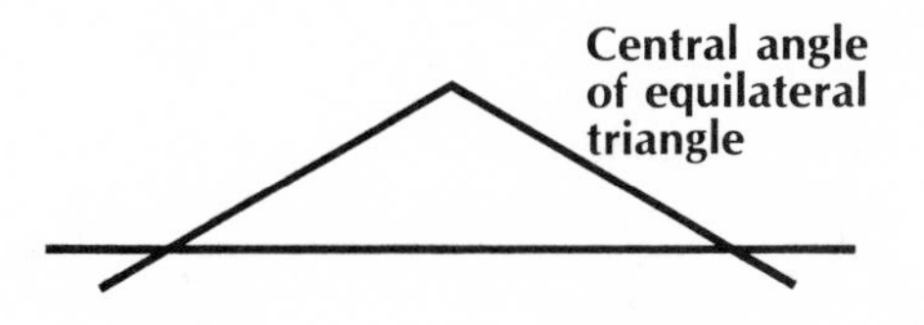

Now draw another line segment and adjust the mirrors as accurately as you can to form as large a square as possible. Again trace the angle formed by the mirrors with your pencil and label it appropriately.

Draw more line segments and trace the angles formed by the mirrors for each of the other regular polygons through the octagon. Now take a protractor and measure the central angles* you have drawn. The central angle of the equilateral triangle should measure about 120°.

3. Write your results in a table like this:

Regular polygon	Triangle	Square	Pentagon	Hexagon	Heptagon	Octagon
Central angle	120°	▒▒▒	▒▒▒	▒▒▒	▒▒▒	▒▒▒

4. What happens to the central angle of a regular polygon as the number of sides increases?

5. Can you state a rule for calculating its size exactly? (Remember that one revolution contains 360°.)

6. If you have thought of a rule, what would be the measure of the central angle of a regular polygon with 90 sides?

---

*If you are not sure of how to do this, look on page 495.

Set III

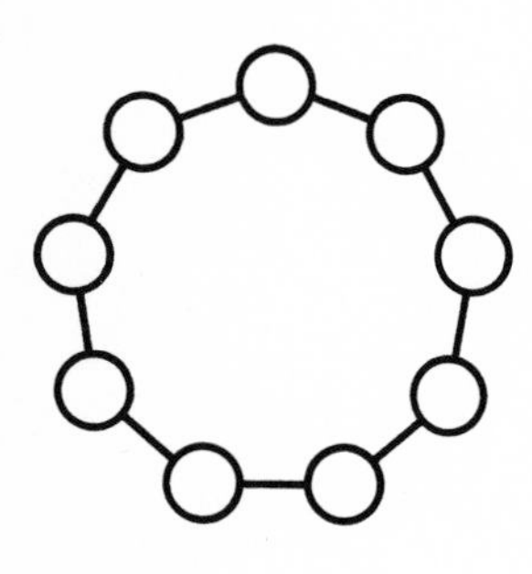

Here is a game for two players that has a winning strategy based on symmetry. A penny is placed on each corner of a regular polygon, say a nonagon, for example.

The two players take turns removing either one penny or two pennies that are next to each other. The player who picks up the last penny wins.

The game isn't fair, because the player who goes second can always win. Suppose the first player picks up two pennies, leaving the board as shown at the left below. The second player should then take the penny exactly opposite, so that the board looks like the diagram at the right below.

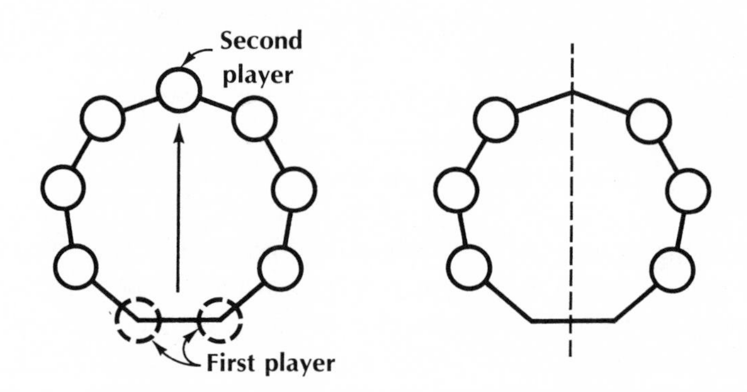

1. What should the second player do from this point on to be sure of winning the game?
2. If the first player picks up only one penny at the start, what should the second player's original move be?
3. Suppose the game were played on a regular polygon with an even number of sides instead of an odd number. Would this make any difference in the second player's strategy?

INTERESTING READING

The Ambidextrous Universe, by Martin Gardner, Basic Books, 1964.

Lesson 2

Some Regular Polygon Constructions

MATHEMATICS was one of the hobbies of the French general and emperor Napoleon. He was especially interested in geometry, and some of the problems he studied were about regular polygons.*

Since regular polygons possess so much point symmetry, they have a special relationship with the circle. The circle is the most symmetric of all geometric figures since it can be turned about its center through any angle and still coincide with its original position. One of the relationships between regular polygons and the circle is that a circle can be drawn so that it passes through every corner of a regular polygon. Such a circle is said to be *circumscribed* about the polygon and the polygon is said to be *inscribed* in the circle.

A regular heptagon inscribed in a circle.

One way to draw a regular polygon is to inscribe it in a circle. The traditional tools for doing this are the *straightedge,* for drawing straight lines, and the *compass,* for drawing circles and parts of circles called arcs. These tools were used by the Greeks in their geometry—it is said that Plato, the famous Greek philosopher, chose them because of their simplicity—and they have

* In 1797, Napoleon explained the solutions to some geometry problems that he had learned about while in Italy to a couple of world famous mathematicians in Paris who had never even heard of them. One of them told Napoleon: "General, we expect everything of you, except lessons in geometry!"

been the means ever since of making geometric constructions.

► **Geometric constructions** are drawings made with only two tools: a straightedge and a compass.

Here is an example of a simple geometric construction. To bisect a line segment means to divide it into two line segments having equal lengths. This can be done with a straightedge and compass in the following way. First, put the metal point of the compass on point A and draw an arc like the one shown in the first figure. Next, put the metal point on point B and draw an equal arc that crosses the first arc at a point above and a point below the line segment. Finally, draw a line through the two points where the arcs cross. This line bisects the line segment.

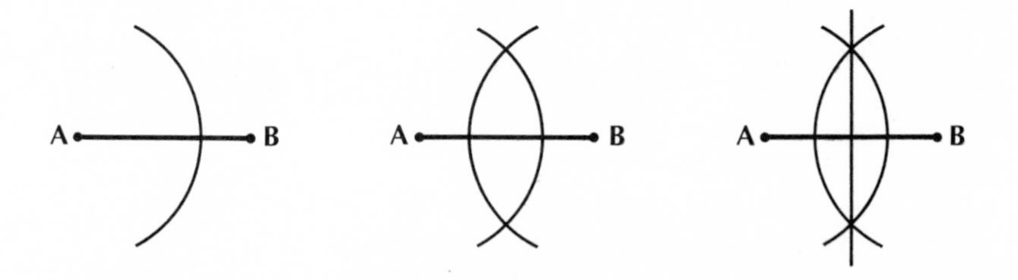

Of all the regular polygons, the regular hexagon is perhaps the easiest to inscribe in a circle, followed closely by the equilateral triangle and the square. Drawing a regular pentagon with straightedge and compass is rather tricky and the construction of a regular heptagon or regular nonagon with these tools is *impossible!* * You might think that it should be possible to solve every problem in mathematics if a person is clever enough and keeps trying. The Seabees of the United States Navy had a motto: "The difficult we do now, the impossible takes a little longer." But there are mathematical problems that are impossible now and always will be, no matter what someone may think of in the future. How do we know? Because these problems have been *proved* impossible to solve by the method in mathematics that guarantees its conclusions: *deductive reasoning.*

Perhaps the most famous impossible problem in mathematics is that of the *trisection* of an angle: dividing an angle with two lines into three equal angles, using only a straightedge and compass. It was not until 1837 that it was proved that this seemingly simple problem can not be done with only these instruments.

* Of course it *is* possible to draw these figures *if* you use a protractor or some other instrument.

EXERCISES

Set I

1. Draw a circle and inscribe a regular hexagon in it. This is easy to do because the *sides* of the hexagon are equal to the *radius* of the circle. Adjust your compass to the radius of the circle and divide the circle into 6 arcs. Then connect the 6 points.

2. Inscribe an equilateral triangle in a circle. Notice that this can be done by dividing the circle into 6 arcs as if you were going to draw a hexagon, and then connecting every other point.

3. One of the problems that interested Napoleon was: is it possible to mark the corners of a square on a circle using *only a compass?* The answer is yes, and the construction works like this.

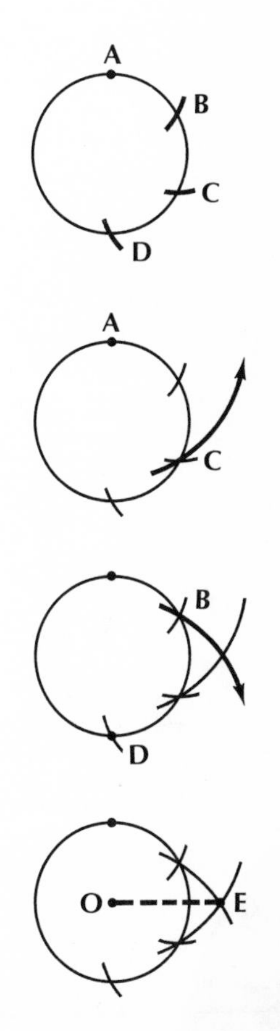

a) Draw a circle, mark a point on the top of it, and, leaving the distance between the metal point and the pencil point of the compass equal to the radius of the circle, draw 3 arcs like those shown in the first figure. Label the points A, B, C, and D.

b) Adjust the compass so that the metal point is on A and the pencil point is on C and draw an arc like the one shown in the second figure.

c) Now put the metal point on D and the pencil point on B and draw another arc crossing the previous one as shown in the third figure.

d) Label the point where the arcs cross E. Adjust the compass so that the metal point is at the center of the circle, O, and the pencil point is on E. The distance between these 2 points is equal to the length of the side of the square.

e) Now pick the compass up and, starting at A, draw 4 arcs around the circle, as shown in the fifth figure. The points where they intersect the circle are the corners of the square and we used just a compass to find them. Of course, to draw the sides of the square you will need to use your straightedge.

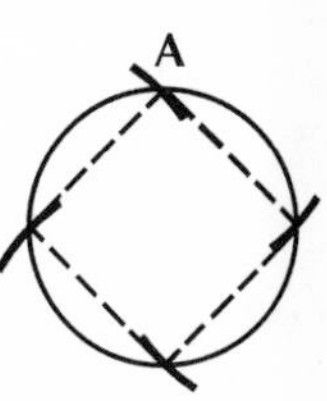

4. To inscribe an octagon in a circle, we can use the square we have just constructed to make things easier.

   a) First, draw a circle with the same radius as the circle with the square and then copy the 4 arcs from the first circle on to the second with your compass, as shown in the first figure below.

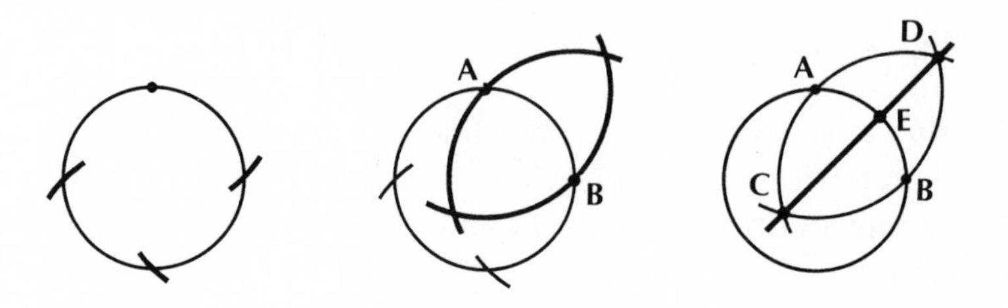

   b) Now label 2 of the points A and B and draw 2 equal arcs with centers at A and B, as shown in the second figure.

   c) Label the 2 points where the arcs you drew intersect C and D and draw a line through them. Label the point where this line intersects the circle E; this point bisects the arc AB (it divides it into 2 equal arcs).

   d) The distance between A and E is equal to the length of the side of the octagon and if you copy it around the circle you should get 8 equal arcs. Join the points with line segments and the octagon is complete.

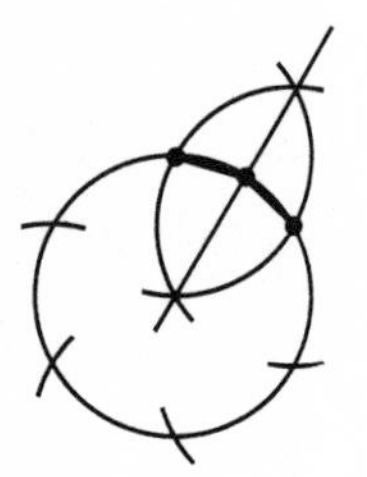

5. Draw a large circle (radius about 2 inches) and use the same method to inscribe a regular dodecagon in it. A *dodecagon* is a polygon with 12 sides, so the first thing to do is to find the corners of a regular hexagon on the circle and then bisect each of the 6 arcs.

Set II

The construction of a regular pentagon with a straightedge and compass takes quite a bit of patience. There are several ways to do it, and we will try one of the shortest of them.

1. First, draw a circle with a radius of about 1.5 inches and draw a diameter in it, as shown in the first figure at the

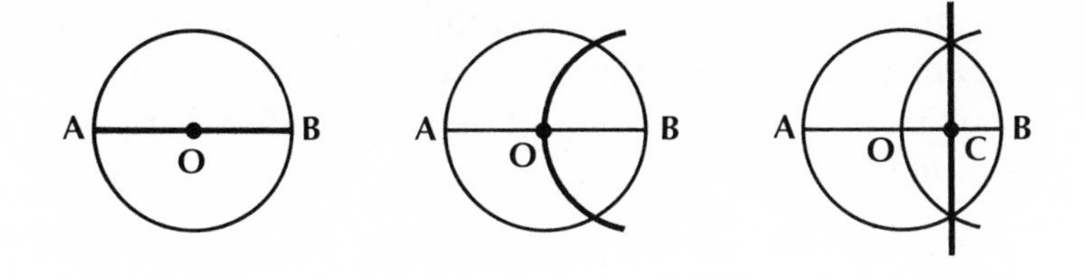

top of this page. Label the center of the circle O and the endpoints of the diameter A and B.

2. Place the metal point of your compass at B and the pencil point on O and draw an arc like the one shown in the second figure above.
3. Draw a line through the 2 points where the arc crosses the circle, and label the point where it intersects the diameter point C. This point is midway between points O and B.

To make the remaining diagrams easier to understand, some of the lines have been omitted. Do not erase, however, any lines on your own paper.

4. Put the metal point of the compass at A and the pencil point about halfway between C and B, and draw an arc as shown in the first figure below.

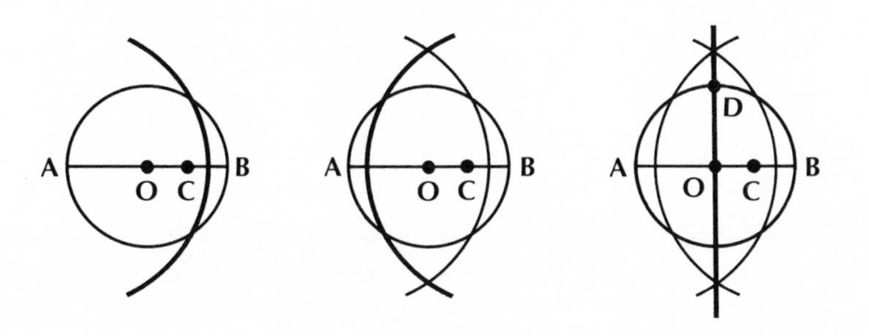

5. Keeping your compass at the same width, put the metal point at B and draw an equal arc which crosses the other arc at points above and below the circle, as shown in the second figure above.
6. Draw a line through the 2 points where the arcs cross. It should pass through the center of the circle, O; notice that it is perpendicular to the diameter AB. Label the point where it crosses the top of the circle D.

7. Put the metal point at C and the pencil point on D. Draw an arc to the left that crosses the diameter of the circle as shown in the figure below. Label the point where it meets the diameter E.

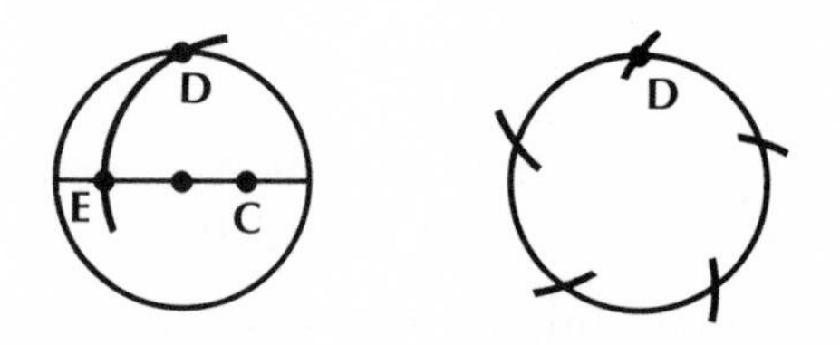

8. The distance between D and E is equal to the length of a side of the pentagon we have been going through all of this work to construct. Put the metal point at D and adjust the compass so that the pencil point falls on E. Now if your work has been very accurate, you should be able to draw 5 equal arcs around the circle, starting and ending at D. If these points are joined with line segments, a regular pentagon will be formed.

You are probably wondering how anyone thought of all of this in the first place and why it works. To explain the answers to these questions would require some algebra which is beyond the scope of this course. By carrying out this construction, you have some idea of how tricky geometric constructions with just a straightedge and compass can be.

Set III

Here is an interesting puzzle. It is called a *dissection* puzzle and the problem is to cut a regular dodecagon up into pieces that can be rearranged to form a square. Some dissection puzzles were discovered by the Greeks, and in the 10th century a Persian mathematician wrote an entire book about them. Another book has been written recently by the world's current expert on dissection puzzles, a man who lives in Australia.*

Place a sheet of paper underneath the large dodecagon you constructed in Set I and poke a hole with the metal point of your compass at each of the 12 corners. Now join the 12 holes on your new sheet of paper to form another dodecagon and then

**Geometric Dissections*, by Harry Lindgren, Van Nostrand 1964.

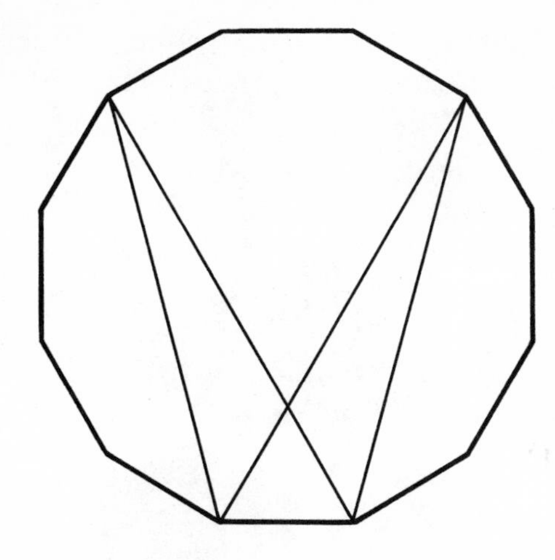

draw in 4 more line segments like those shown in the figure at the top of this page to divide it into 6 parts. Notice that the resulting figure has line symmetry and that one of the 6 parts seems to be an equilateral triangle.

Cut the 6 pieces apart with scissors and try to rearrange them to form a square. If you can, make a sketch to show what the arrangement of the pieces looks like.

INTERESTING READING

The Borders of Mathematics, by Willy Ley, Pyramid Books, 1967: Chapter 2, "The Seven-Spoked Wheel," and Chapter 3, "Enter Carl Friedrich Gauss."

Lesson 3

Mathematical Mosaics

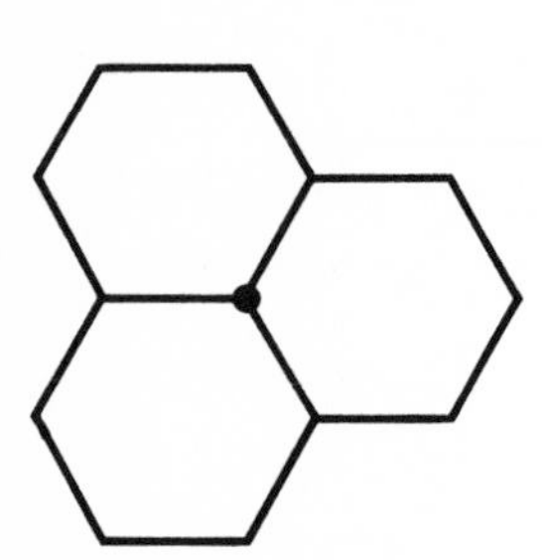

DID you know that bees are expert mathematicians when it comes to building honeycombs? The Greek mathematician, Pappus, who lived in the fourth century A.D., said: "Though God has given to men the best and most perfect understanding of wisdom and mathematics, He has allotted a partial share to some of the unreasoning creatures as well. . . . This instinct is specially marked among bees. They prepare for the reception of the honey the vessels called honeycombs, with cells all equal, similar and adjacent, and hexagonal in form." *

Pappus had observed that the bees use the regular hexagon exclusively for the shape of the cells in the honeycomb. The photograph shows what the arrangement of the cells looks like. Each corner point in the honeycomb is surrounded by exactly 3 hexagons, and there is no space wasted between the cells.

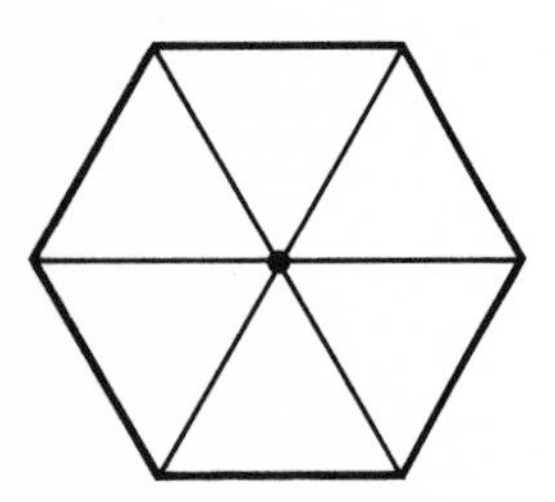

Is this hexagonal arrangement of cells the most efficient one? Or would other regular polygons work just as well? Let's try some of them and see.

The simplest regular polygon is the equilateral triangle. The angles of an equilateral triangle are smaller than those of a regular hexagon and so more than 3 equilateral triangles will fit around a point. In fact, there is room for exactly 6.

---

* *Selections Illustrating the History of Greek Mathematics,* by Ivor Thomas, Harvard University Press, 1939.

If the bees used equilateral triangles, their honeycomb would look like this.

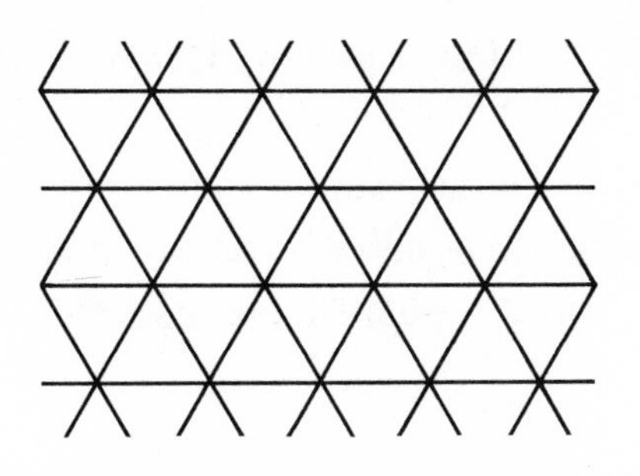

This arrangement, however, wastes wax, because if the cells are to be the same size as in the hexagonal honeycomb, more material is needed to form them.

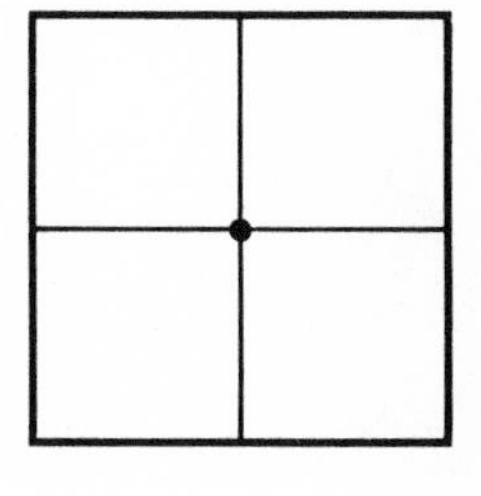

Another possibility is the square. Each angle of a square is a right angle and exactly 4 will fit around a point. A honeycomb having square cells requires, as does the one with triangular cells, more material for the walls and so it also would waste wax.

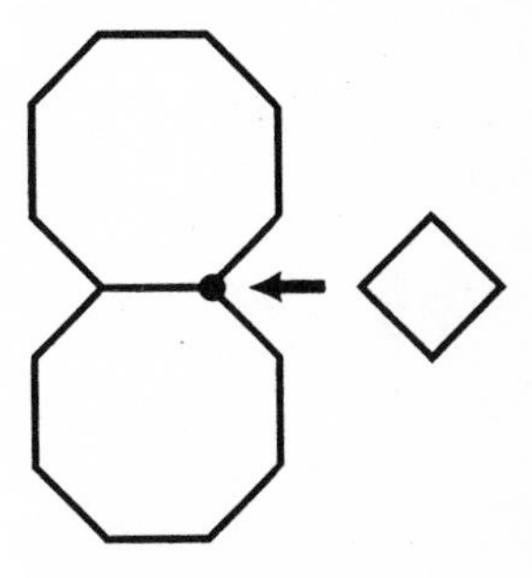

Suppose the bees used a regular polygon having *more* sides than a hexagon instead of fewer sides. How would, say, regular octagons work? The angles of a regular octagon are larger than those of a hexagon, so that there is room for only 2 around a point, leaving some space unfilled. However, there is just the right amount of room for a square in this space and so 2 regular octagons and 1 square surround a point exactly. If this pattern is repeated, the honeycomb would look like this.

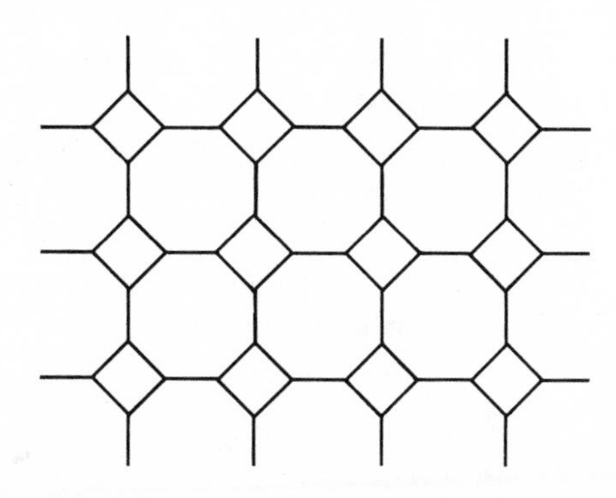

This design has the disadvantage of being more complicated than the simple hexagonal pattern and it also would require more wax.

► Each of the designs we have considered might be called a **mathematical mosaic.** By a mathematical mosaic, we mean an arrangement of regular polygons in which every corner point is surrounded exactly by the same number of polygons of each kind.

To represent the honeycomb arrangement, we will use the symbol 6-6-6 to indicate that each point is surrounded by 3 regular hexagons (6-sided polygons). The symbols for the other mosaics we have considered are:

3-3-3-3-3-3 (6 equilateral triangles surround each point),
4-4-4-4 (4 squares surround each point), and
4-8-8 (1 square and 2 regular octagons surround each point).

EXERCISES

Set I

Experiment. Mathematical Mosaics

You have seen four different ways of covering a surface with regular polygons to form a mathematical mosaic. What other ways are possible? To answer this, we will experiment with regular polygons of several different shapes. You may be provided with the necessary polygons or you can make a set yourself out of stiff paper as follows.

The figure on the facing page contains the 6 shapes you will need. Trace the figure onto a piece of tracing paper. Put a sheet of stiff paper underneath the tracing paper and poke a hole with the metal point of a compass at each of the 12 corners of the dodecagon. Then remove the paper, use a straightedge to draw the sides of the polygon, and carefully cut it out. Repeat this procedure until you have:

2 regular dodecagons,
2 regular octagons,
2 regular hexagons,
3 regular pentagons,
2 squares,
4 equilateral triangles.

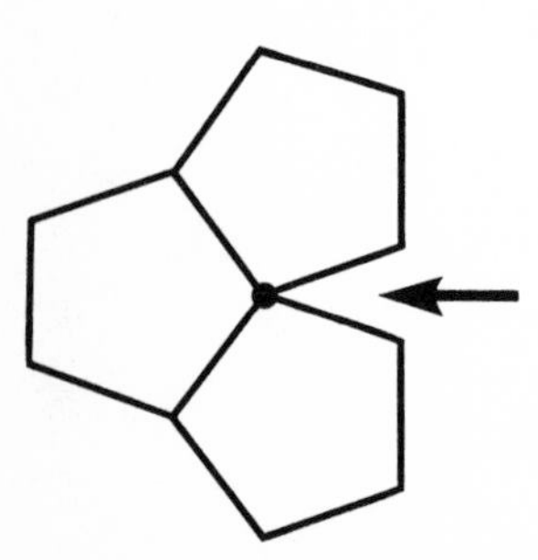

Now, draw a point on your paper and try to surround it with the 3 pentagons. The result should look something like the figure at the left. When the pentagons are moved so that 2 pairs of sides touch, there is a gap between the third pair of sides. The gap is

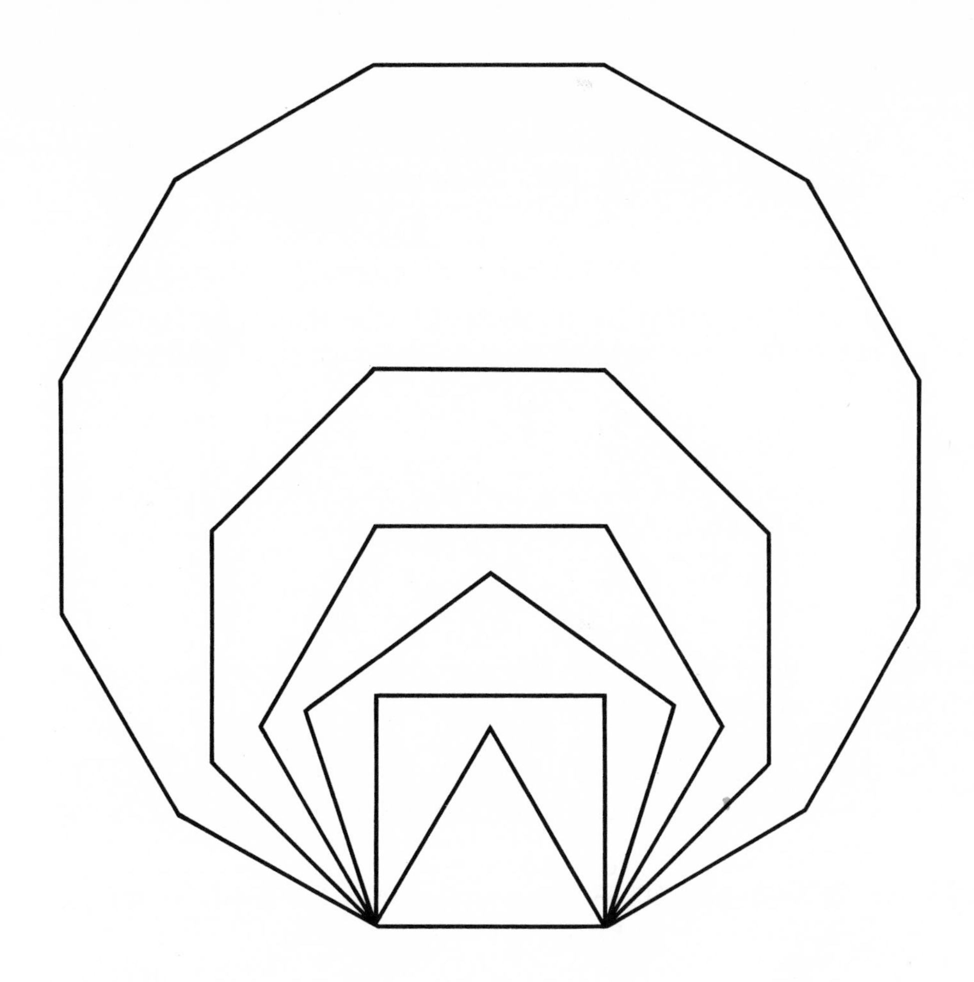

too small for another pentagon to fit, so it is impossible to make a mosaic with regular pentagons only.

Try surrounding the point with 2 dodecagons and an equilateral triangle. This time the fit is perfect, so we can record this mosaic with the symbol 3-12-12.

There are several other possible arrangements for forming mathematical mosaics. See how many of them you can find, and keep track of your discoveries by adding their symbols to this list of the ones we have found so far:

3-3-3-3-3-3
4-4-4-4
6-6-6
4-8-8
3-12-12.

Set II

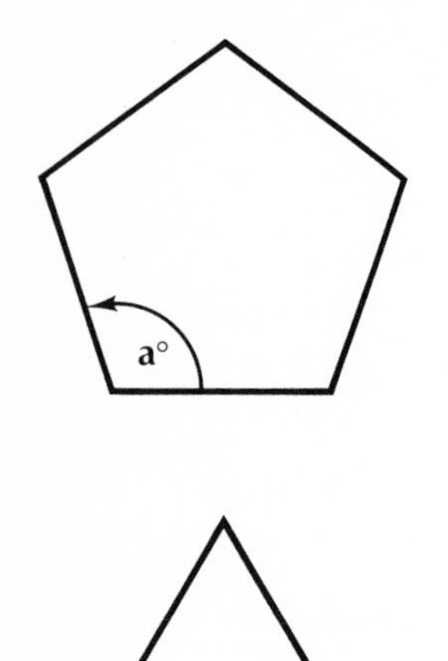

The size of an angle of a regular polygon is a function of the number of sides. A formula for this function is

$$a = 180 - \frac{360}{n},$$

where a represents the measure of one of the angles in degrees and n represents the number of sides. For an equilateral triangle,

$$a = 180 - \frac{360}{3} = 180 - 120 = 60°.$$

1. Copy the following table for this function and use the formula to fill in the missing numbers.

n	3	4	5	6	8	12
a	60°					

2. What happens to a, the size of the angles, as n, the number of sides, increases?

Exactly 360 angles, each with a measure of 1°, will completely surround a point. We can check that the first two entries in our list of mosaics are correct like this:

3-3-3-3-3-3	60° + 60° + 60° + 60° + 60° + 60° = 360°
4-4-4-4	90° + 90° + 90° + 90° = 360°.

3. Check each of the remaining entries in your list in the same way.
4. In some cases, you may have gotten results other than 360°, such as 363°. What would a number like this mean: that there was a slight *gap* between angles, or that there was a slight *overlap* that you didn't notice?

Set III

We have defined a mathematical mosaic as an arrangement of *regular polygons* that completely cover a surface. It is possible to fill a surface with a repeating pattern of other shapes as well. The clever Dutch artist, Maurits Escher, whose drawing *Waterfall* is reproduced on page 19 of this book, has invented many drawings using picture shapes to fill a surface without leaving any open spaces. A remarkable example of these drawings is the

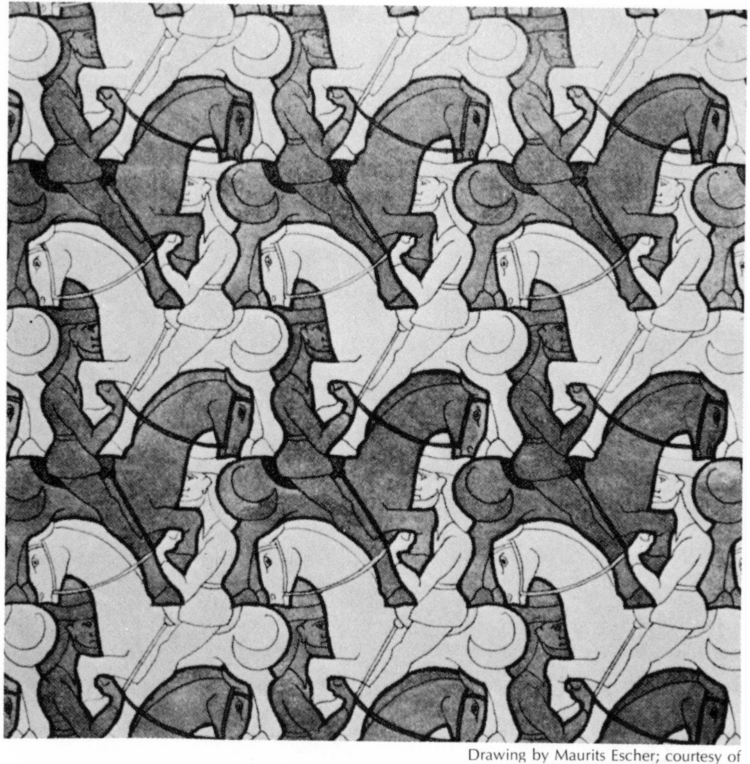

Drawing by Maurits Escher; courtesy of G. W. Breughel, Zwolle, Netherlands.

one of knights on horseback shown above. Needless to say, a design like this is extremely difficult to work out.

Although only 3 regular polygons can be used to form mosaics containing just one shape, four-sided figures of many different shapes will work! Make a copy on a 3 × 5 card of the quadrilateral shown here. Cut it out and use it as a pattern to cut out about 15 copies. If you make them carefully, you can get 4 copies out of one 3 × 5 card.

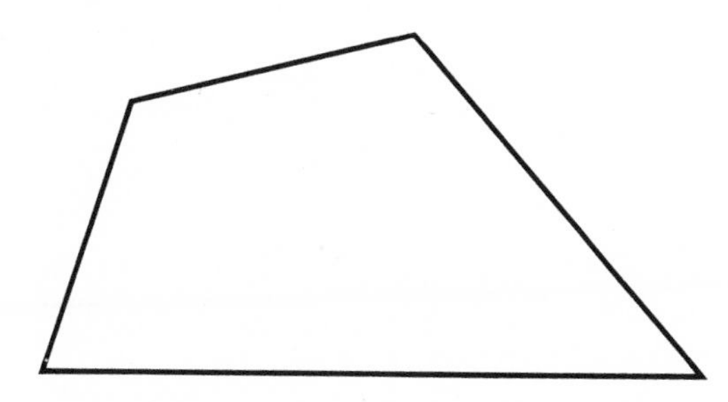

Now slide these around on another sheet of paper until you can get them to fit together to cover the surface without any gaps. Pairs of *equal sides* should be *touching* and every corner should be surrounded by 4 quadrilaterals. When you find the pattern, trace part of it on the paper with your pencil.

INTERESTING READING

The Graphic Work of M. C. Escher, by M. C. Escher, Meredith, 1967.

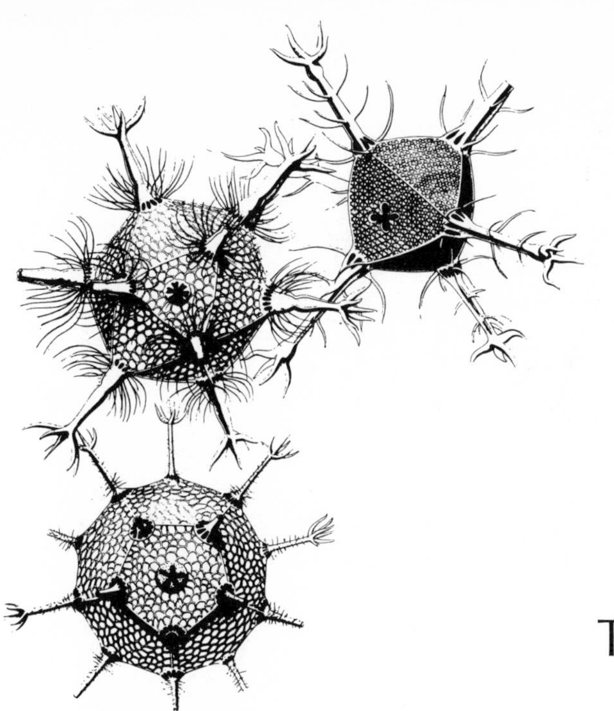

Lesson 4

The Regular Polyhedra

SINCE we have found regular hexagons in the bees' design of the honeycomb, perhaps it is not too surprising that other regular polygons also appear in nature. The skeletons of some tiny sea creatures called radiolaria, so tiny that their structure can only be seen under a microscope, are shown here. These one-celled animals are found floating near the surface of warm ocean water. Although their surfaces are not flat, the skeletons of radiolaria appear to be built of equilateral triangles and regular pentagons; more than that, each one has the shape of a *regular polyhedron.*

► A **regular polyhedron** * is a solid with faces (surfaces) in the shape of regular polygons. All of its faces, edges, and corners are identical.

The regular polyhedra were first studied by a group of Greek mathematicians under the leadership of Pythagoras more than 2,000 years ago. It is thought that the famous geometry book by Euclid, the *Elements,* may have been written primarily to deal with these geometric solids.

The simplest regular polyhedron is the regular **tetrahedron,** ** a solid with 4 equilateral triangles for its faces. Three triangles meet at each corner of a tetrahedron. The figure at the top of the following page shows a tetrahedron and the three triangles that meet at one of its corners.

---

*The plural of *polyhedron* is *polyhedra.*

**The Greek word for "four" was "tetra."

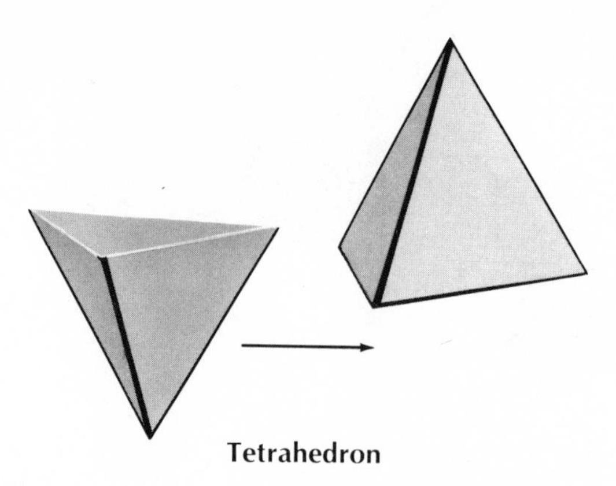

Tetrahedron

If there were 4 triangles at each corner instead of 3, we would have a solid with 8 faces, called a regular **octahedron.**

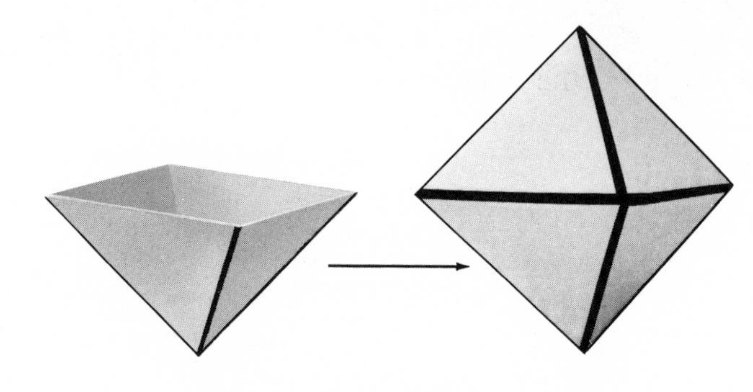

Octahedron

Five triangles at each corner result in a solid called a regular **icosahedron.** An icosahedron has 20 faces in all.

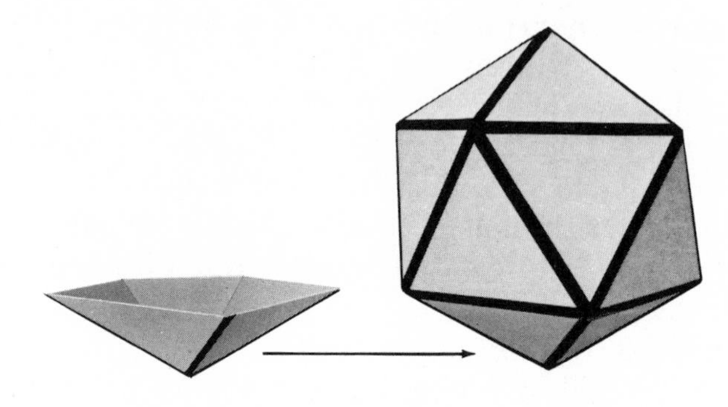

Icosahedron

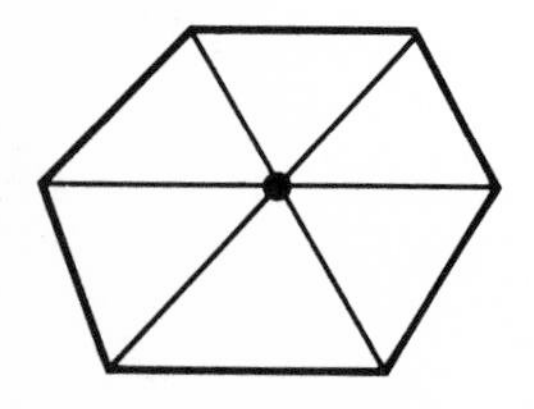

It is impossible to build a regular polyhedron having 6 triangles at each corner because 6 equilateral triangles meeting at one point lie flat ($6 \times 60° = 360°$). Since there must be at least 3 faces to form a corner, the only regular polyhedra having triangles for faces are the three we have just considered: the tetrahedron, the octahedron, and the icosahedron.

The most familiar regular polyhedron is the one whose faces are squares. Three squares meet at each corner of a **cube**; since it has 6 faces altogether, what do you think the Greek name for a

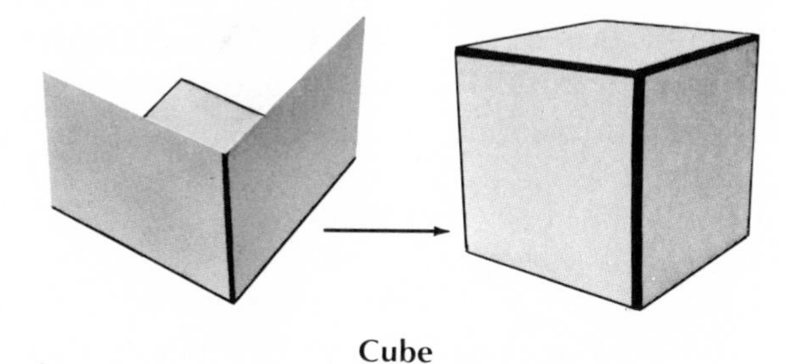

Cube

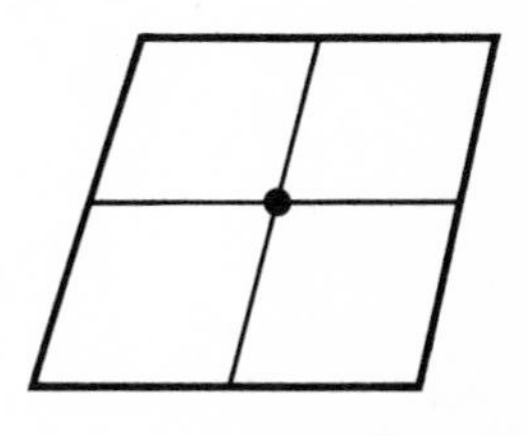

cube was? Because 4 squares meeting at one point also lie flat ($4 \times 90° = 360°$), the cube is the only regular polyhedron with square faces.

You will recall that when we put 3 regular pentagons together around a point, they left a gap. If we fold them upward to close the gap, we get a corner of a regulár **dodecahedron,** a solid having 12 faces in all. Since 3 regular hexagons around one point

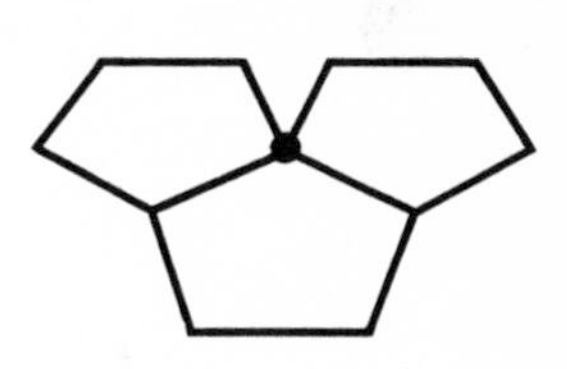

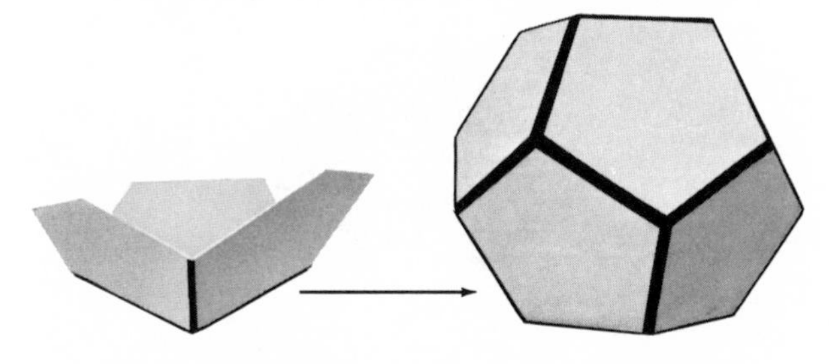

Dodecahedron

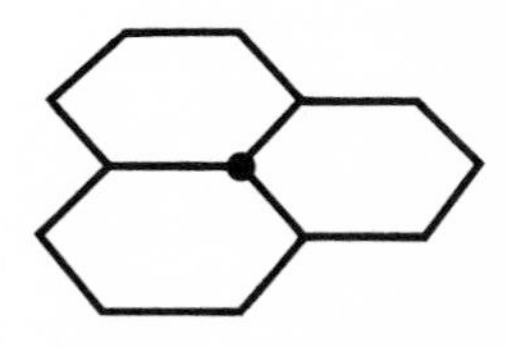

already lie flat, no regular polyhedron is possible with faces having 6 or more sides. This means that there are only *five regular polyhedra* in all;* in fact, Lewis Carroll once called them "provokingly few in number."

The regular polyhedra are sometimes referred to as the *Platonic solids,* named for Plato, who claimed that the atoms of the elements had these shapes. The Greeks thought that there were 4 elements: fire, earth, air, and water. Plato said that atoms of fire had the shape of tetrahedrons, atoms of earth, the shape of cubes,

---

*You will recall that the number of different *regular polygons* is unlimited.

atoms of air, octahedrons, and atoms of water, icosahedrons. The philosophers said that the universe was in the shape of a dodecahedron.

Now, these ideas seem silly, and yet Plato was on the right track in his attempt to explain the shape of matter! Although chemists have long known that atoms do not have the shape of regular polyhedra, it is true that many crystals do, because of the arrangement of their atoms! The cube and tetrahedron appear as the shape of sodium chlorate crystals; the octahedron in crystals of chrome alum. It looks as if these crystals were cut into these shapes, but this is not the case; they *grew* that way!

EXERCISES

Set I

We have been representing mathematical mosaics with numerical symbols to show how many regular polygons of what kind surround each corner point. The same kind of symbols can be used to represent the regular polyhedra. For instance, the symbol for a tetrahedron would be 3-3-3 (to show that 3 triangles meet at each corner), and the symbol for an octahedron would be 3-3-3-3.

1. Write symbols for the cube, dodecahedron, and icosahedron.

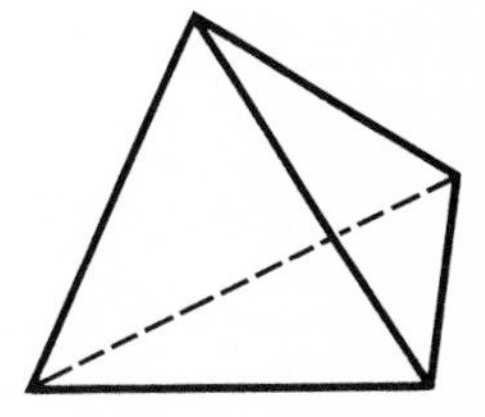

A tetrahedron has 4 faces (surfaces), 4 corners, and 6 edges (line segments), as this "transparent" diagram (right) shows.

2. Use the diagram below to figure out how many faces, corners, and edges a cube has.

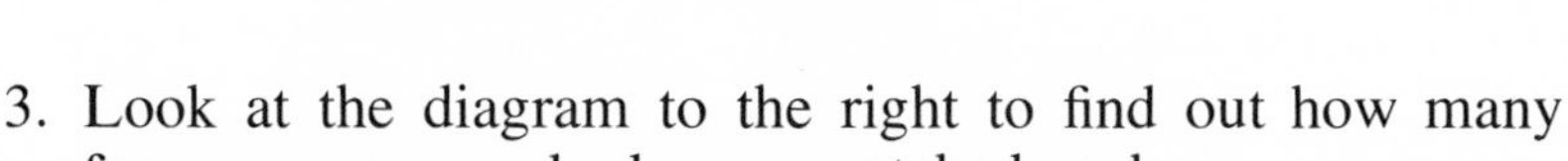

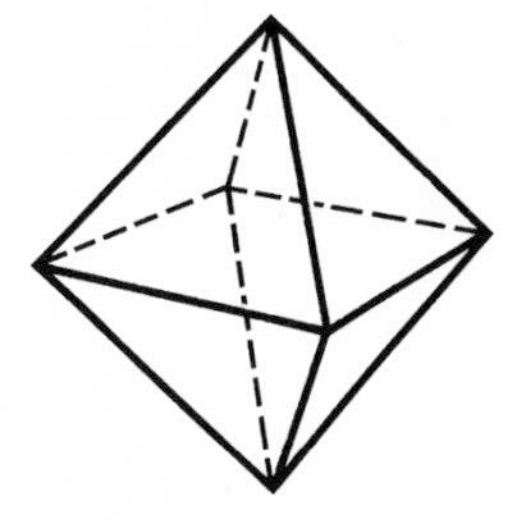

3. Look at the diagram to the right to find out how many faces, corners, and edges an octahedron has.

It would be rather difficult to find the number of corners and edges of a dodecahedron or icosahedron by counting, so we will use deductive reasoning instead.

A dodecahedron has *12 faces,* each of which is a regular pentagon.

4. How many corners does a pentagon have?
5. How many corners do 12 separate pentagons have?
6. In a dodecahedron, each *corner* is a point where 3 pentagons meet (3 corners "overlap"), so if you divide your answer to the last question by 3, you will have the number of corners in a dodecahedron.
7. How many sides does a pentagon have?
8. How many sides do 12 separate pentagons have?
9. In a dodecahedron, each *edge* is a place where 2 pentagons meet (2 sides "overlap"). How many edges does a dodecahedron have?

An icosahedron has *20 faces,* each of which is an equilateral triangle.

10. How many corners do 20 separate triangles have?
11. How many triangles meet at each corner of an icosahedron?
12. How many *corners* does an icosahedron have?
13. How many sides do 20 separate triangles have?
14. How many triangles meet at each edge of an icosahedron?
15. How many *edges* does an icosahedron have?
16. Copy and complete the following table. Use your results in the previous exercises to fill in the missing numbers.

Regular polyhedron	No. of faces	No. of corners	No. of edges
Tetrahedron	4	4	6
Cube	6		
Octahedron	8		
Dodecahedron	12		
Icosahedron	20		

17. Can you find some patterns in your table that are illustrated by the models of regular polyhedra shown here?

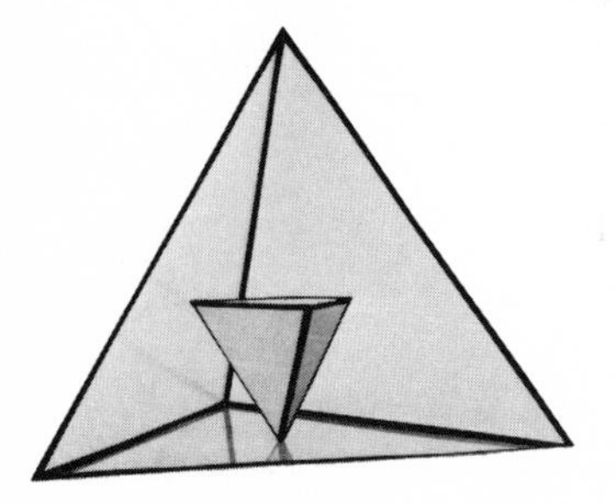
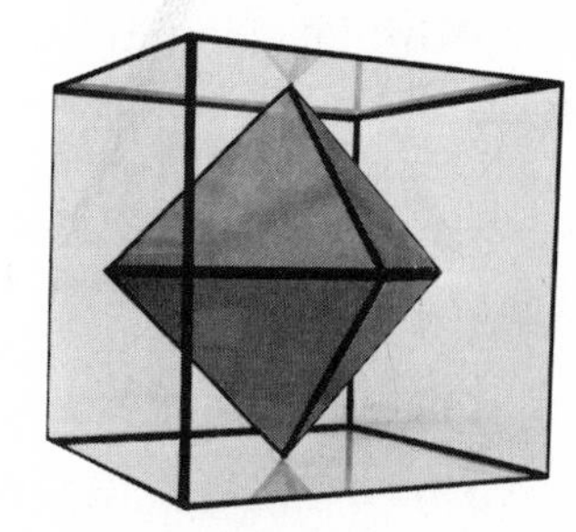

18. At the beginning of this lesson, you saw some pictures of sea animals called radiolaria; can you identify the shape of each one shown?

Set II

Experiment. Models of Two Regular Polyhedra

Part 1. A Model of an Icosahedron We will construct this model by threading drinking straws together to form the edges.

First, cut the straws into 30 *equal* lengths (about 3 inches is convenient). Thread 3 of these on a long length of thread, form a triangle, and tie a knot at the 2 ends of the thread like this. Cut off the extra thread.

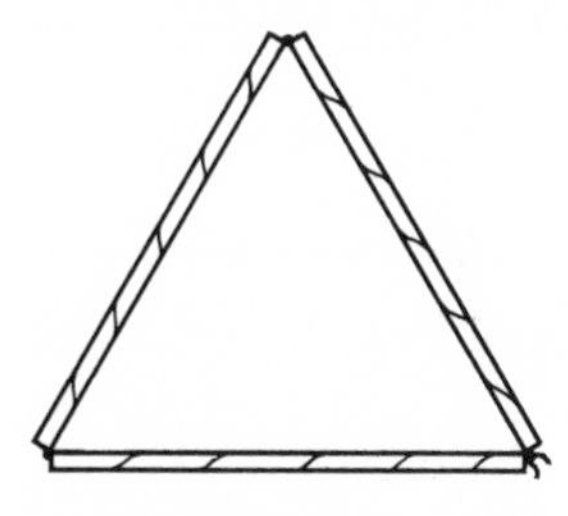

Now thread again through one of the straws of the triangle, add 2 more straws, and tie together to form a second triangle that shares one of its sides with the first triangle.

Continue in the same way, *making sure that 5 triangles are formed at each corner*. The icosahedron will not hold its shape very well until most of the 30 straws have been threaded into place. Your finished model should look something like this.

Part 2. A Model of a Dodecahedron It would be impossible to build a dodecahedron by threading straws together because the faces of a dodecahedron are pentagons; *triangles* are the only polygons that are rigid and hold their shape. We will use another method instead. The large figure at the top of page 216 is a pattern for our model. Make 2 copies of this figure on tracing paper. Then put a piece of stiff paper underneath the tracing paper and poke a hole with the metal point of a compass at each of the 20

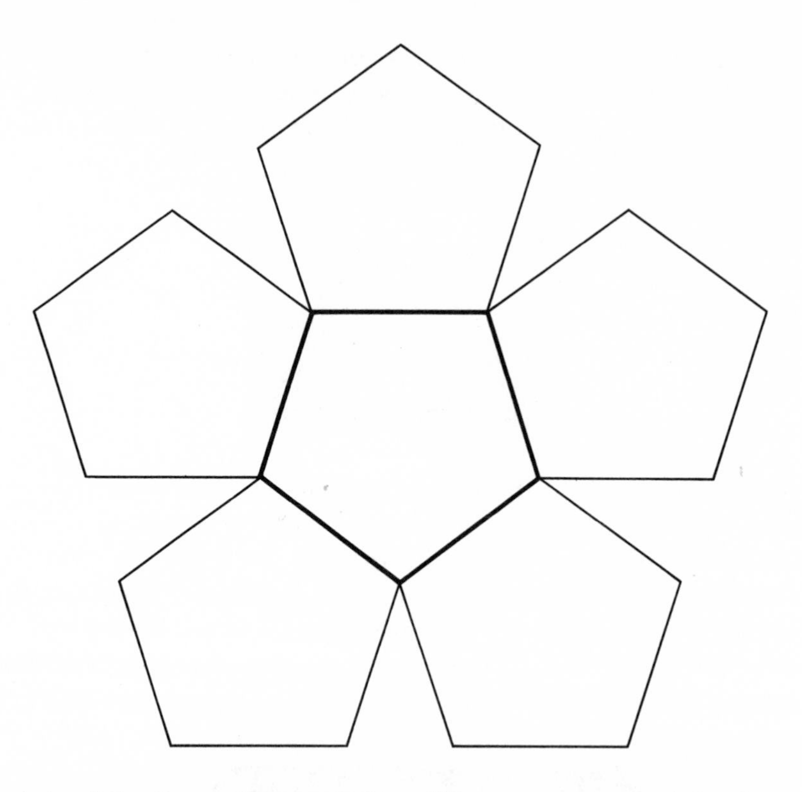

points of the figure. Remove the stiff paper, use a straightedge to draw all of the line segments between the 20 holes, and cut the figure out. Score each of the 5 sides of the pentagon in the center of each copy (the heavy line segments on the pattern) with your scissors, and fold each of the 5 surrounding pentagons up to make a bowl-shaped figure.

Now hold the 2 "bowls" facing each other and press them together so that they flatten out. Take a rubber band and weave it alternately above and below the corners as shown in this figure,

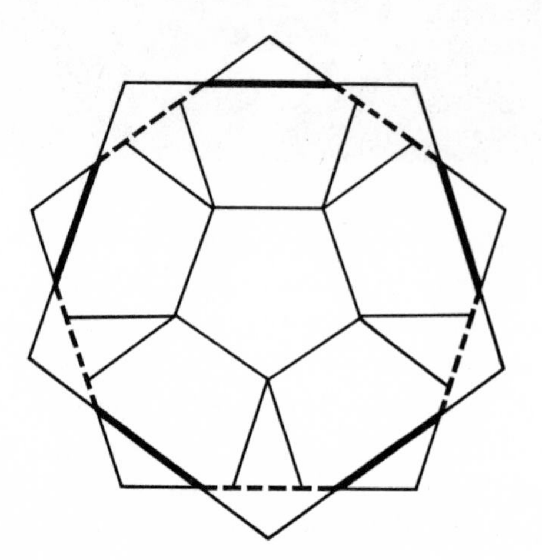

while holding the 2 pieces flat.

If you carefully let go, a dodecahedron will pop into shape!

Set III

Alexander Graham Bell, the inventor of the telephone, was fascinated by the regular tetrahedron. He invented and flew a number of large kites made from networks of tetrahedrons several years before the Wright brothers built their first airplane. The architecture of a museum of Bell's inventions in Nova Scotia is based on the tetrahedron shape.

Here is a puzzle with a tetrahedron which is currently available in some stores and which you can easily make yourself. Make 2 copies of the pattern shown below on a couple of 4×6

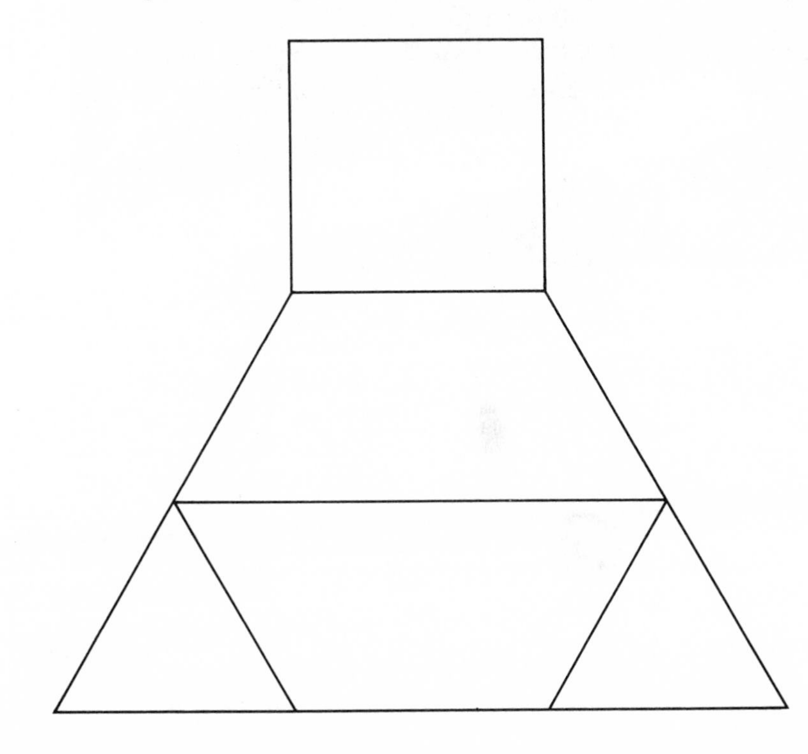

cards and cut them out. Fold each one along the lines and tape the edges of each together to form a pair of identical solids.

The puzzle is to put the two solids together to form a tetrahedron. Can you do it?

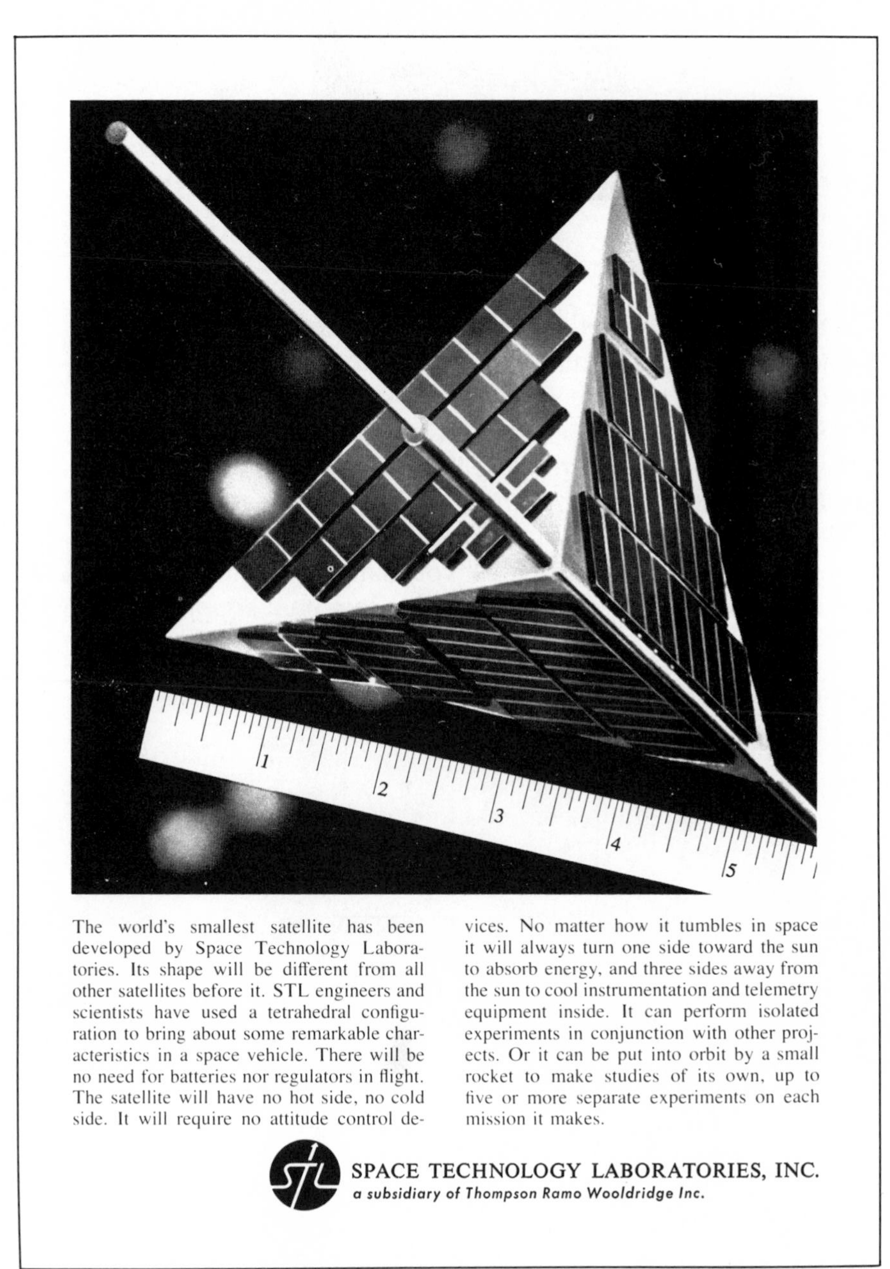

Courtesy of TRW Inc.

The tetrahedron used in the design of a space satellite.

INTERESTING READING

The Architecture of Molecules an illustrated book by Linus Pauling and Roger Hayward, W. H. Freeman and Company, 1964.

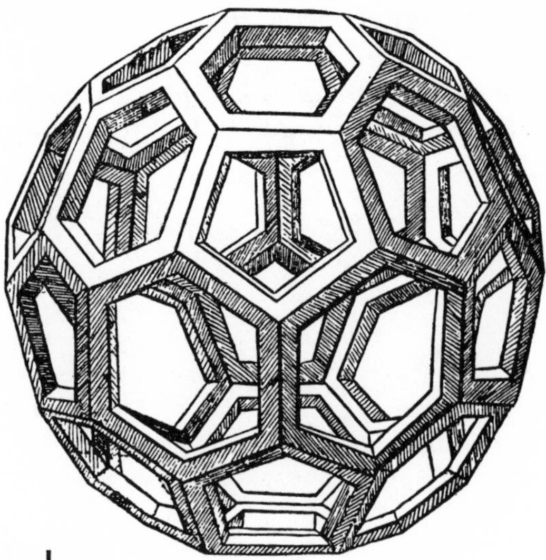

Lesson 5

Some Semi-Regular Polyhedra

LEONARDO DA VINCI, the great Renaissance artist, scientist, and inventor, began one of his books with these words: "Let no one who is not a mathematician read my works." The painter of the *Mona Lisa* and the *Last Supper* was, like Napoleon, interested in geometry, and devoted much study to the geometric solids called polyhedra.

You have learned that there are only five *regular polyhedra,* or Platonic solids: solids with faces in the shape of *regular polygons of one kind* and with the same number of polygons at each corner. Another set of geometric solids are the **semi-regular polyhedra**, solids with faces in the shape of more than one kind of regular polygon, yet with every corner the same. Among the semi-regular polyhedra are thirteen solids called the **Archimedean solids.** They are named after Archimedes, who wrote a book about them. Leonardo da Vinci made models of the Archimedean solids by using wooden sticks for their edges; his drawing of one of these models, a "truncated* icosahedron," is shown above. The faces of this polyhedron have two different shapes, and at each corner one regular pentagon and two regular hexagons meet. We have used numerical symbols to represent the pattern of regular polygons in mathematical mosaics

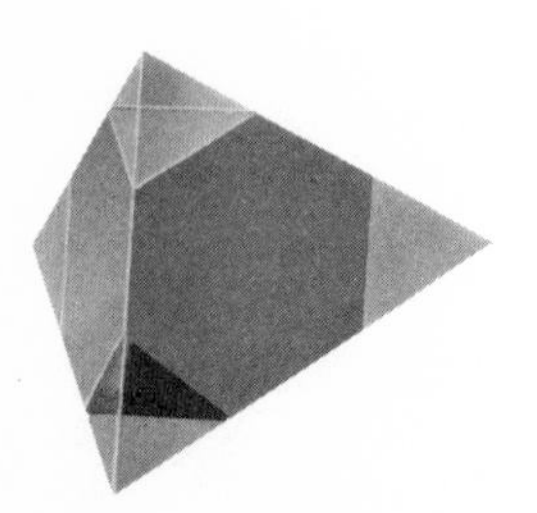

* "Truncated" means that the corners have been "cut off." Here, a regular tetrahedron has been cut at the corners to form a "truncated tetrahedron."

and the regular polyhedra; the symbol for the truncated icosahedron is 5-6-6. It is interesting to know that a soccerball has the pattern of a truncated icosahedron on its surface.

Here is a drawing of another Archimedean solid, called a "cuboctahedron." Its symbol is 3-4-3-4, since two equilateral triangles and two squares surround each corner. The cuboctahedron, like all of the regular and semi-regular polyhedra, is very symmetrical. In fact, such a polyhedron can be put inside a sphere so that each corner touches the sphere; the polyhedron is said to be "inscribed" in the sphere.

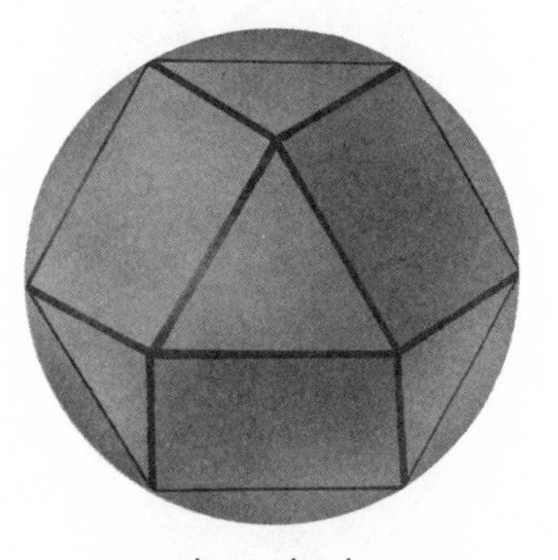

A cuboctahedron inscribed in a sphere.

EXERCISES

Set I

The names (some of them are real tongue-twisters) and photographs of models of the 13 Archimedean solids are shown on the following 4 pages.

Count the number and kind of regular polygons that surround one corner of each solid and write the appropriate symbols.

1. Truncated tetrahedron

2. Truncated cube

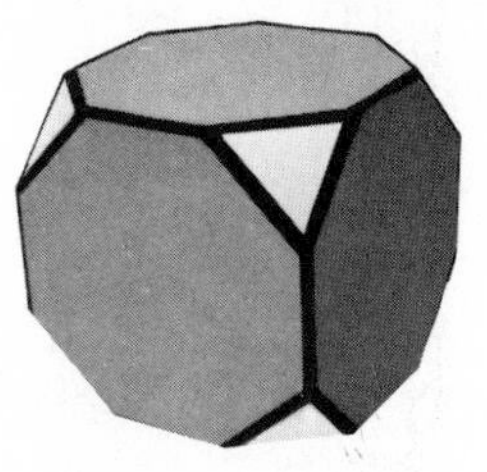

3. Truncated octahedron

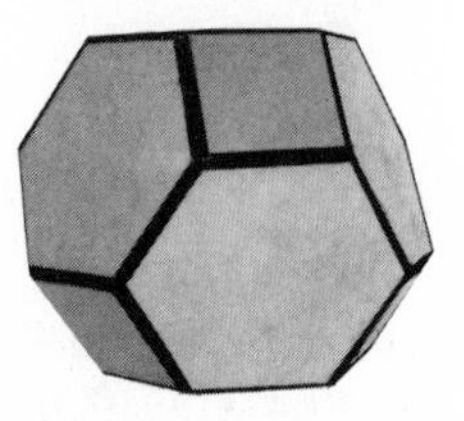

4. Truncated dodecahedron

5. Truncated icosahedron

6. Cuboctahedron*

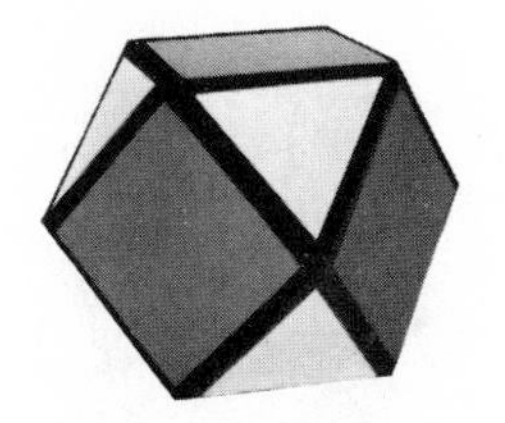

7. Rhombicuboctahedron

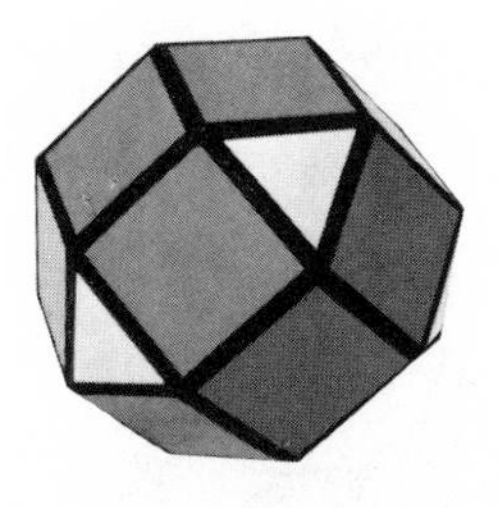

8. Great rhombicuboctahedron

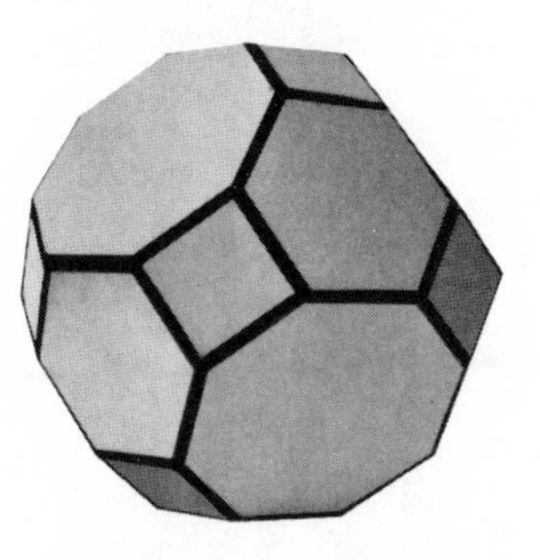

---

*Which of the following symbols do you think is more appropriate for this solid: 3-3-4-4 or 3-4-3-4?

9. Icosidodecahedron

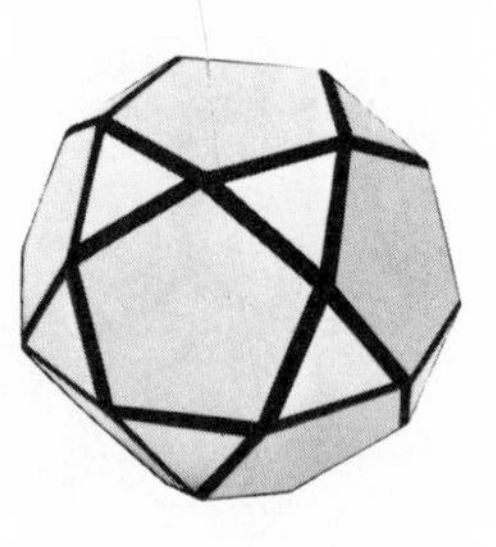

10. Rhombicosidodecahedron

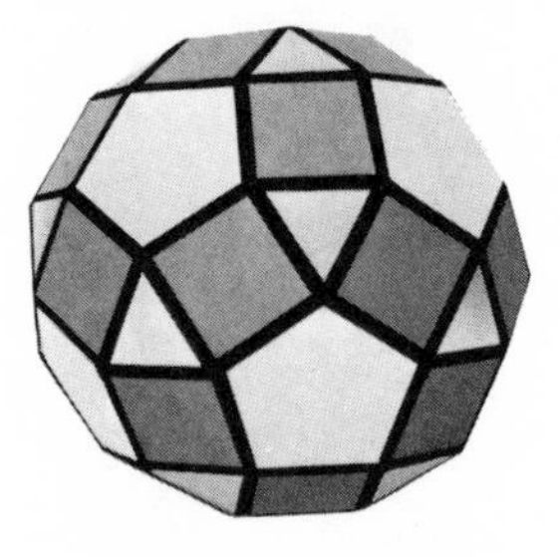

11. Great rhombicosidodecahedron

12. Snub cube

13. Snub dodecahedron

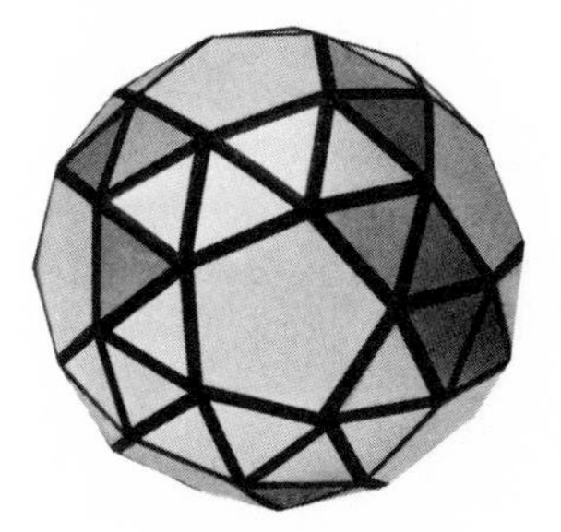

Set II

A truncated tetrahedron has 8 faces, of which 4 are triangles and 4 are hexagons. How many corners and how many edges does it have? To answer these questions without trying to count them, we can reason like this:

A triangle has 3 corners, so 4 separate triangles have 12 corners. A hexagon has 6 corners, so 4 separate hexagons have 24 corners. Adding these, we get 36 corners. Since 3 faces meet at each corner of the solid (3 corners "overlap"), the solid has $36/3 = 12$ corners in all.

A polygon always has the same number of sides as it has corners, so we also have 36 sides (the 4 triangles have 12 sides and the 4 hexagons have 24 sides). Since 2 sides meet at each edge of the solid, the solid has $36/2 = 18$ edges in all.

Use the following information and the same method of reasoning to find the number of corners and the number of edges in each of these solids. Refer to the figures in Set I.

1. A truncated octahedron has 14 faces, of which 6 are squares and 8 are hexagons. (Figure 3)
2. A truncated icosahedron has 32 faces, of which 12 are pentagons and 20 are hexagons. (Figure 5)
3. A great rhombicuboctahedron has 26 faces, of which 12 are squares, 8 are hexagons, and 6 are octagons. (Figure 8)
4. A rhombicosidodecahedron has 62 faces, of which 20 are triangles, 30 are squares, and 12 are pentagons. (Figure 10)

5. A snub dodecahedron has 92 faces, of which 80 are triangles and 12 are pentagons. (Figure 13)

6. Use your results to the previous exercises to complete the following table.

Polyhedron	No. of faces	No. of corners	No. of edges
Truncated tetrahedron	8	12	18
Truncated octahedron	14		
Truncated icosahedron	32		
Great rhombicub-octahedron	26		
Rhombicosidode-cahedron	62		
Snub dodecahedron	92		

7. Do you notice any patterns in this table?

8. A snub cube has 38 faces and 24 corners. How many edges do you think it has?

Set III

Experiment. A Model of a Snub Cube

We will construct this model by folding and gluing together two copies each of pattern A (on page 227) and B (on page 228).

Copy pattern A onto a sheet of tracing paper. Then put a sheet of construction paper underneath the tracing paper and poke a hole with the metal point of a compass at each of the circled points of pattern A.

Remove the paper and use a straightedge to draw all of the line segments. The pattern will be folded along each of the lines within it, so score each line with a pair of scissors.

Cut the pattern out and fold up along each line, so that the pencil marks will be *inside* the finished model. Glue* each of the four flaps to the next panel as indicated by the arrows, so that a bowl is formed like the one shown at the bottom of page 227.

All of the directions to this point should be carried out twice, so that you have two identical bowls.

* Rubber cement works very well. Apply it to both surfaces, let them dry, and then press together.

Pattern A

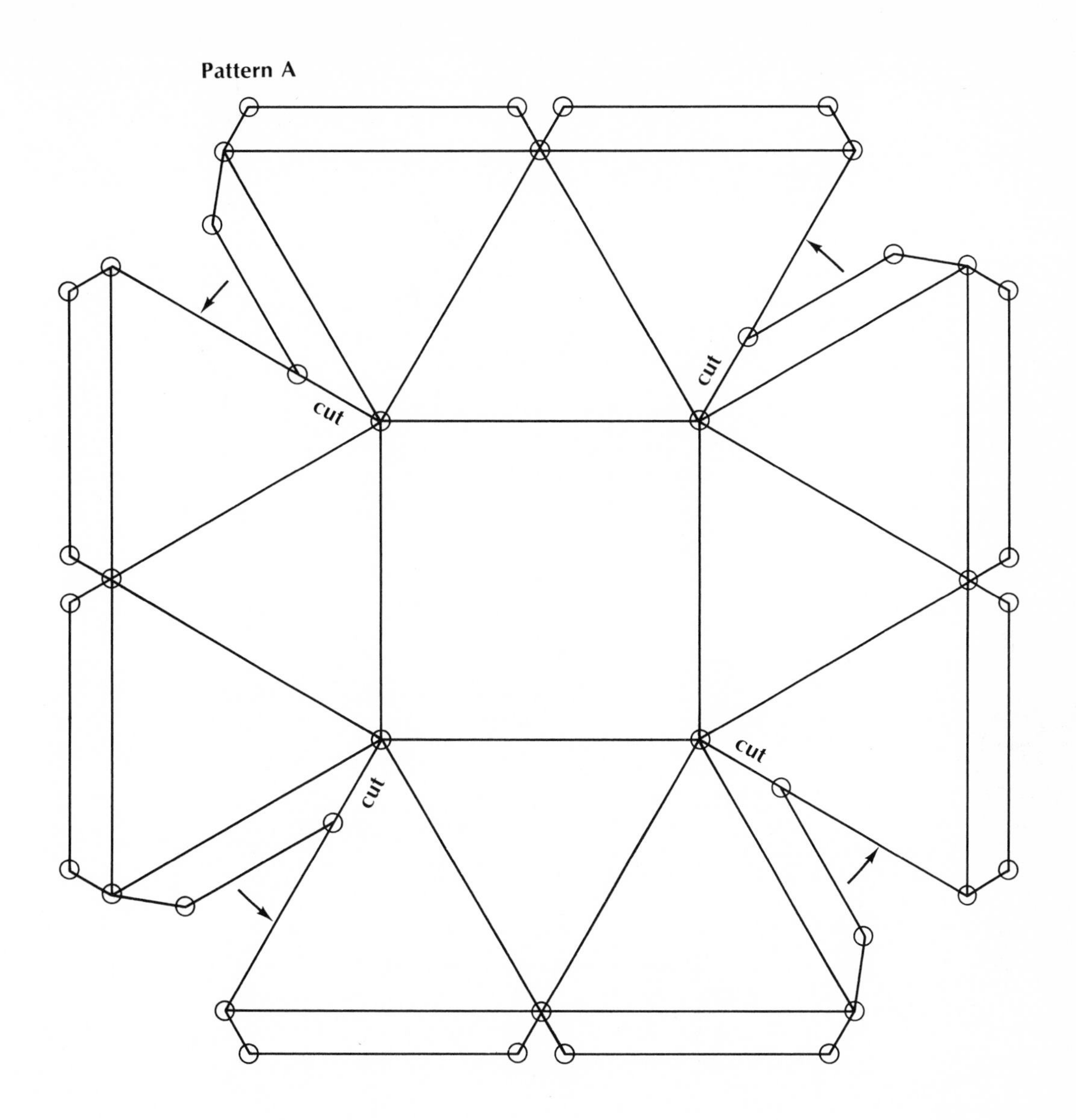

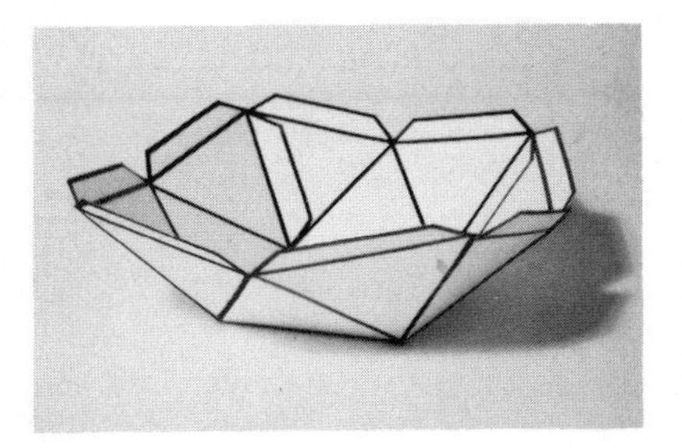

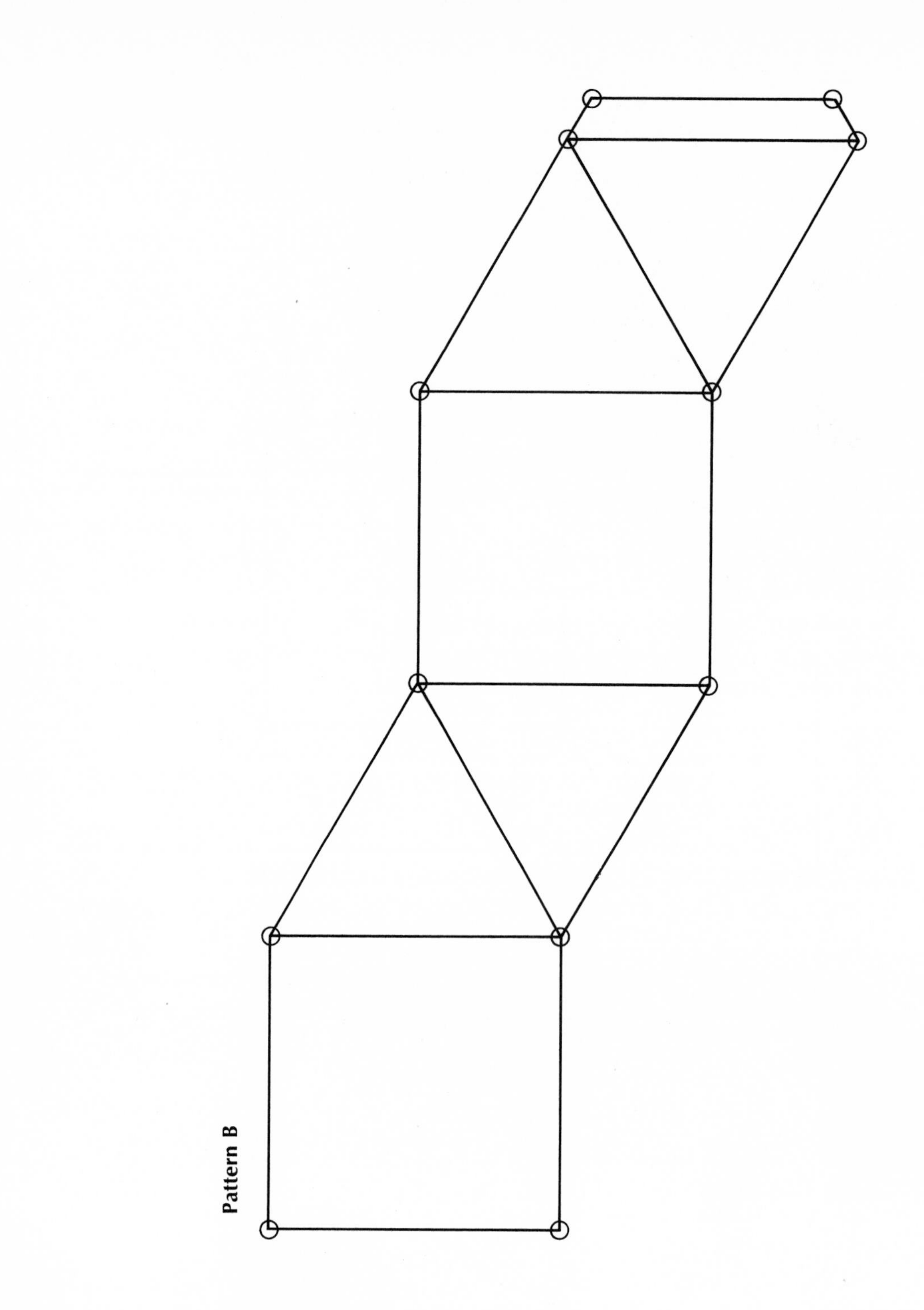
Pattern B

Now make two copies of pattern B in the same way. Glue them together to make a band like the one shown here.

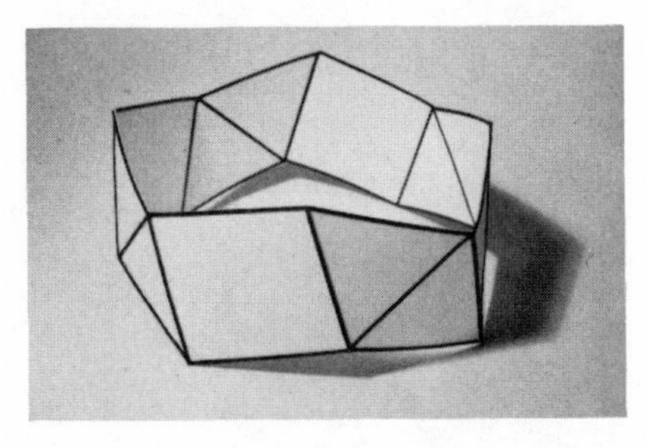

Use the flaps on one bowl to glue it to one side of the band (they should be inserted inside the band), and then glue the other bowl to the other side.

The model will look more attractive if you do the following as well. Cut out six squares with sides of length $1\frac{3}{4}$ inches from construction paper of a different color and glue them over the six square faces. When finished, your snub cube should look something like Figure 12 on page 224.

Lesson 6

Photograph from Moody Institute of Science.

Regular Pyramids and Prisms

THE pyramids of ancient Egypt were built more than four thousand years ago as tombs for the pharoahs. About eighty of them still stand, and of these the largest one, called the Great Pyramid, was once considered one of the "seven wonders of the world." It was built in about 2600 B.C. under the direction of King Khufu and was put together from more than two million stone blocks. These blocks weigh between 2 and 150 tons each and were moved into place by thousands of men using only ropes and log rollers. The shape of the base of the Great Pyramid is a square and the other four faces have the shape of triangles; these triangles have the same *size and shape* and are said to be *congruent.* The Great Pyramid is one member of an unlimited family of geometric solids called *regular pyramids.*

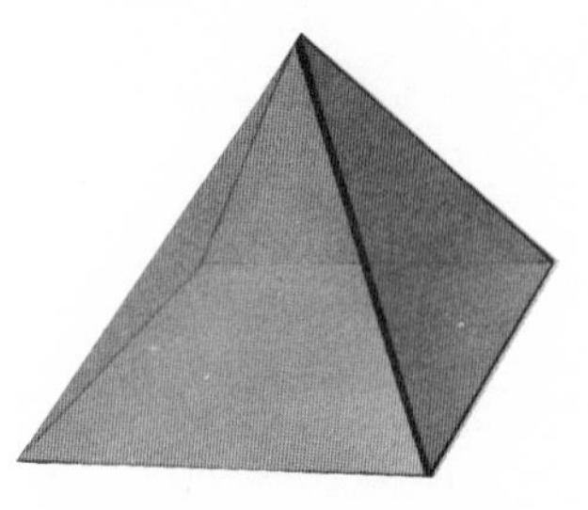

► A **regular pyramid** is a polyhedron that has a regular polygon for its base and congruent triangles for the rest of its faces.

The base of a pyramid is considered to be one of its faces and pyramids are named according to the shape of their base. Here are some more examples of regular pyramids.

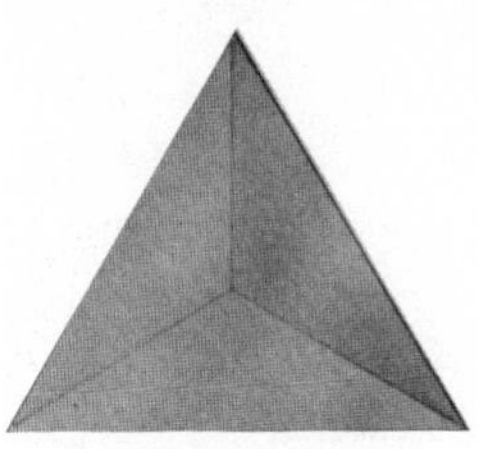

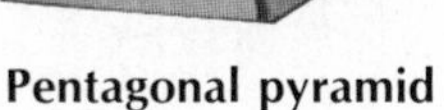

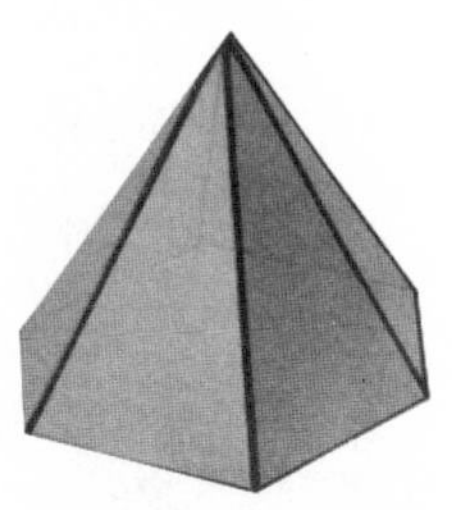

Triangular pyramid **Pentagonal pyramid** **Hexagonal pyramid**

Notice that one of the regular polyhedra, the tetrahedron* is also a triangular pyramid. Another, the octahedron,* might be considered to be a pair of square pyramids whose bases have been put together.

Another family of geometric solids are the *regular prisms.*

► A **right regular prism** is a polyhedron that has a pair of congruent regular polygons for its bases and congruent rectangles for the rest of its faces. (Remember that congruent figures have the same size and shape.)

The bases of a prism are always parallel to each other, and prisms, like pyramids, are named according to their shape. Some right regular prisms are shown here.

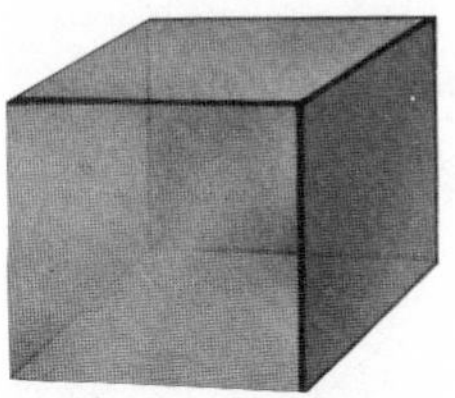

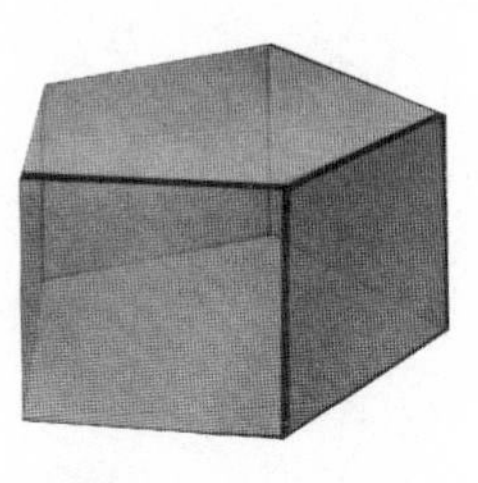

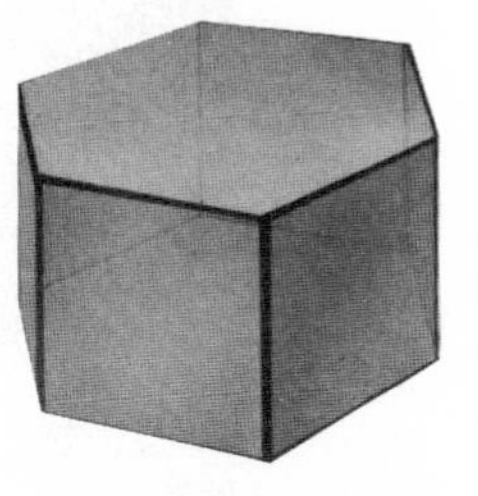

Square prism **Pentagonal prism** **Hexagonal prism**

The cube, another regular polyhedron, is an example of a square prism.

You have seen many man-made regular pyramids and prisms. These kinds of polyhedra occur in nature, too. Quartz crystals are tiny prisms. An awesome display of natural prisms is the

---

*The tetrahedron and octahedron are pictured on page 210.

Courtesy of the National Park Service.

Devil's Post Pile

Devil's Post Pile National Monument in California. The Devil's Post Pile is a set of tall columns of rock, some of them 60 feet high. Many of these columns have the shape of right regular prisms, most of them hexagonal and others pentagonal.

So even the geologist has discovered mathematics in his study of the earth.

EXERCISES

Set I

1. Draw an octagonal pyramid. The base, in perspective, might look something like this.

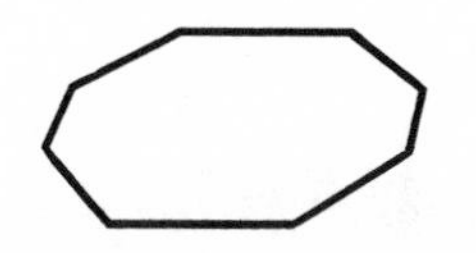

2. Draw a right triangular prism. (Its bases are in the shape of an equilateral triangle which, in perspective, looks something like this.)

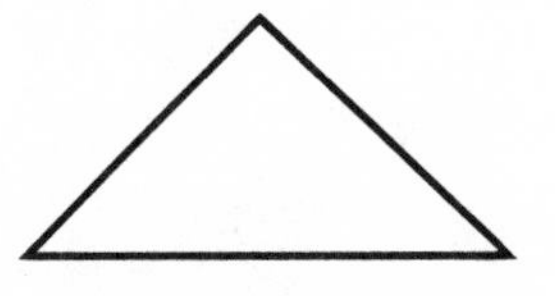

3. What geometric figure does a regular polygon become more and more like as the number of its sides increases indefinitely?
4. What happens to the number of faces of a pyramid if the number of sides in its base is increased?
5. Draw a "pyramid" whose base is a circle. The circle, in perspective, would look something like this.

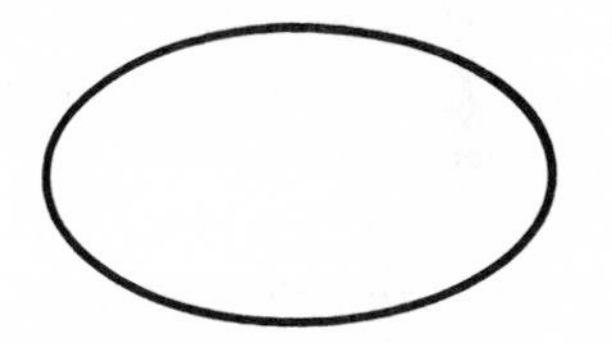

6. Notice that instead of having many narrow flat triangular faces, the "pyramid" you have drawn has one curved surface. Do you know the name for such a geometric solid? (Hint: If you turn it upside-down and take off the base, it makes a good container for ice cream.)
7. What happens to the number of faces of a prism if the number of sides in its bases is increased?
8. Draw a "right prism" whose bases are circles.

9. The "right prism" you have drawn has one curved surface. How many flat surfaces does it have?
10. Do you know the name for such a geometric solid? (Hint: It also starts with the letter "c" and is the name of an important part of an automobile engine.)

"Actually, what I had in mind was something a little larger."

Set II

1. Copy the following table and then look back at the drawings of the pyramids on page 231 to help you fill in the missing numbers. Check to see that you agree with the numbers on the first line.

Pyramid	No. of sides in base	No. of faces	No. of corners	No. of edges
Triangular	3	4	4	6
Square				
Pentagonal				
Hexagonal				

2. Study the pattern in your table. Then add a line for a decagonal pyramid (whose base is a decagon) and guess what its numbers would be.

If we let n represent the number of sides of the base of a pyramid, its number of faces can be represented as $n + 1$.

3. How would you represent its number of corners?
4. How would you represent its number of edges?
5. Add a line to your table to show these results for a pyramid with a base having n sides:

	No. of sides in base	No. of faces	No. of corners	No. of edges
Base of n sides	n	$n + 1$		

Letting F represent the number of faces, C represent the number of corners, and E represent the number of edges, the formula

$$F + C = E + 2$$

should fit every pyramid listed in your table.

6. Substitute the numbers involving n on the last line of the table into the formula to show that it fits all pyramids.
7. Look back at the drawings of the prisms on page 231 to help in completing the following table.

Prism	No. of sides in a base	No. of faces	No. of corners	No. of edges
Triangular	3	5	6	9
Square				
Pentagonal				
Hexagonal				

8. Add a line of numbers to your table for a decagonal prism.

If we let n represent the number of sides of one base of a prism, its number of faces can be represented as $n + 2$.

9. How would you represent its number of corners?
10. How would you represent its number of edges?
11. Add a line to your table of prisms to show these results as illustrated at the top of the next page.

	No. of sides in a base	No. of faces	No. of corners	No. of edges
Base of n sides	n	$n + 2$	▯▯▯▯	▯▯▯▯

12. Substitute the numbers involving n on the last line into the formula,

$$F + C = E + 2,$$

to show that it fits all prisms.

Set III

This star-shaped polyhedron was discovered by the German astronomer and mathematician, Johann Kepler, in about 1620. It is called a "small stellated dodecahedron" and can be constructed by adding 12 pentagonal pyramids to the faces of a dodecahedron. The drawing below shows 5 of the pyramids being added.

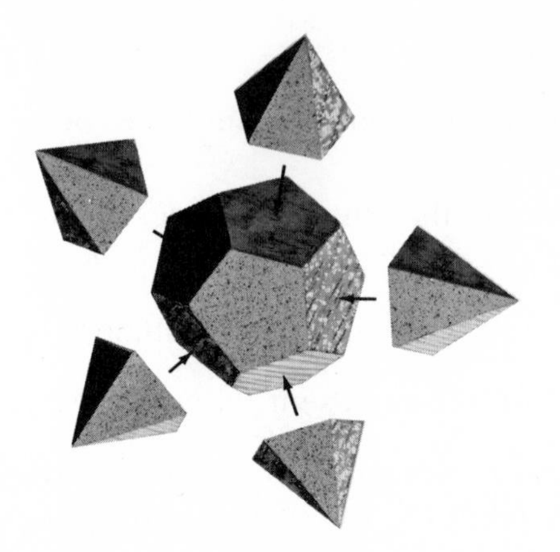

Kepler discovered another star-shaped polyhedron, shown below, at the same time. It can be formed by adding pyramids to the faces of another regular polyhedron.

1. What kind of pyramids are added?
2. To what regular polyhedron do you think they are added?
3. How many points does the star have altogether?

INTERESTING READING

Ancient Egypt, by Lionel Casson, a book in Time-Life's "Great Ages of Man" series, 1965: "The Pyramid Builders," pp. 129–139.

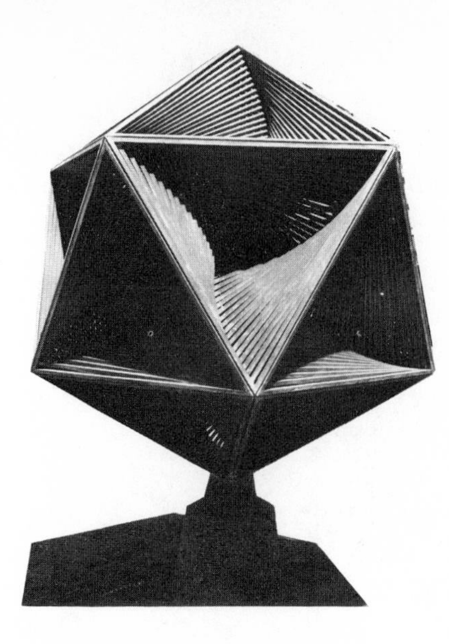

Icosaspirale, a sculpture by Charles Perry at the Alcoa Golden Gate Plaza in San Francisco.

Chapter 5 / Summary and Review

In this chapter we have studied regular polygons and their relationships on a flat surface and in space.

Regular polygons and symmetry *(Lesson 1)* A *regular polygon* is a figure all of whose sides have the same length and all of whose angles are equal.

A figure has *line symmetry* if it can be folded in half along a line so that the two halves coincide.

A figure has *point symmetry* if it can be rotated less than 360° about a point so that it coincides with its original position.

Some regular polygon constructions *(Lesson 2)* Every regular polygon can be inscribed in a circle. A polygon is inscribed in a circle if every one of its corners lies on the circle.

Geometric constructions are made with a straightedge and compass.

Mathematical mosaics *(Lesson 3)* A *mathematical mosaic* is an arrangement of regular polygons in which every corner point is surrounded by the same number of polygons of each kind.

The regular polyhedra *(Lesson 4)* A *regular polyhedron* is a solid with faces in the shape of regular polygons. All of its faces, edges, and corners are identical.

There are only five regular polyhedra: the tetrahedron, cube, octahedron, dodecahedron, and icosahedron.

Some semi-regular polyhedra *(Lesson 5)* A *semi-regular polyhedron* is a solid with faces in the shape of more than one kind of regular polygon, yet with every corner the same.

The 13 *Archimedean solids* are semi-regular polyhedra.

Regular pyramids and prisms *(Lesson 6)* A *regular pyramid* is a solid that has a regular polygon for its base and congruent triangles for the rest of its faces.

A *right regular prism* is a solid that has a pair of congruent regular polygons for its bases and congruent rectangles for the rest of its faces.

Pyramids and prisms are named according to the shape of their bases.

EXERCISES

Set I

1. If a knot is tied in a strip of paper and then pressed flat, a regular polygon is formed. Which one is it?

2. What must be true if a regular polygon is inscribed in a circle?
3. What regular polygon inscribed in a circle has sides equal in length to the radius of the circle?
4. Name a regular polygon that it is impossible to inscribe in a circle using only a straightedge and compass.

5. This photograph was taken through a microscope and is of ice crystals. What shape do ice crystals have?

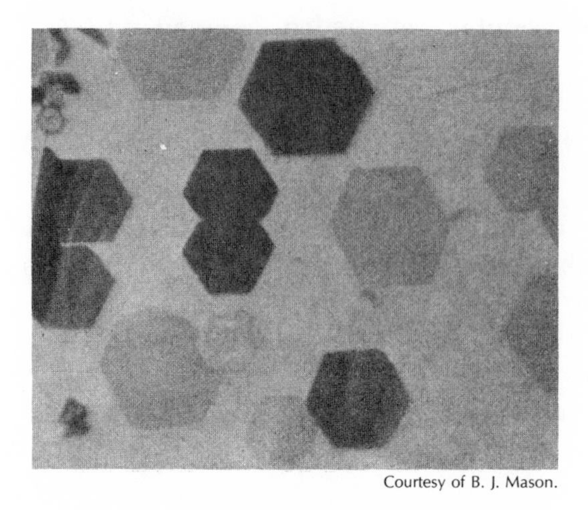

Courtesy of B. J. Mason.

6. It has been recently discovered that certain viruses have the shape of one of the regular polyhedra. Can you guess from this microphotograph which polyhedron it is?

Courtesy of Kenneth M. Smith and Robley C. Williams.

7. Look at Thor's wheel in the cartoon on page 186. What kind of geometric solid is it? (Ignore the hole in the center.)

8. At the time the astronomer Kepler lived, there were six known planets in the solar system. Kepler tried to explain the relative distances between them by means of the regular polyhedra. You may recall that a regular polyhedron

can be put *inside* a sphere so that each *corner* touches it. A regular polyhedron can also *surround* a sphere so that each *face* touches it.

Kepler imagined a series of six spheres, one for each planet, with the sun at their center. These spheres were separated by the five regular polyhedra, as this drawing by Kepler shows. Can you identify all five, naming them in order from largest to smallest?

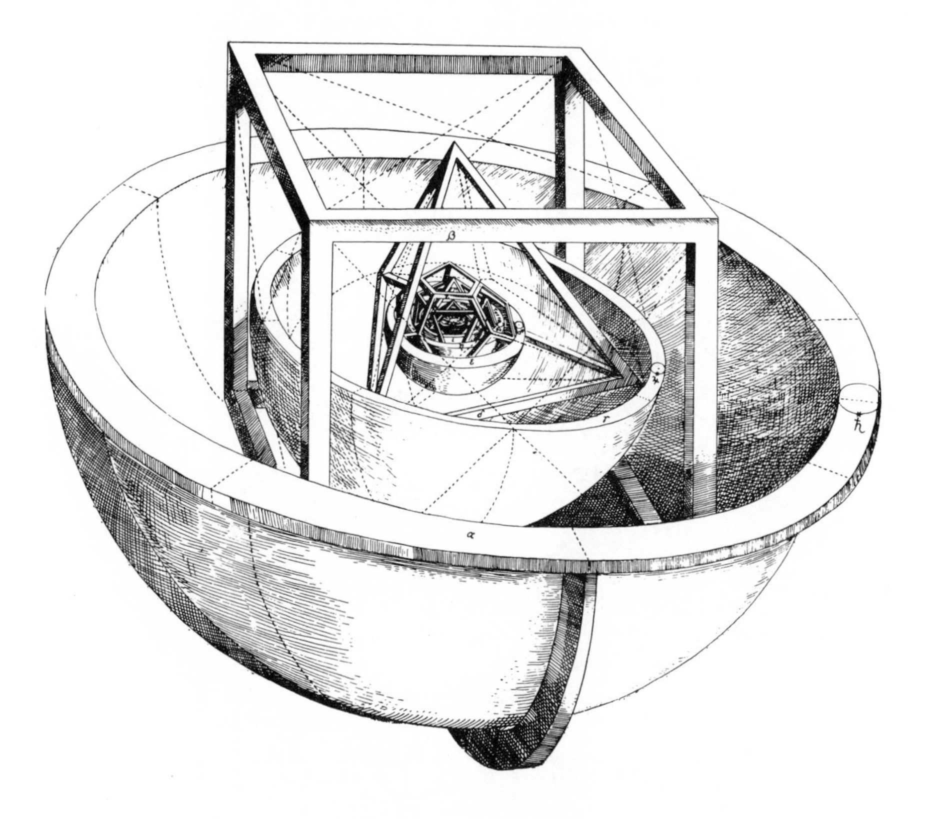

Set II

1. The four symbols used on playing cards are symmetrical. What kinds of symmetry does each symbol have?

Heart **Spade** **Diamond** **Club**

2. There are three simple ways of planting an orchard so that the distance between each tree and each of its closest neighbors is the same. Two of them are shown below. Make a sketch showing the third way. (Hint: It involves another regular polygon.)

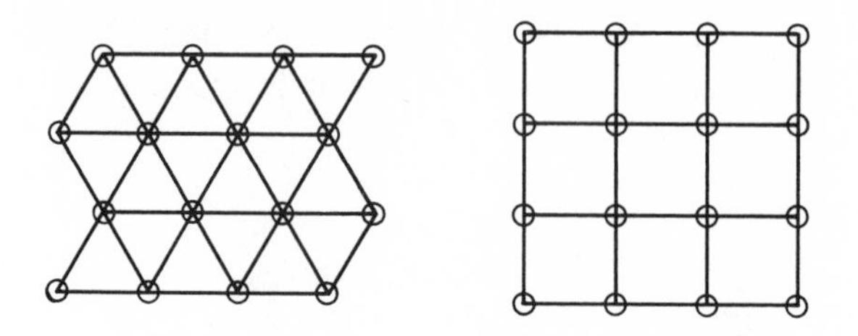

3. In what way are the two mathematical mosaics shown here alike? Write the symbol for each.

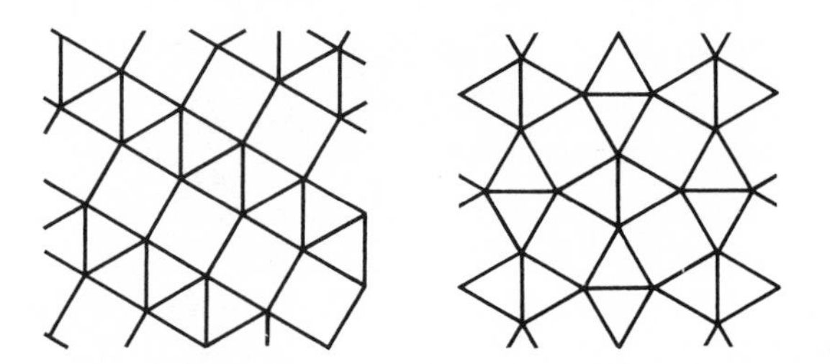

The geometric solid shown below is called a "compound" polyhedron.

4. Name the regular polygons that you see in it.
5. The solid consists of two intersecting regular polyhedra. In this model one is white and the other is gray. Can you identify both of them?
6. What kinds of pyramids can you find in this solid?

Set III

Most drums make sounds with no definite pitch. The kettledrum, however, can be tuned to play notes of different pitches by changing the tension on the skin that is its head.

The head of a kettledrum has the shape of a circle. Suppose that it could have the shape of a regular polygon instead and that we had a set of five kettledrums with heads of the following shapes: equilateral triangle, square, regular hexagon, regular octagon, and circle. If the heads of these drums all have the same area and the same tension, then their pitches (or frequencies) will be different, because of their different shapes.

If the circular kettledrum is tuned to a pitch of 130 vibrations per second, the following table shows the pitches of the other four drums:

Shape of head	Pitch (in vibrations per second)
Equilateral triangle	146
Square	136
Regular hexagon	132
Regular octagon	131

1. Which drum of the five has the highest pitch (most vibrations per second)?
2. Complete the following statement: As the number of sides of the head of a "regular polygon" drum increases, its pitch ▥▥▥▥.

Do you think that a drum could be built with a head in the shape of a regular polygon, and having the same area and tension as the drums shown in the table, so that it had the following pitch? Explain each of your answers.

3. 140 vibrations per second. (Hint: Is there a regular polygon with more sides than a triangle and fewer sides than a square?)
4. 134 vibrations per second.
5. 125 vibrations per second.
6. What do you think the pitch of a drum like these would be if it had a head in the shape of a regular polygon having 20 sides?

Chapter 6

MATHEMATICAL CURVES

Lesson 1

The Circle and the Ellipse

THE fisherman in this cartoon is in for a big surprise. He seems to have some competition at cutting a circle in the ice.

► The **circle** is the simplest mathematical curve. It is the set of *all points in a plane* (a flat surface) that are *at the same distance* from a fixed point in the plane. The fixed point is called the center of the circle.

The circle is the most symmetrical of all mathematical curves since every line drawn through its center is a line of symmetry. Its center is an unusual point of symmetry because a circle can be turned on its center point through *any* angle and still coincide with its original position.

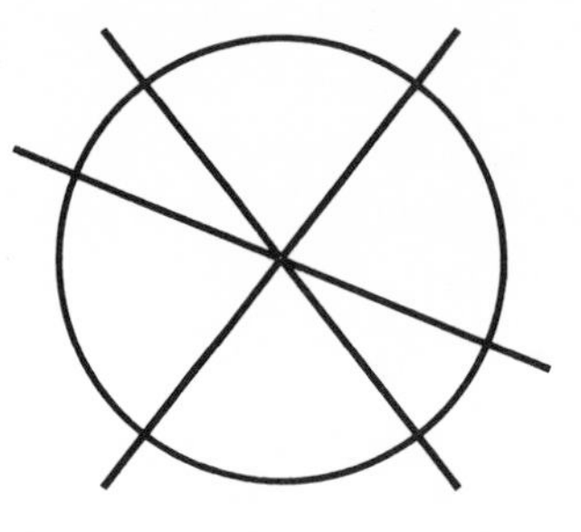

Although the circle is a very common curve in the world around us, it is not the one we most often see. This is because a circle seen from an angle appears to be another mathematical curve, called an **ellipse.** Unlike the circle, whose shape never changes, an ellipse can have many different shapes, from nearly round to long and narrow. Here are some drawings of ellipses of various shapes.

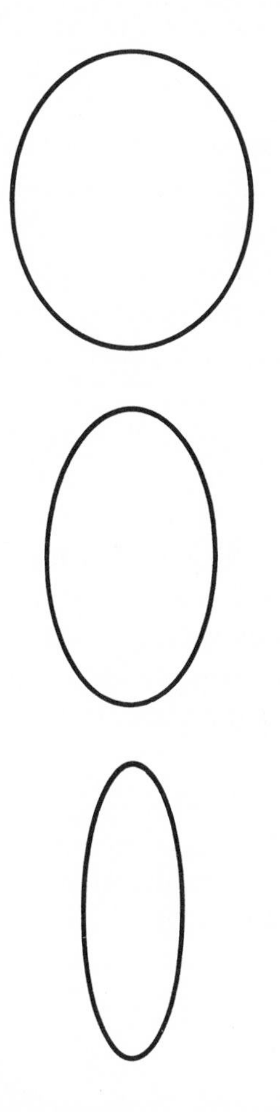

An ellipse has two lines of symmetry, which are perpendicular to each other. These lines contain two segments called the *major axis* and *minor axis* of the ellipse; the major axis is always the longer of the two. An ellipse has point symmetry, because it can be rotated 180° to coincide with its original position.

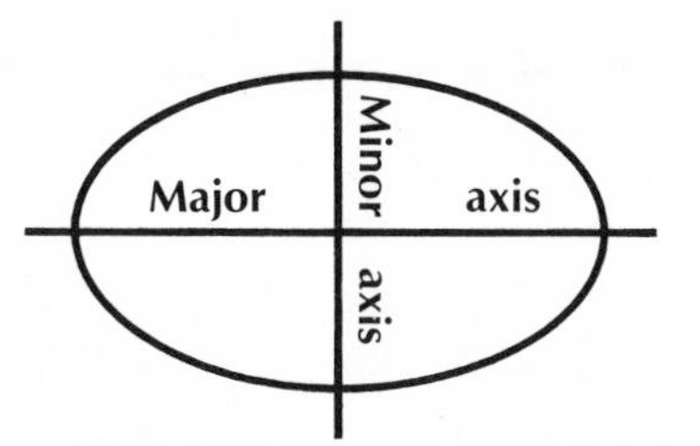

Mathematics is a subject in which new ideas are often studied without any regard for their practical value. The Romans did not create any mathematics of importance, because they looked down upon learning for which they did not see any immediate use. In fact, Cicero said that mathematics was good only for measuring and counting.

The Greeks, on the other hand, didn't care whether or not what they were studying seemed useful and they invented much mathematics merely because it was interesting or beautiful. They made a thorough study of the ellipse and became well acquainted with its properties, even though such knowledge seemed to be completely useless at the time. The discovery that the ellipse had practical applications in astronomy did not come until centuries later.

The early Greek astronomers thought that the planets moved in circular orbits, since the circle is the simplest mathematical curve. In the sixteenth century, the great Polish astronomer Nicholas Copernicus, who claimed that the sun, rather than the earth, was the center of the solar system, also tried to explain the motions of the planets by means of circles. Johann Kepler, a century later, invented the complicated model shown on page 241 in an attempt to explain the solar system. After deciding that the model was wrong, he eventually discovered that the planets do not travel around the sun in circles, but in elliptical orbits instead. The orbits of artificial satellites of the earth are also ellipses. Notice the nearly elliptical orbit of Apollo 11 shown in this diagram of its historic trip to the moon.

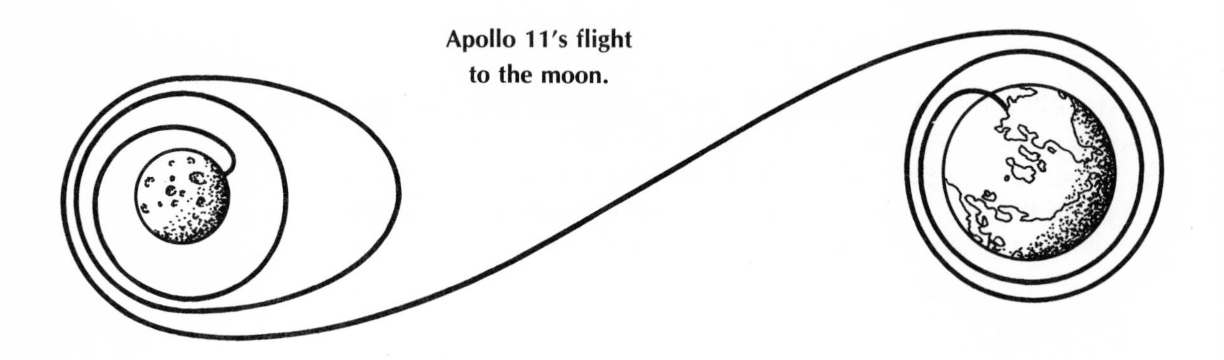
Apollo 11's flight to the moon.

► An **ellipse** can be defined as the set of *all points in a plane such that the sum of the distances of each point from two fixed points is the same.*

The ellipse below shows what this means. The two fixed points are called the *foci** of the ellipse; in this figure they are labeled F_1 and F_2. Do you see that the sum of the distances of point A from points F_1 and F_2 is 3.0 inches and that the sum of the distances of point B from these same two points is also 3.0 inches? This is true not for just points A and B but for *every* point of the ellipse. We will make use of this property to draw some ellipses of different shapes.

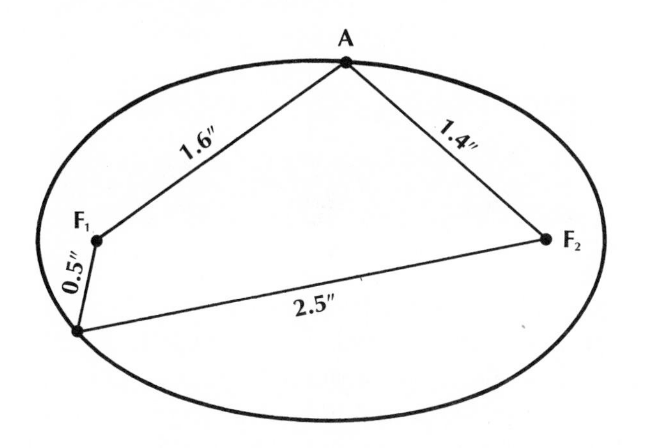

EXERCISES

Set I

Experiment. An Easy Way to Draw Ellipses that Has "Strings Attached"

On a sheet of graph paper (4 units per inch), draw a pair of perpendicular axes with their origin at the center of the paper. Label the axes *x* and *y* and the origin O. Tack the graph paper on a square piece of corrugated cardboard (see top figure, page 250).

Mark the two points on the *x*-axis that are 6 units to the left and 6 units to the right of the origin with your pencil. Put a tack in each point (the tacks should not go all the way into the cardboard).

**Foci* (pronounced "fō-sī") is the plural of *focus*.

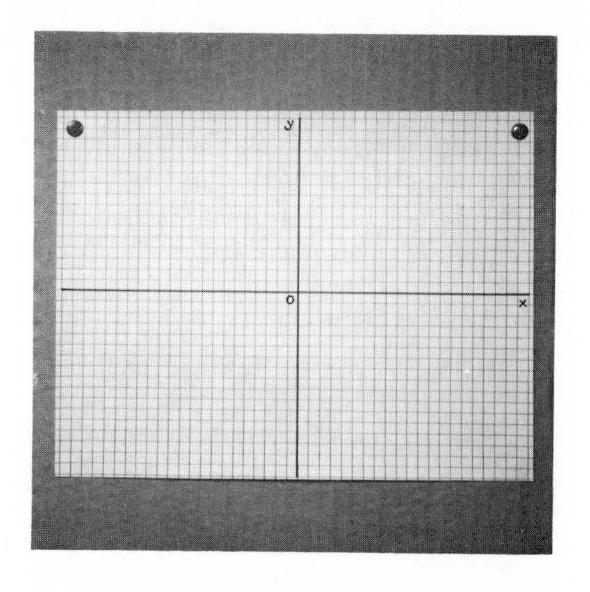

Now take a string about 10 inches long and, as accurately as you can, tie a knot in it to make a loop exactly 8 inches around. Place the loop on the board around the tacks and use your pencil to pull it taut, forming a triangle. Now move the pencil around the paper, keeping the string taut, and an ellipse will be drawn.

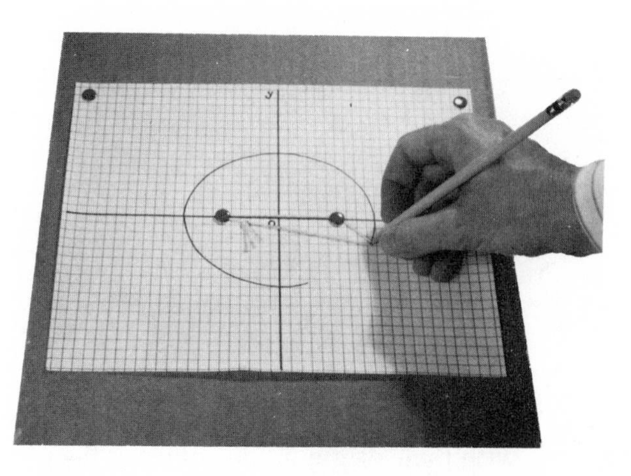

Label the ellipse by writing the letter A on it at some point.

1. The major axis of the ellipse should be about 20 units long. Check this on your paper. How long is its minor axis?
2. The two points where you put the tacks are the foci of the ellipse. The coordinates of one focus of ellipse A are (6, 0). What are the coordinates of the other focus?

Mark the two points on the x-axis that are 3 units to the left and right of the origin and move the tacks to these points. Use the same loop of string to draw another ellipse. Label it B.

3. The foci of ellipse B are closer together than those of ellipse A. How do the lengths of the major and minor axes of ellipse B compare to those of A?

Now move the tacks together again so that the foci coincide at the origin (you need only one tack since they are the same point). Draw another curve, being careful to keep the string taut. Label it C.

4. How do the lengths of the major and minor axes of curve C compare to each other?
5. What kind of curve is curve C?
6. What happens to the shape of an ellipse as its foci move closer and closer together?

Remove the sheet of graph paper from the cardboard and tack a new sheet in place. Again draw and label a pair of axes. Mark the two points on the x-axis that are 12 units to the left and right of the origin and place tacks in these points.

Take a longer piece of string and make, as accurately as possible, a loop that is 16 inches around. Put the loop around the tacks, draw an ellipse, and label it D.

7. The major axis of ellipse D should be about 40 units long. Check this. How long is its minor axis?
8. What are the coordinates of the foci of this ellipse?

Move the tacks farther apart so that the foci are at $(-15, 0)$ and $(15, 0)$. Draw another ellipse and label it E.

9. What are the lengths of the major and minor axes of ellipse E?
10. What happens to the shape of an ellipse as its foci move farther and farther apart?
11. Is it possible for two ellipses to have different sizes and yet have the same shape? (Look at the five curves you have drawn to decide.)

12. Can you give the lengths of the major and minor axes of an ellipse having the same shape as ellipse E, but which is smaller?

Set II

Every mathematical curve can be associated with an equation.* For example, curve C, which you drew in Set I, has the equation

$$x^2 + y^2 = 256,$$

and curve A has the equation

$$64x^2 + 100y^2 = 6{,}400.$$

To get a better idea of the relationship between curves and equations, we will draw graphs for a couple of equations and see what they look like.

What curve has the equation

$$x^2 + y^2 = 25?$$

To answer this, we will plot some points whose x and y coordinates "fit" the equation. For example, (5, 0) is one of these points, because

$$5^2 + 0^2 = 25,$$

and (−3, −4) is another of these points, because

$$(-3)^2 + (-4)^2 = 25.$$

To save you the trouble of doing a lot of arithmetic, some more points have been worked out for you.

1. The table below includes, in addition to the two points already given, other points that also fit the equation, $x^2 + y^2 = 25$. (Some of the points fit only approximately.) Draw a graph for the equation by first plotting the points and then joining them with a smooth curve.

x	0	1	2	3	4	5	4	3	2	1	0	−1
y	5	4.9	4.5	4	3	0	−3	−4	−4.5	−4.9	−5	−4.9

x	−2	−3	−4	−5	−4	−3	−2	−1
y	−4.5	−4	−3	0	3	4	4.5	4.9

*Several illustrations of this are shown on page 130.

2. What kind of curve did you get?

One of the five curves you drew in Set I has the complicated looking equation

$$64x^2 + 289y^2 = 18,496.$$

Which one is it? To find out, we will again plot some points whose x and y coordinates fit the equation. For example, (0, 8) is one of these points, because

$$64(0)^2 + 289(8)^2 =$$
$$0 \quad + 289(64) = 18,496.$$

3. Turn a sheet of graph paper sideways and draw and label a pair of axes like those you drew in Set I. Plot the points in the table below, which fit the above equation, and join them with a smooth curve.

x	0	3	6	9	12	15	16	17	16
y	8	7.9	7.5	6.8	5.7	3.8	2.7	0	−2.7

x	15	12	9	6	3	0	−3	−6	−9
y	−3.8	−5.7	−6.8	−7.5	−7.9	−8	−7.9	−7.5	−6.8

x	−12	−15	−16	−17	−16	−15
y	−5.7	−3.8	−2.7	0	2.7	3.8

x	−12	−9	−6	−3
y	5.7	6.8	7.5	7.9

4. What kind of curve did you get?

5. Which one of the curves that you drew in Set I matches this one; in other words, which one has the equation

$$64x^2 + 289y^2 = 18,496?$$

6. Do you think that the graph of the equation, $x^2 + y^2 = 100$, is a circle or an ellipse?

7. Is the graph of the equation, $4x^2 + 25y^2 = 100$, a circle or an ellipse?

Set III

Several years ago, pool tables in the shape of an ellipse became available for sale in New York City. These tables have only one pocket, which is located at one focus. Because of the elliptical shape of the table, a ball that is hit in *any* direction from the focus opposite the pocket will end up in the pocket! This drawing shows some of the paths that the ball might take.

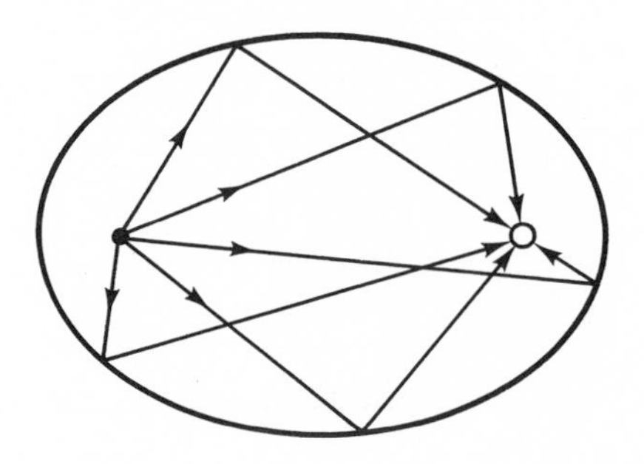

As the ball strikes the cushion of the table, it rebounds from it at the same angle. Since the ellipse is not straight, but curved, we must imagine a straight line just touching it at the point where the ball hits in order to see the equal angles. Such a line is called a *tangent* to the curve.

The tangent can touch the ellipse at *any* point and it will *always* form equal angles with the lines from that point to the foci. This is why a ball hit from the other focus will always land in the pocket.

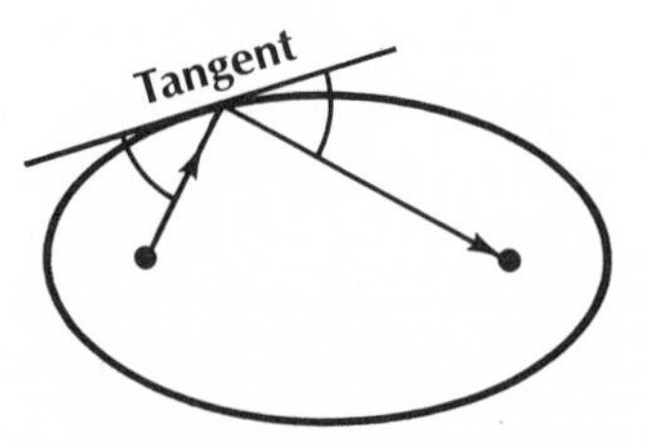

At the beginning of this book, it was mentioned that Lewis Carroll thought of the possibility of playing billiards on a circular table.

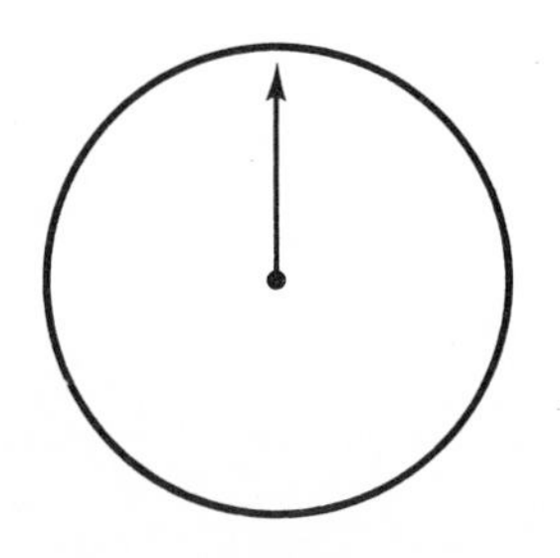

1. If the ball were hit from the center of the table, can you figure out what its path would be?
2. What relationship would the path have to the tangent drawn to the circle at the point where the ball hits the cushion of the table?

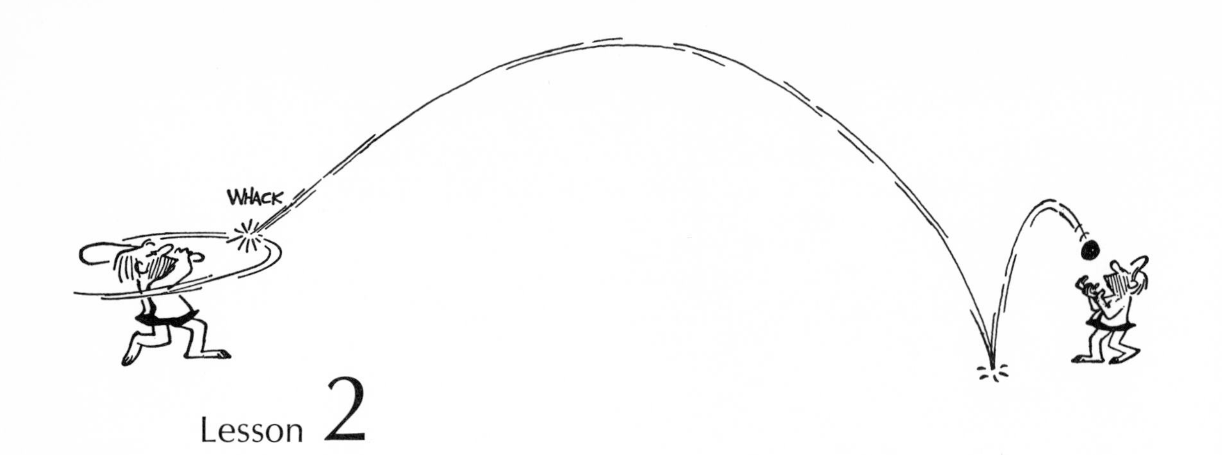

Lesson 2

The Parabola

WHEN a baseball is hit into the air, it moves along a path in the shape of a mathematical curve called a *parabola.* If the ball is hit at a 45° angle with the ground, its path looks like the one in the cartoon above.

It was Galileo who proved that the path of an object thrown through space is this curve. You probably recognize the parabola since you have drawn several graphs of functions with its shape (see Chapter 3, Lesson 4).

► A **parabola** can be defined as the set of *all points in a plane such that the distance of each point from a fixed point is the same as its distance from a fixed line.*

To see just what this means, look at the figure below.

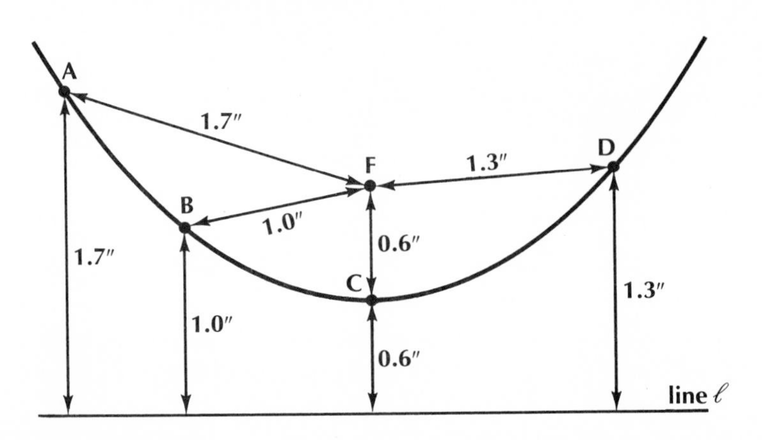

By permission of Johnny Hart and Field Enterprises, Inc.

The distance from point A on the parabola to the point F, 1.7 inches, is equal to the distance from point A to the line ℓ, 1.7 inches. Points B, C, and D on the parabola are also the same distance from point F as they are from line ℓ, and this is true for every point on the curve. Point F is called the *focus* of the parabola. Every parabola has one focus; you will recall that an ellipse has two foci.

A parabola has one line of symmetry which passes through its focus. It is easy to see that if the parabola is folded along this line, its two halves will coincide.

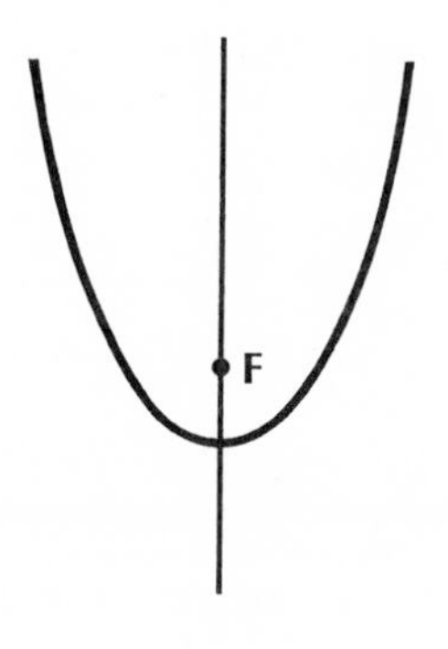

Since it has no ends and cannot be measured, a parabola is called an *infinite* curve. An ellipse, on the other hand, is a *finite* curve because it has a definite length.

Courtesy of San Francisco Convention and Visitors Bureau.

If cables are hung between two towers and used to suspend a bridge, each one forms a curve that is very close to the shape of a parabola. The Golden Gate Bridge in San Francisco is a world-famous example of this mathematical curve.

The parabola is also used in automobile headlights. The mirror in each headlight has a curved surface formed by rotating a parabola about its axis of symmetry. If a light is placed at the focus of the mirror, it is reflected in rays parallel to the axis. In this way a straight beam of light is formed. The opposite principle is used

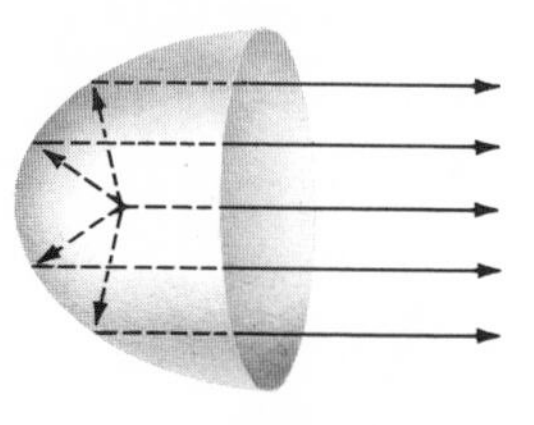

in the giant mirrors in reflecting telescopes and in antennas used to collect light and radio waves from outer space:* the beam comes *toward* the parabolic surface and is brought into focus at the focus point.

*See the photograph on page xv of the antenna of the National Radio Observatory at Green Bank, West Virginia.

EXERCISES

Set I

You have already drawn some graphs for functions which turned out to be parabolas, and in the previous lesson you drew some ellipses. What does the graph of each of the following functions look like? To find out, make a table including the x numbers indicated and draw a graph for each on a separate pair of axes. For each graph, use the scales shown in the figure below.

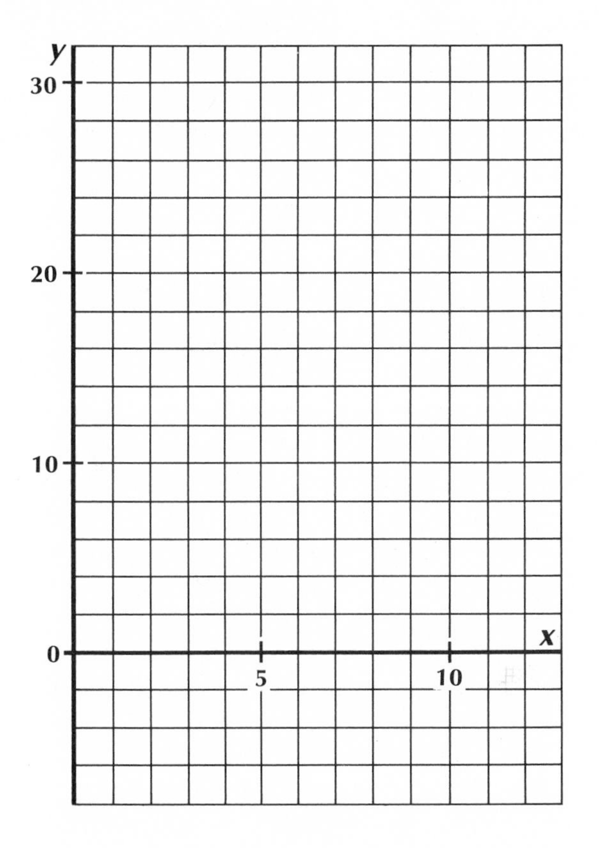

1. $y = x^2 + 16 - 8x$ $\qquad$ x from 0 to 10

   Here are some examples to remind you how to figure out the y numbers.

   If $x = 0$, $y = 0^2 + 16 - 8(0) = 0 + 16 - 0 = 16$.
   If $x = 1$, $y = 1^2 + 16 - 8(1) = 1 + 16 - 8 = 9$.
   If $x = 2$, $y = 2^2 + 16 - 8(2) = 4 + 16 - 16 = 4$, etc.

2. $y = x^2 + 29 - 10x$ $\quad$ x from 0 to 10

3. $y = 15 + 6x - x^2$ $\quad$ x from 0 to 8

4. $y = x^2 + 43 - 14x$ $\quad$ x from 1 to 12

5. $y = 7 + 4x - x^2$ $\quad$ x from 0 to 6

6. What do you notice about these graphs?

Set II

Experiment. A Parabola Computer

In this experiment, you will draw a graph that can be used to multiply numbers by drawing lines between points on a parabola.

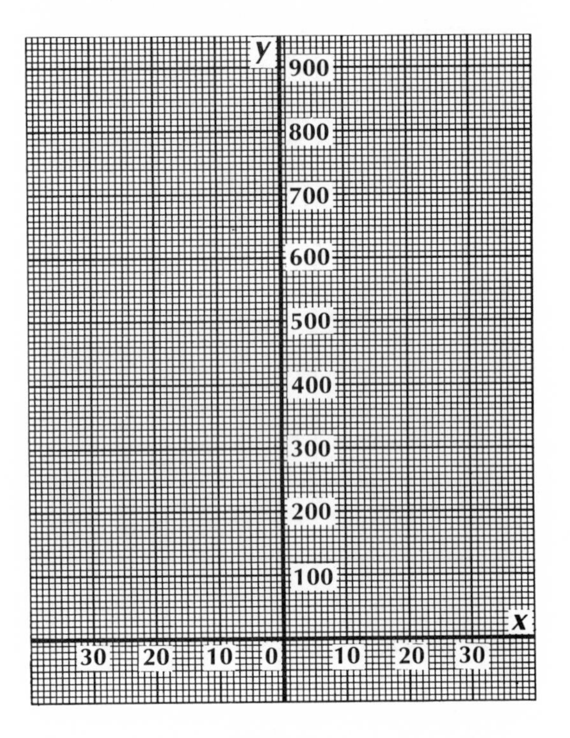

Draw a pair of axes on a sheet of graph paper, which is ruled 10 units per inch, as in the diagram here. Number the axes as shown, letting 1 inch on the x-axis represent 10 and 1 inch on the y-axis represent 100. Notice that the numbers on the x-axis to the left of the origin have been written as if they were positive.

Now use the following table of the function, $y = x^2$, to plot a

Table for $y = x^2$

x	y	x	y
0	0	16	256
1	1	17	289
2	4	18	324
3	9	19	361
4	16	20	400
5	25	21	441
6	36	22	484
7	49	23	529
8	64	24	576
9	81	25	625
10	100	26	676
11	121	27	729
12	144	28	784
13	169	29	841
14	196	30	900
15	225		

series of points to the left and right of the y-axis as accurately as you can. Notice that, except for (0, 0) which corresponds to the origin, each pair of numbers in this table corresponds to two points on the graph. Part of the graph, showing some of the points, is pictured here.

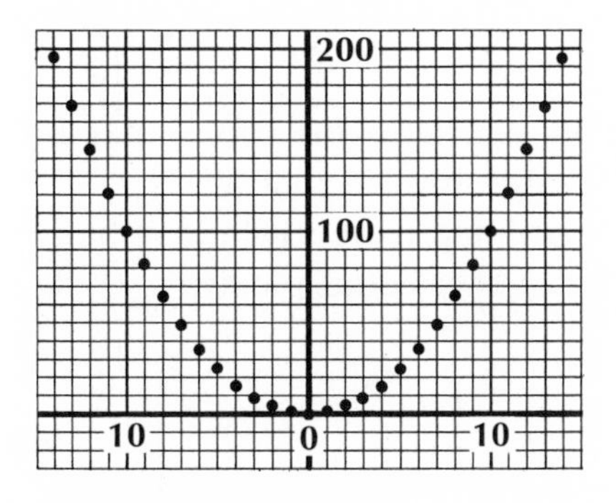

After you have plotted all of the points (61 altogether), join them with a smooth curve.

Your parabola is now ready for multiplying numbers. Here is an example of how it works. To multiply 5 and 8, find these numbers on the x-axis, one to the left and the other to the right of the origin. Locate the two points on the curve *directly above* these points and use a straightedge to join them with a line. The point where this line crosses the y-axis indicates the answer, 40.

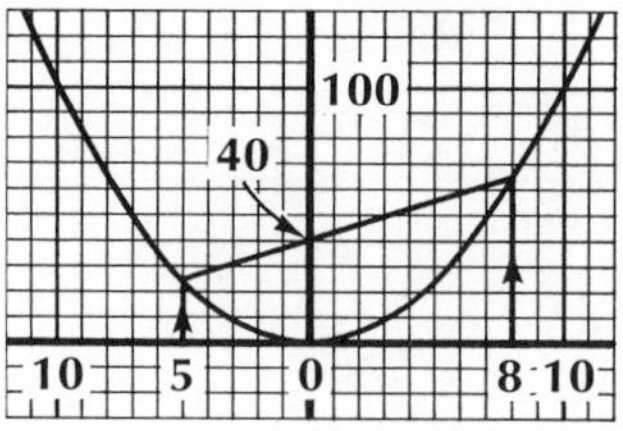

Now, following the same procedure, use your graph to find approximate answers to these problems by drawing the appropriate lines.

1. 10×8
2. 12×17
3. 19×19
4. 21×26
5. 26×27
6. 30×12
7. 27.5×29.5

Set III

Experiment. Parabolas by Folding Paper

In this experiment, you will form some parabolas without plotting points or drawing any lines at all!

Take a sheet of unlined paper, fold it in half like this, and tear it along the fold into two equal pieces.

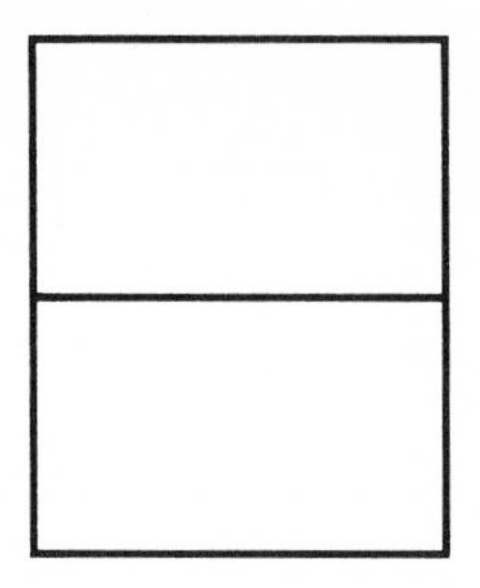

Take one piece and mark a point on it about one inch above the center of the lower edge. Fold your paper and make a sharp crease, as shown here, so that the lower edge falls along the point. Open the paper flat and fold again at a different angle, being careful that the lower edge again comes to the marked point. Repeat this about 20 times, folding the paper in a different direction each time, and a parabola should appear. Trace it with your pencil. The marked point is its focus.

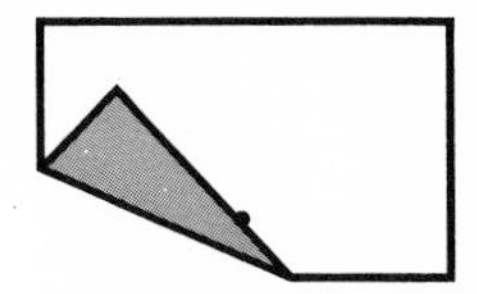

Take the other sheet of paper, mark a point in about the center of it and do the same as before. Trace the curve that results.

Do you think that the parabola that appears this time has the same shape as the first one?

Lesson 3

The Hyperbola

WHEN an airplane flies faster than the speed of sound (about 770 miles per hour), it creates a shock wave heard as a "sonic boom." This shock wave has the shape of a cone and it intersects the ground in part of a mathematical curve called a **hyperbola.**

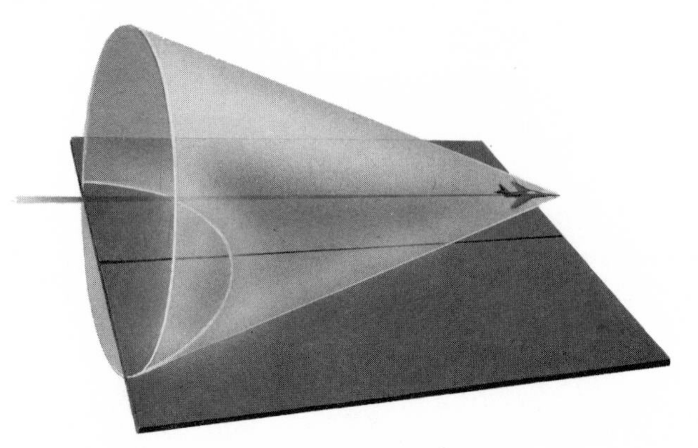

The sonic boom hits every point on this curve at the same time, so that people in different places along the curve on the ground all hear it at once. No sound is heard outside of the curve, but the boom eventually covers every place inside it.

The early Greek mathematicians knew all about the hyperbola, the "sonic boom" curve. In fact, they gave the curve its name. The Greeks were equally familiar with the parabola (the "baseball" curve), as well as the ellipse. How did they happen to become acquainted with these curves so long ago? A Greek named Apollonius, who lived in the third century B.C., wrote a book in which he showed that these curves could be produced by slicing a cone in different directions. If the slice is in the same direction as the side of the cone, the curve that results is the parabola.

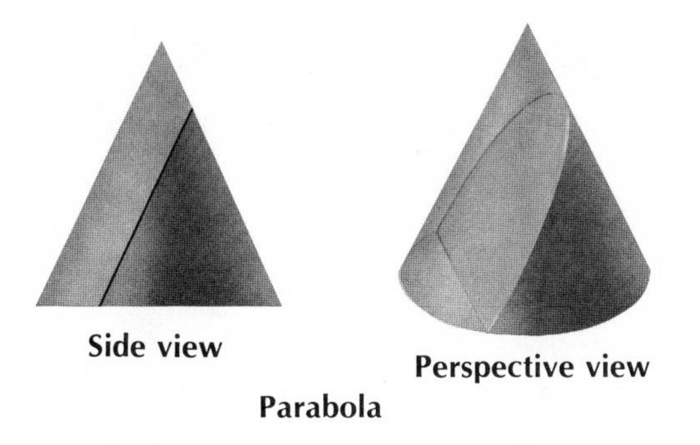

Side view

Perspective view

Parabola

If the slice is tilted from the direction of the side of the cone toward the horizontal, the curve is an ellipse instead. If the slice

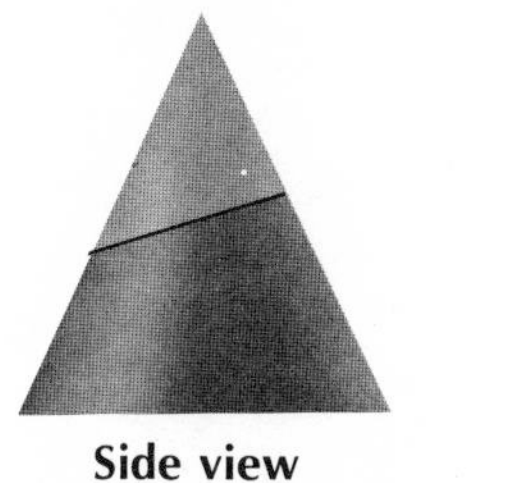

Side view

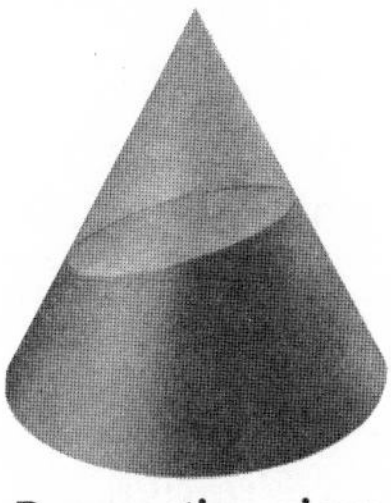

Perspective view

Ellipse

is tilted in the other direction, that is, away from the direction of the side of the cone and toward the vertical, part of a hyperbola is formed.

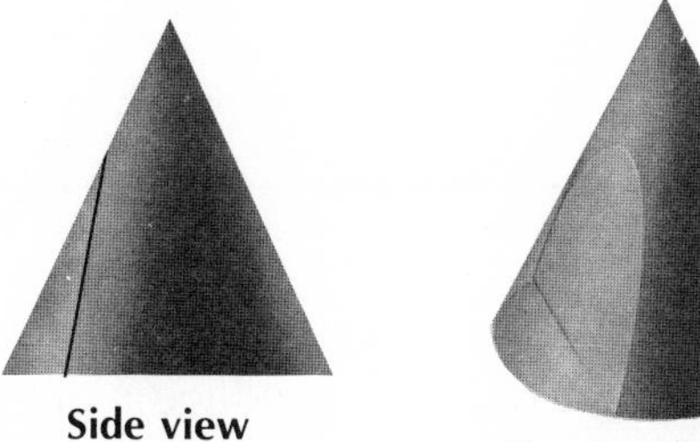

Side view

Perspective view

Hyperbola

Look at the curve in the diagram on page 116. It is part of a hyperbola; two more hyperbolas are shown on pages 117 and 118. A hyperbola consists of two "opposite" curves, called its "branches." To see these branches as slices of a cone, imagine a pair of cones put together so that they touch at their points. A slice cutting through the two cones forms a pair of curves which are the branches of a hyperbola.

Since the parabola, ellipse, and hyperbola are formed when a cone is sliced into sections, they are called **conic sections.** Apollonius named his book *The Conics* and it has since been translated into many different languages. One of these translations was by the English astronomer Edmund Halley, for whom Halley's Comet is named. This comet travels along a path in the shape of an ellipse and Halley used this conic section to predict the time of its return.

EXERCISES

Set I

1. The circle is also a conic section. In what direction should a cone be sliced in order to produce a circle? To answer this, draw a side view of a cone, showing the direction of the slice. Also make a perspective drawing.

Photograph from the Mount Wilson and Palomar Observatories.

Halley's Comet photographed in 1910.

2. Comets travel in orbits that are conic sections. Those that are part of our solar system, including Halley's Comet, have elliptical orbits. Other comets with greater speeds follow parabolic and hyperbolic orbits. Do you think that comets traveling in any of these orbits ever pass near the earth more than once? Which ones?

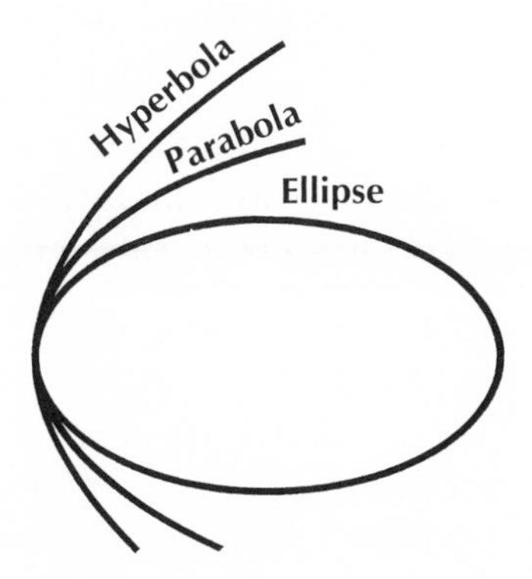

3. Is a hyperbola a *finite* or an *infinite* curve?

4. A hyperbola, like an ellipse, has two foci; they are the points labeled F_1 and F_2 in the diagram below. Three points on this hyperbola have been labeled A, B, and C.

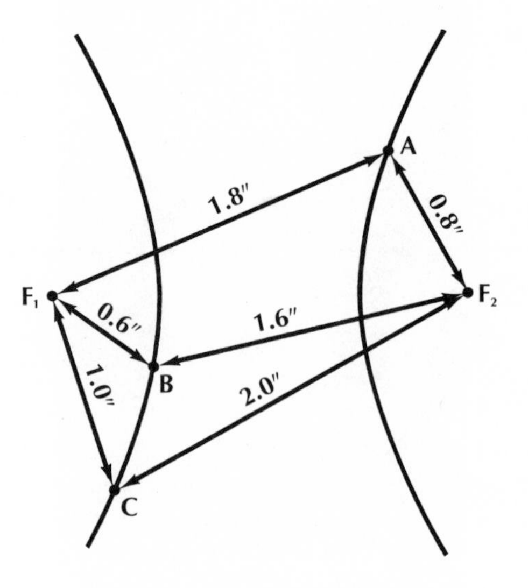

What is the difference between the distance from A to F_1 and the distance from A to F_2? The difference between the distances from B to F_1 and from B to F_2? The difference between the distances from C to F_1 and from C to F_2?

5. Copy and fill in the missing word in the following statement: "A **hyperbola** can be defined as the set of *all points in a plane such that the* ______ *between the distances of each point from two fixed points,* called the foci, *is the same.*" (Compare this definition with the definition of an ellipse on page 248.)

You have learned that every mathematical curve can be associated with an equation and you have already drawn graphs of each of the conic sections by plotting points and joining them with smooth curves.

The graph of the equation

$$16x^2 - 25y^2 = 400$$

is one of the conic sections. One of the points whose x and y

coordinates fit this equation is (5, 0) since

$$16(5)^2 - 25(0)^2 =$$
$$16(25) - 25(0) =$$
$$400 - 0 = 400.$$

6. Draw and label a pair of axes as shown below.

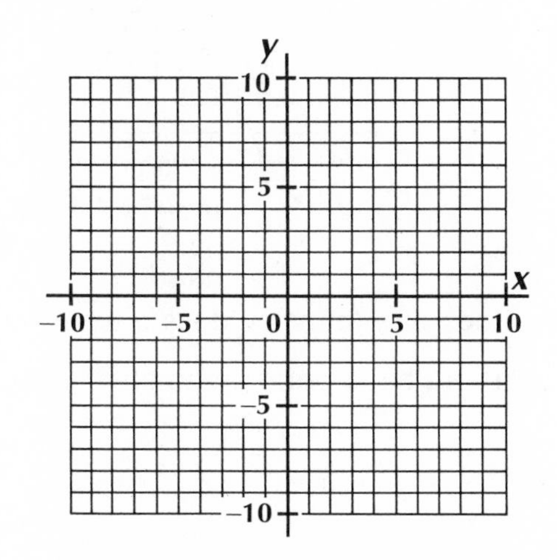

Plot the points in the table below, whose coordinates approximately fit the equation $16x^2 - 25y^2 = 400$.

x	10	9	8	7	6	5	6	7
y	6.9	6.0	5.0	3.9	2.7	0	−2.7	−3.9

x	8	9	10
y	−5.0	−6.0	−6.9

x	−10	−9	−8	−7	−6	−5	−6
y	6.9	6.0	5.0	3.9	2.7	0	−2.7

x	−7	−8	−9	−10
y	−3.9	−5.0	−6.0	−6.9

7. Along what kind of curve do you think these points lie? Draw it.

8. The graphs of these two equations

$$x^2 - 4y^2 = 4 \qquad y^2 - 4x^2 = 4$$

are shown on the following page. What do you notice about the graphs and their equations?

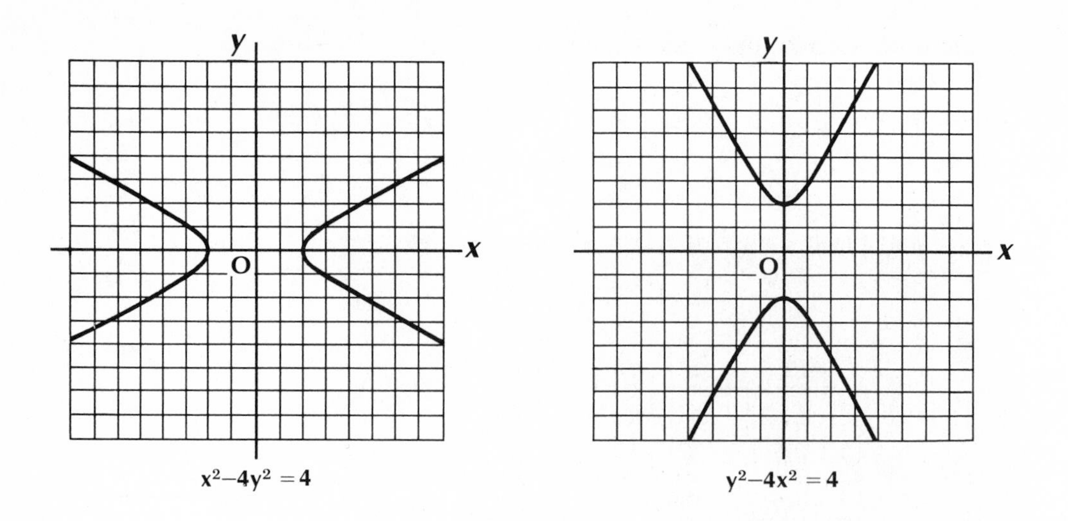

9. How many lines of symmetry does a hyperbola have?
10. Does a hyperbola have point symmetry? Explain.

Set II

Experiment. A Hyperbola by Folding Paper

Draw a circle with a diameter of about 3 inches on a sheet of tracing paper like this. Mark a point on the paper about 1 inch below the circle.

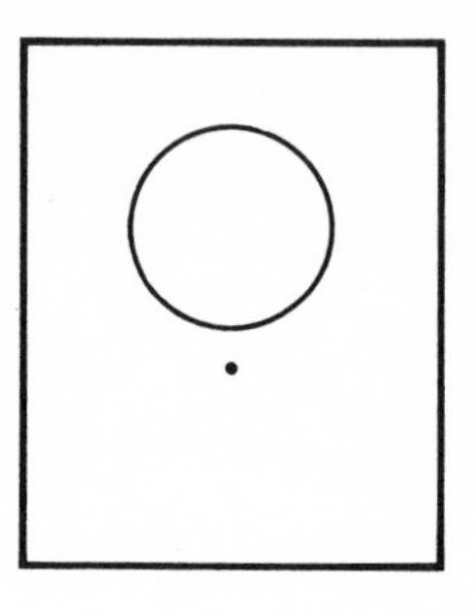

Now fold the paper so that the point falls on the circle and make a sharp crease. Open the paper flat and fold in a different direction so that the point again falls on the circle. Continue in this way until the point has been folded into different positions *all the way around* the circle.

You should be able to find both branches of a hyperbola when you are finished. Trace them in on the paper.

The point you marked is one focus of the hyperbola. Where is the other focus?

Courtesy of William G. Johnson, Concert Recording.

Set III

The pipes in a large pipe organ range in size from 32 feet long to as short as a couple of inches. The longer the pipe, the lower the pitch of the note it produces. An open 32-foot pipe produces a note with a frequency of 16 vibrations per second.

Here is a table showing the lengths of some pipes and their pitches.

Note	Pitch	Length of Pipe
C	16	32.0
G	24	21.3
C	32	16.0
E	40	12.8
G	48	10.7
C	64	8.0
E	80	6.4
G	96	5.3
C	128	4.0

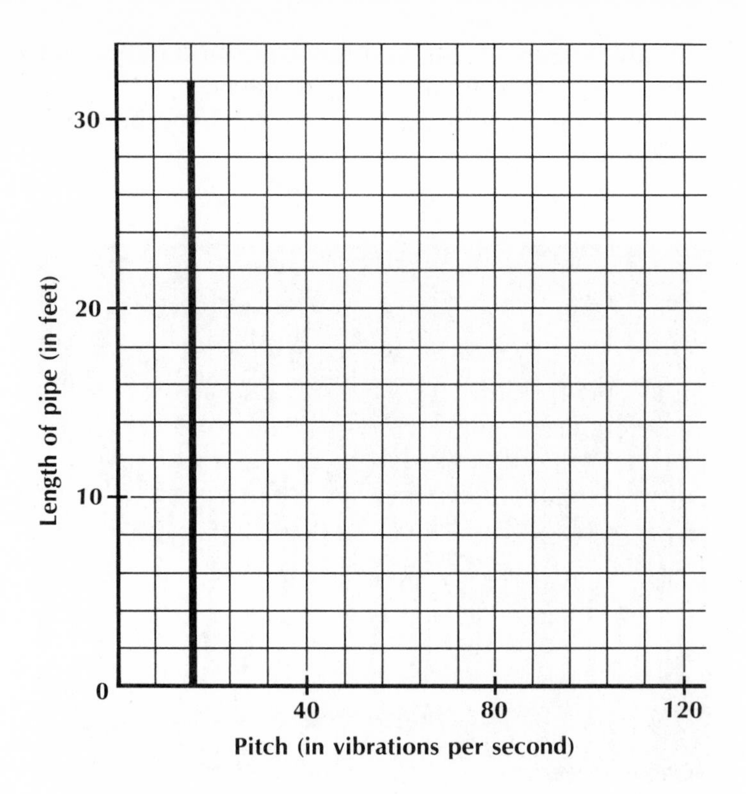

1. Draw a pair of axes on a sheet of graph paper and label and number them as shown above. Notice that each unit on the pitch axis represents 8 vibrations per second, and that each unit on the pipe-length axis represents 2 feet.

   Now draw a series of line segments on the graph to represent the pipes included in the table on page 271. The first one has been drawn in the diagram as an example.

2. Draw a smooth curve through the upper points of the line segments you have drawn.

3. What do you notice?

The pitches of the notes named C* are part of a familiar number sequence:

16 32 64 128

4. What kind of number sequence is it?

5. What do you notice about the lengths of the pipes that

---

* Look back also at the drawing of the piano keyboard on page 176.

sound these C notes? (Look in the table on page 271.)

6. What do you think would be the pitch of an organ pipe having a length of 2 feet?

A formula expressing the pitch, p, of an organ pipe as a function of its length, l, looks like this:

$$p = \frac{\square}{l}$$

7. This is the formula of the curve that you drew through the upper points of the line segments on your graph. Can you figure out the missing number in it? (Hint: Multiply.)

INTERESTING READING

Sound and Hearing, by S. S. Stevens and Fred Warshofsky, a book in Time-Life's "Life Science Library" series, 1965: "The Unwanted Sounds," pp. 170–180.

The Universe, by David Bergamini, a book in Time-Life's "Life Nature Library" series, 1962: "Planets, Meteorites and Comets," pp. 63–83.

Lesson 4

"That's a note I don't hit very often."

The Sine Curve

WHEN musical sound waves are changed into visual images by an electronic instrument called an oscilloscope, they have a regular pattern, which repeats itself many times each second. Here is a diagram showing one of these repeating patterns, produced by a trumpet.

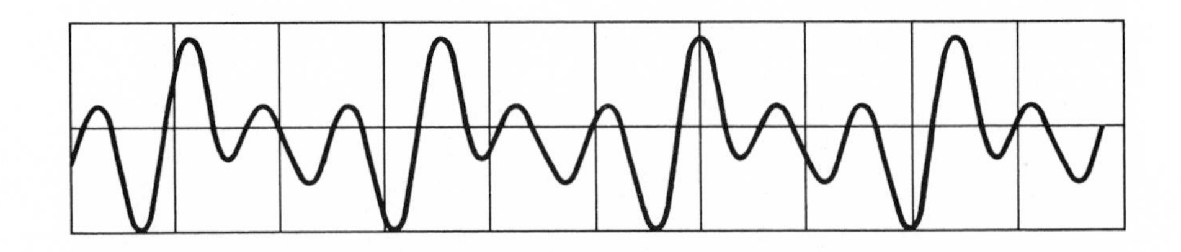

A French mathematician named Joseph Fourier, who lived during the time of Napoleon, showed that every sound wave that repeats itself like this one is related to a mathematical curve called the **sine curve.**

To understand what a sine curve is and why it has the shape that it has, imagine a clock whose hour hand always points at 3 and whose minute hand moves backwards. Suppose that we first look at the clock when the minute hand points in the same direction as the hour hand. As the minute hand moves counterclockwise, the angle that it makes with the hour hand becomes larger. Some angles made during one revolution are on the facing page. In each revolution the two hands form angles varying in measure from 0° to 360°. When the hands point in the same direction, it is convenient to think of them as forming a 0° angle.

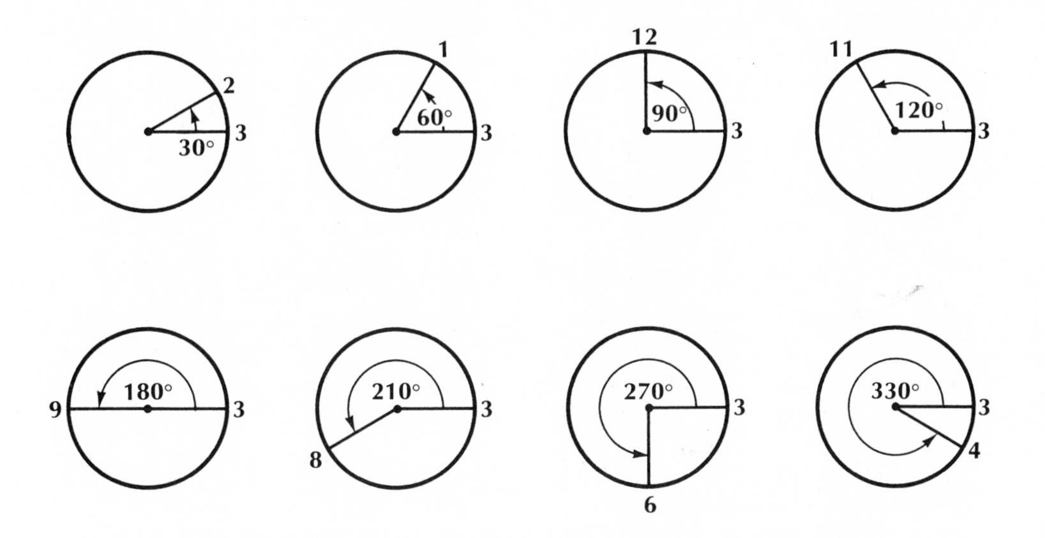

Now suppose that the minute hand of the clock is one unit long and that the clock is divided in half by a horizontal line through its center. If we let x represent the measure of the angle formed by the clock's hands at any moment, then the vertical distance y from this horizontal line to the tip of the minute hand is called the "sine" of the angle x.

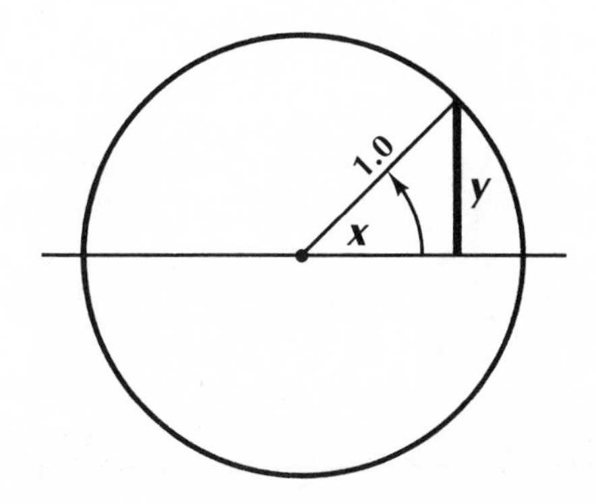

Diagrams of approximate sines of different angles are given at the top of the following page. In the last two, the sines are negative numbers; this is because the distances from the hori-

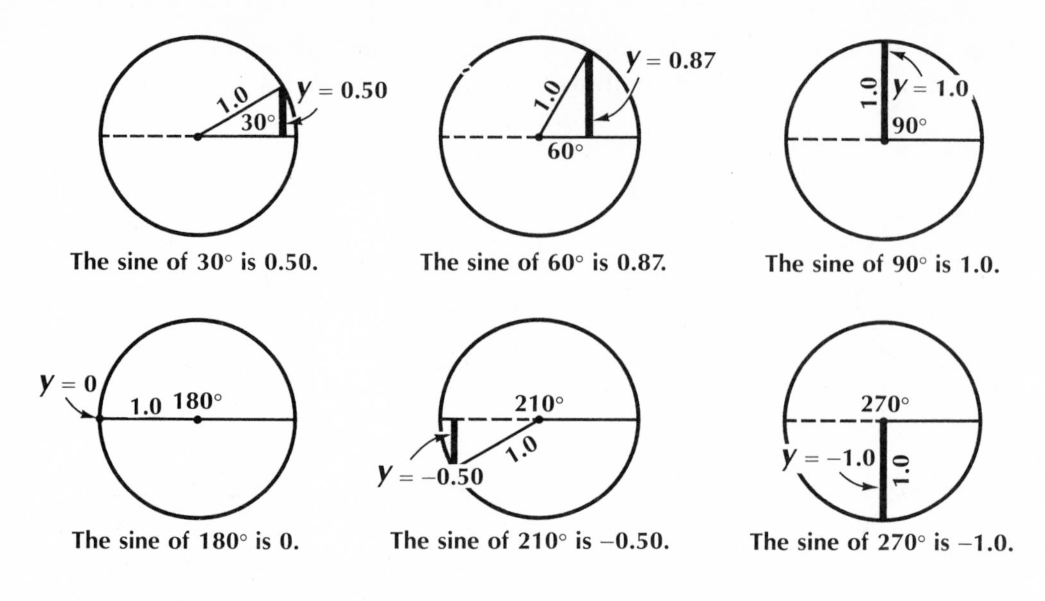

The sine of 30° is 0.50. The sine of 60° is 0.87. The sine of 90° is 1.0.

The sine of 180° is 0. The sine of 210° is −0.50. The sine of 270° is −1.0.

zontal line to the tip of the minute hand are measured downward instead of upward.

The sine of an angle is a function of the measure of the angle, and a graph of the "sine function" is shown below.

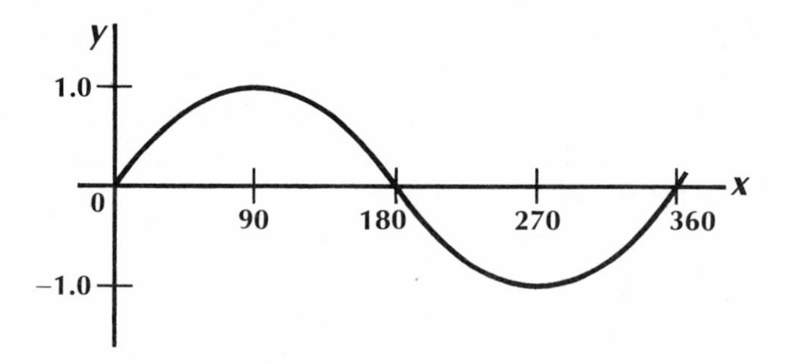

If the graph is continued, the curve is repeated over and over. This curve, called the *sine curve,* is important in many fields other than the study of sound. For instance, it shows how the strength of an alternating electric current changes with time.

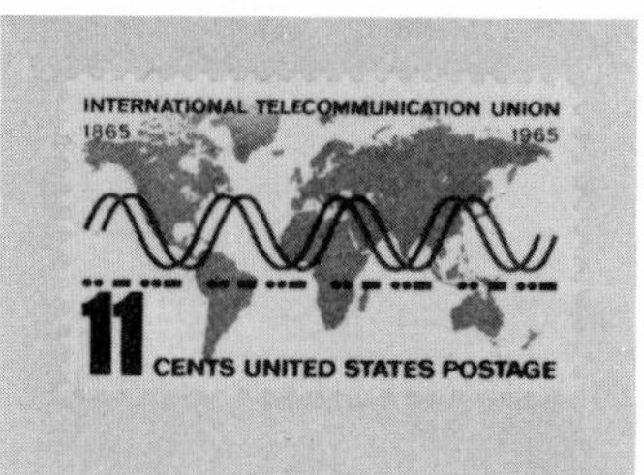

The sine curve on a United States postage stamp.

EXERCISES

Set I

Here is a large diagram from which you can estimate the sines of some angles to the nearest hundredth.

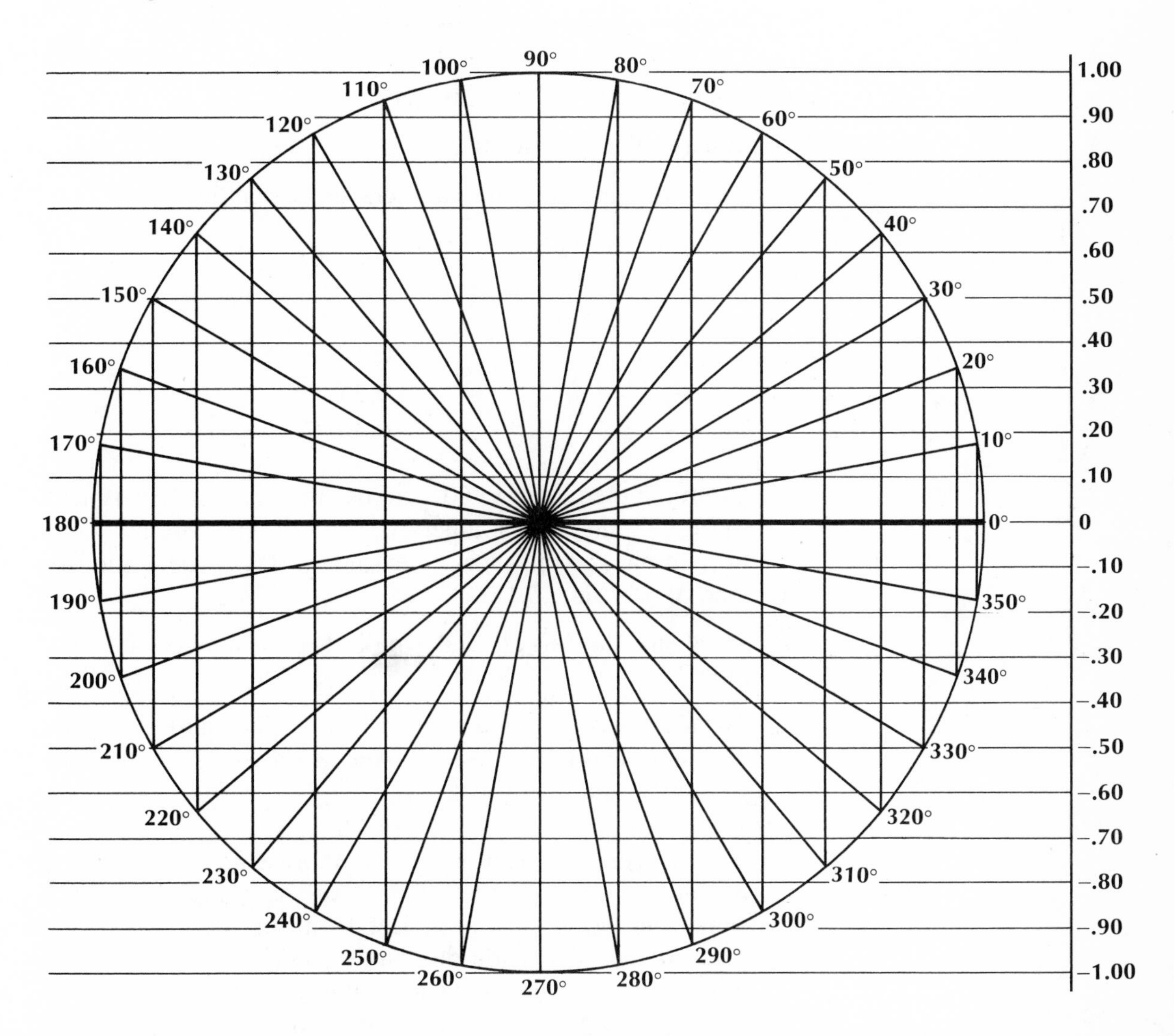

For example, the sine of 10° is approximately 0.17

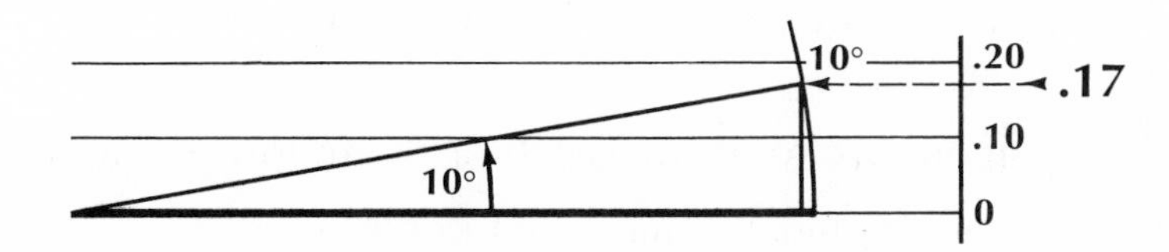

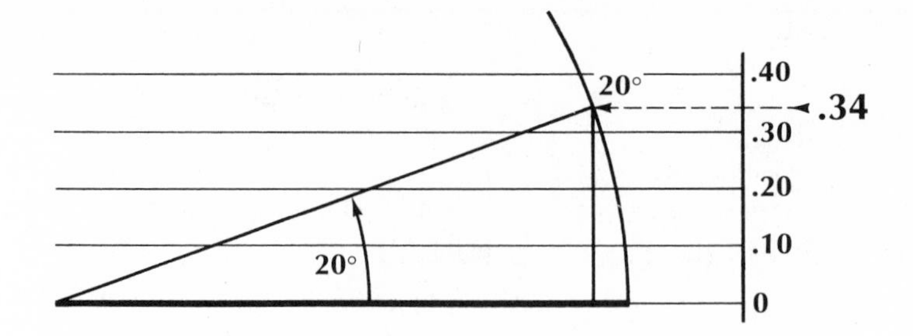

and the sine of 20° is approximately 0.34.

1. What is the sine of 30°?

2. Copy the table of sine numbers shown below.

Angle x	y = sine of x
0°	0
10°	0.17
20°	0.34
30°	0.50

   Continue the table to 360° and use the diagram to estimate the remaining sine numbers. Here are two more numbers to help check your work: the sine of 100° is 0.98 and the sine of 190° is −0.17.

3. What are the largest and smallest sine numbers in your table?

4. Turn a sheet of graph paper (4 units per inch) sideways and draw and label a pair of axes as shown in the figure at the top of the facing page.

   The measure of each angle x is represented on the horizontal axis and the sine of each of these angles is represented on the vertical axis. Let 1 unit on the x-axis represent 10° and 1 unit on the y-axis represent 0.1. Use your table to plot a point for each angle from 0° through 360° and join the points with a smooth curve.

5. What kind of symmetry does the part of the sine curve you have drawn have?

Use your sine curve to answer each of the following questions.

6. What angles have a sine equal to 0?

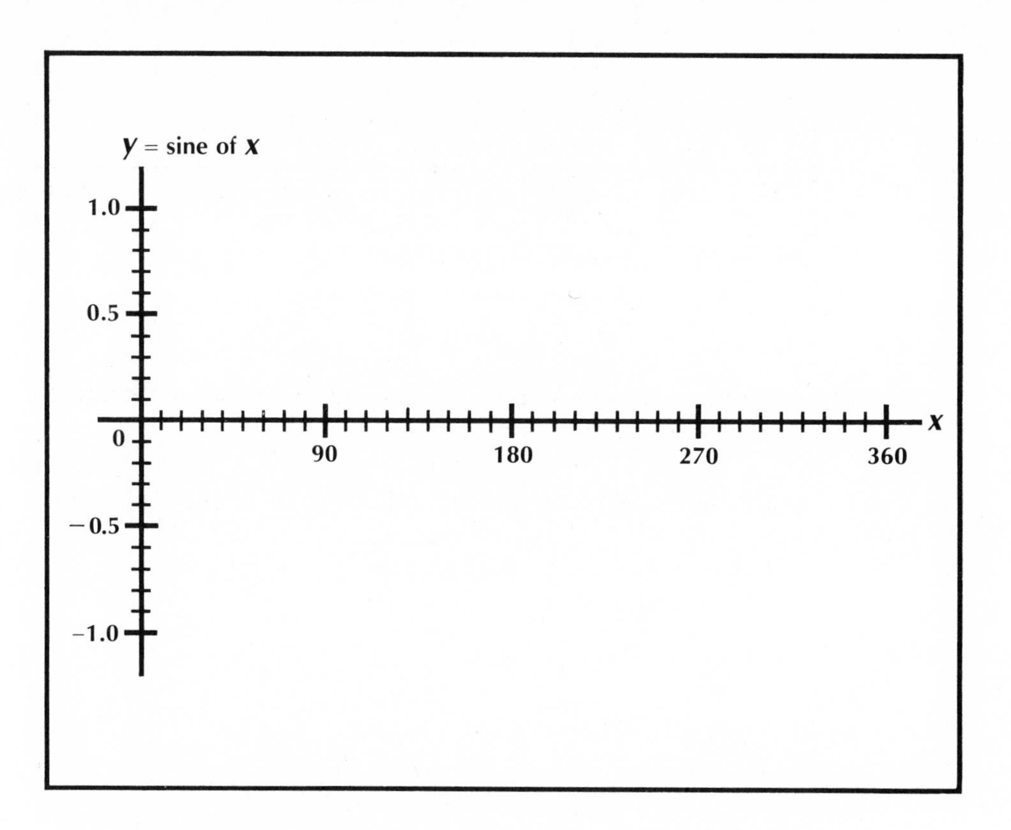

7. Two angles shown on your curve have a sine of $\frac{1}{2}$. What angles are they?
8. What angle do you think has the same sine as an angle of 5°?
9. What two angles shown on your curve have a sine of $-\frac{1}{2}$?
10. What angle do you think has the same sine as an angle of 185°?

Set II

The **amplitude** of a sine curve is its maximum distance from the x-axis. The **period** of a sine curve is the distance along the x-axis of one wave length (the smallest part of the curve that is repeated).

The equation for the curve shown at the top of page 280 is:

$$y = \text{sine } x.^*$$

---

*This means: y is the sine of angle x.

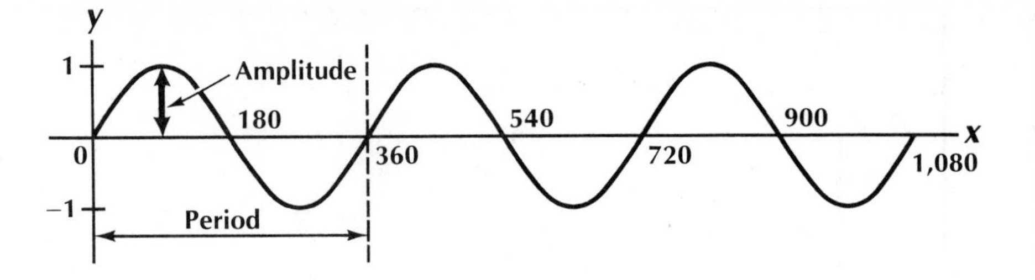

The amplitude of this curve is 1 and its period is 360°; three wave lengths are shown.

What are the amplitudes and periods of each of the following curves?

1. $y = 2 \text{ sine } x$

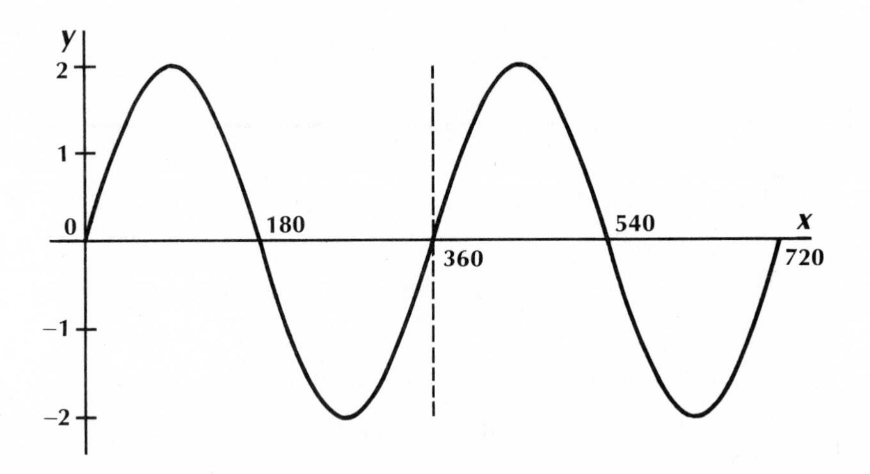

2. $y = \text{sine } 2x$

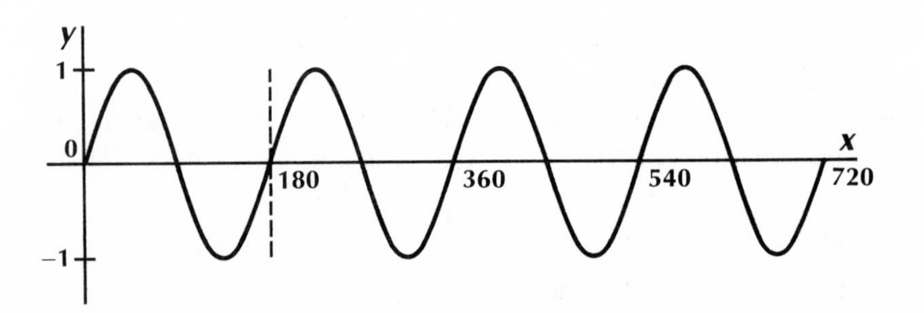

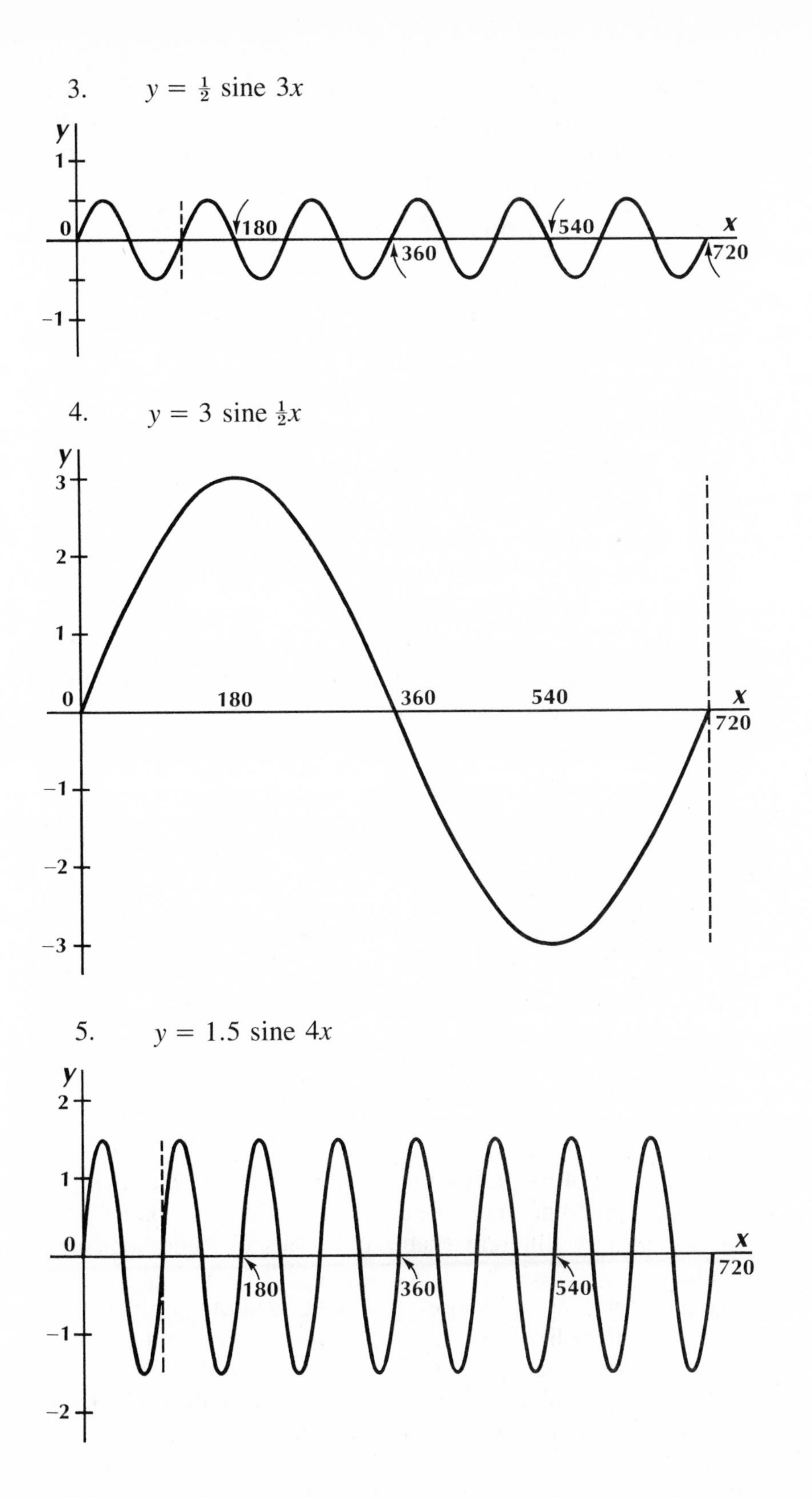
3. $y = \frac{1}{2}$ sine $3x$
y
1
0
−1
x
180
360
540
720
4. $y = 3$ sine $\frac{1}{2}x$
y
3
2
1
0
−1
−2
−3
x
180
360
540
720
5. $y = 1.5$ sine $4x$
y
2
1
0
−1
−2
x
180
360
540
720

6. Compare the amplitude you have written down for each of these curves with its equation. If you see a relationship, what do you think would be the amplitude of a curve with the equation, $y = 5$ sine $2x$?
7. Compare the period you have written for each of the curves with its equation. If you see a relationship, what do you think would be the period of a curve with the equation, $y = 4$ sine $6x$? (Hint: The relationship involves division and 360°.)
8. The *loudness* of a sound depends upon the *amplitude* of its wave. If the sine curves in exercises 1-5 represent sound waves, which one is loudest?
9. The *pitch* or frequency of a sound depends upon its *period*. A high frequency sound has a small period since its wave length is short. Which one of the sound waves shown in exercises 1-5 has the highest pitch?
10. Here is a diagram showing three sound waves at once. Do these sound waves have the same loudness or do they have the same pitch?

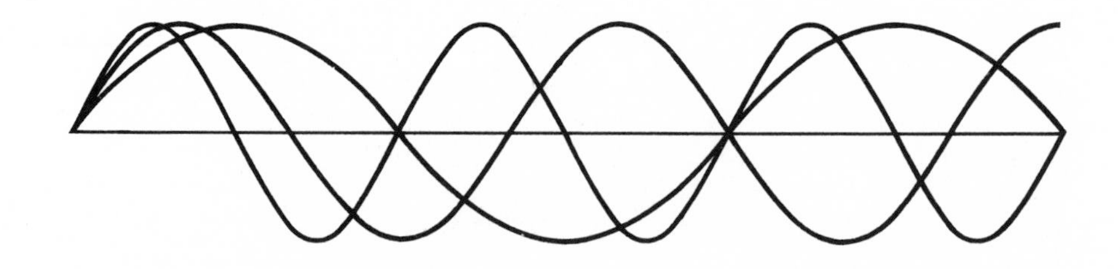

Set III

A different kind of graph from the kind we have been drawing is the *polar* graph. When the sine curve is drawn on a polar graph, it has an entirely different shape and looks like another curve instead.

Use a compass to draw a set of 10 circles all having the same center and with radii of $2\frac{1}{2}$, $2\frac{1}{4}$, 2, $1\frac{3}{4}$, $1\frac{1}{2}$, $1\frac{1}{4}$, 1, $\frac{3}{4}$, $\frac{1}{2}$, and $\frac{1}{4}$ inches. Use a protractor to draw a series of lines through this center, called the "pole" of the graph, at 10° angles with each other. Number the lines 0°, 10°, 20°, 30°, and so forth to 350°.

Now plot a point on each of these lines at a distance from the pole equal to the sine of the angle it represents. We will consider the radius of the largest circle to be 1 unit, so that the radius of the smallest circle is 0.1 unit, the radius of the next circle is 0.2 unit, and so on. Since the sine of 0° is 0, the first point is the pole itself. The sine of 10° is 0.17, so the next point is on the 10°

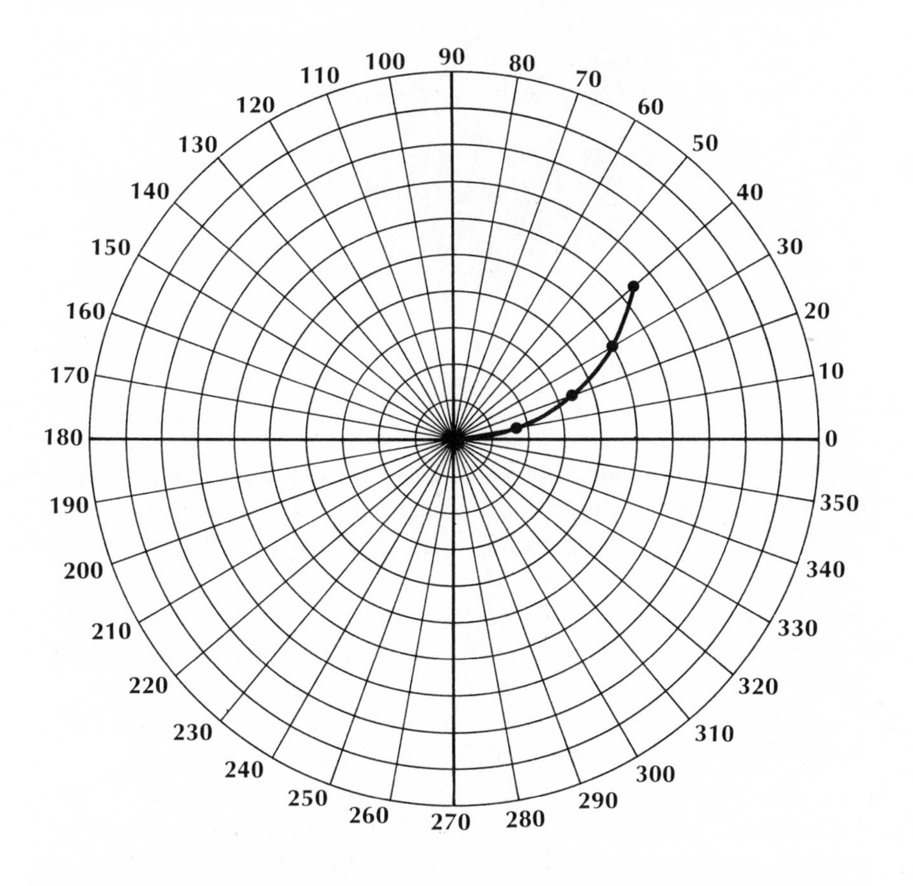

line at a distance of 0.17 units from the pole. The points corresponding to 0°, 10°, 20°, 30°, and 40° have been plotted on this diagram as an example. Plot a point on each line from 0° through *180*° and join them in order with a smooth curve. What curve is the result?

Lesson 5

Spirals

THE chambered nautilus is a sea creature that moves into a series of successively larger compartments as it grows. Each compartment or chamber has the same shape except the last one, when the animal is fully grown. This photograph shows the shell of a chambered nautilus which has been cut in half to reveal these compartments. It has the shape of a mathematical curve called a *logarithmic spiral.*

A **spiral** is a curve traced by a point that moves around a fixed point from which it continually moves away.

There are several kinds of spirals. The logarithmic spiral was discovered by Descartes, the man who invented coordinate geometry. Archimedes wrote a book titled *On Spirals* in which he described another type of spiral which has since been named for him. The groove in a phonograph record is in the shape of an *Archimedean spiral.*

Spirals occur in nature in many different ways. They appear in the heads of daisies, in elephant tusks, and in the webs of certain spiders. The nerve fibers that make up the sound receptors in our ears are in a spiral-shaped arrangement. The most spectacular spirals in nature can be seen only through a telescope—these are the galaxies and nebulas of the universe. Most galaxies have spiral shapes, including the Milky Way, the one in which our sun is a star. Our solar system is about three-fourths of the way from the center of the spiral to the edge.

Spirals that appear in nature—the nautilus shell (left), the spider's web, a galaxy.

Courtesy of The American Museum of Natural History.

Photograph from the Mount Wilson and Palomar Observatories.

A spiral galaxy 50 million light-years away.

EXERCISES

Set I

A person walking from the center of a merry-go-round at a steady speed along a radius of the floor travels, with respect to the ground, along an Archimedean spiral. To draw a picture of his path, it is convenient to use a polar graph.

1. Prepare your paper by following the directions given in the second paragraph of the Set III exercise in the last lesson. Then plot a point on each of the angle lines at a distance from the pole as given in the table at the top of the facing page. (This time we will consider $\frac{1}{4}$ inch to be 1 unit.) The points corresponding to the angles from 0° through 100° have been plotted on the diagram below as an example.

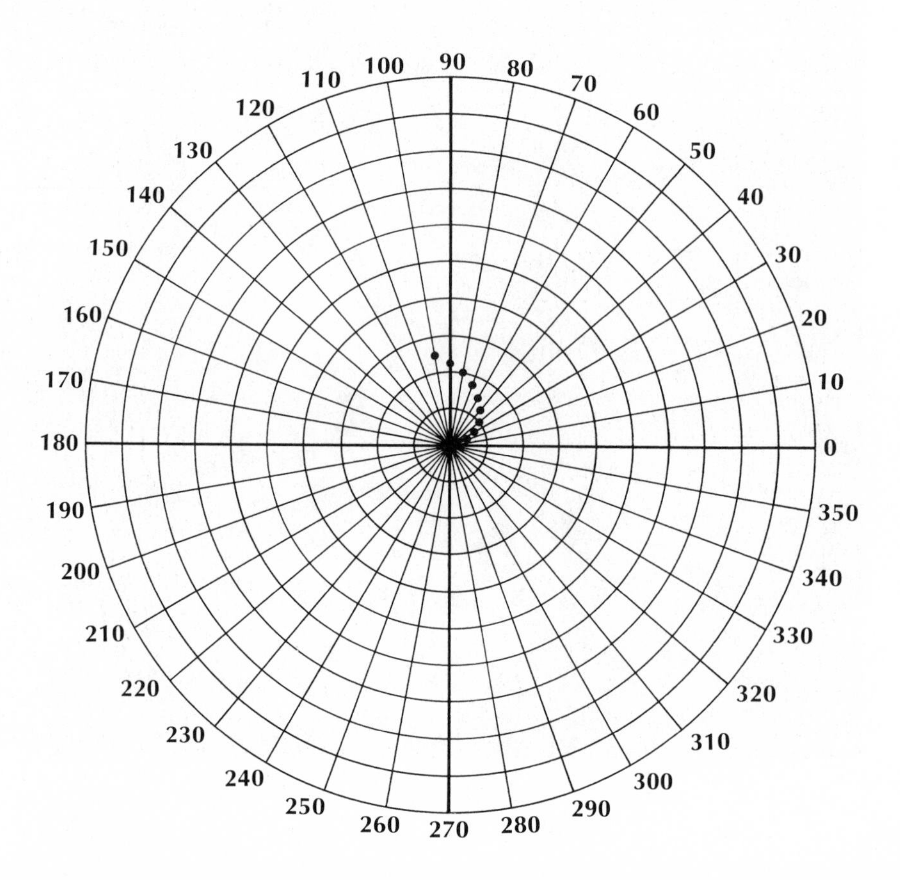

Angle line	Distance from pole
0°	0
10°	0.25
20°	0.50
30°	0.75
40°	1.00
50°	1.25
etc.	

Plot a point on each line from 0° through 360° and then continue until you come to the outermost circle; join the points in order with a smooth curve. The result is an Archimedean spiral.

2. The spiral you have drawn represents the person's path on the merry-go-round with respect to the ground. Through how many degrees has the merry-go-round turned by the time he is *halfway* from the center to the edge?
3. Through how many degrees in all will the merry-go-round have turned by the time he arrives at its edge?
4. The angles listed in the table for this curve form an arithmetic sequence. In what kind of number sequence are the successive distances of the curve from the pole?

If a person's speed while walking from the center to the edge of the merry-go-round is slower or faster, his path with respect to the ground is still an Archimedean spiral, but one of a different shape.

5. On the same polar graph, plot another series of points to show a path corresponding to a speed that is twice as fast. The beginning of the table for this curve is shown below.

Angle line	Distance from pole
0°	0
10°	0.50
20°	1.00
30°	1.50
40°	2.00
etc.	

6. Through how many degrees has the merry-go-round turned

when the person walking along this path arrives at the edge?

7. Is the number sequence of the successive distances of this curve from the pole at each 10° the same kind as that for the other curve?

8. Copy and complete the following statement: "As the angle of an *Archimedean spiral* increases in arithmetic sequence, the successive distances of the curve from the pole form a ▒▒▒▒▒▒ sequence."

9. How many revolutions does a 45-rpm phonograph record make as the needle travels along the groove of a song which is 3 minutes long? (Hint: What does 45-rpm mean?)

10. Through approximately how many *degrees* does this record turn as the needle moves from the edge toward the center? (Hint: How many degrees are in one revolution?)

11. The needle does not travel along the groove at the same speed throughout the record because the loops of the spiral become progressively shorter. Does the needle speed up or slow down as the record plays?

Courtesy of Irv Phillips; by permission of Publishers-Hall Syndicate.

12. The average length of each loop of the groove of a 45-rpm record is about 17 inches. The needle covers one loop of the groove in each revolution. About how long is the groove of a 3-minute song?

Set II

1. The spiral of the chambered nautilus is a logarithmic spiral. To draw a picture of it, plot the points represented in the table below on a polar graph (you can use the same polar graph paper you used in Set I), and join them with a smooth curve.

Angle line	Distance from pole	Angle line	Distance from pole
0°	1.0	190°	3.0
10°	1.1	200°	3.2
20°	1.1	210°	3.4
30°	1.2	220°	3.6
40°	1.3	230°	3.8
50°	1.3	240°	4.0
60°	1.4	250°	4.2
70°	1.5	260°	4.5
80°	1.6	270°	4.7
90°	1.7	280°	5.0
100°	1.8	290°	5.3
110°	1.9	300°	5.6
120°	2.0	310°	6.0
130°	2.1	320°	6.3
140°	2.2	330°	6.7
150°	2.4	340°	7.1
160°	2.5	350°	7.5
170°	2.7	360°	8.0
180°	2.8		

2. The successive distances of the curve from the pole have been rounded off in this table. They form a number sequence which you should be able to recognize by copying

the table below and filling in the missing numbers:

Angle	0°	120°	240°	360°
Distance from pole	1.0	▮▮▮▮	▮▮▮▮	▮▮▮▮

3. What kind of sequence do the distances form?
4. Copy and complete the following statement: "As the angle of a *logarithmic spiral* increases in arithmetic sequence, the successive distances of the curve from the pole form a ▮▮▮▮▮▮▮▮▮▮ sequence."
5. Without drawing the curve, what kind of spiral do you think would have these distances from the pole for the angles in this table?

Angle	0°	100°	200°	300°
Distance from pole	1.0	3.0	9.0	27.0

6. What kind of spiral has this table?

Angle	0°	100°	200°	300°
Distance from pole	1.0	3.0	5.0	7.0

Set III

Experiment. A Logarithmic Spiral in a "Rectangle of Whirling Squares"

You have learned that the Fibonacci number sequence* begins like this:

1 1 2 3 5 8 13 21 34 55 89 . . . .

This sequence, which frequently appears in plant growth, is closely related to the logarithmic spiral.

Turn a sheet of graph paper (ruled 10 units per inch) sideways and draw a rectangle that is 55 units wide and 89 units long. See Diagram A on the facing page.

Draw line segments in the rectangle to divide it into a series of squares whose sides are 55, 34, 21, 13, 8, 5, 3, 2, 1, and 1 respectively, as shown in Diagram B on the facing page.

* See page 72.

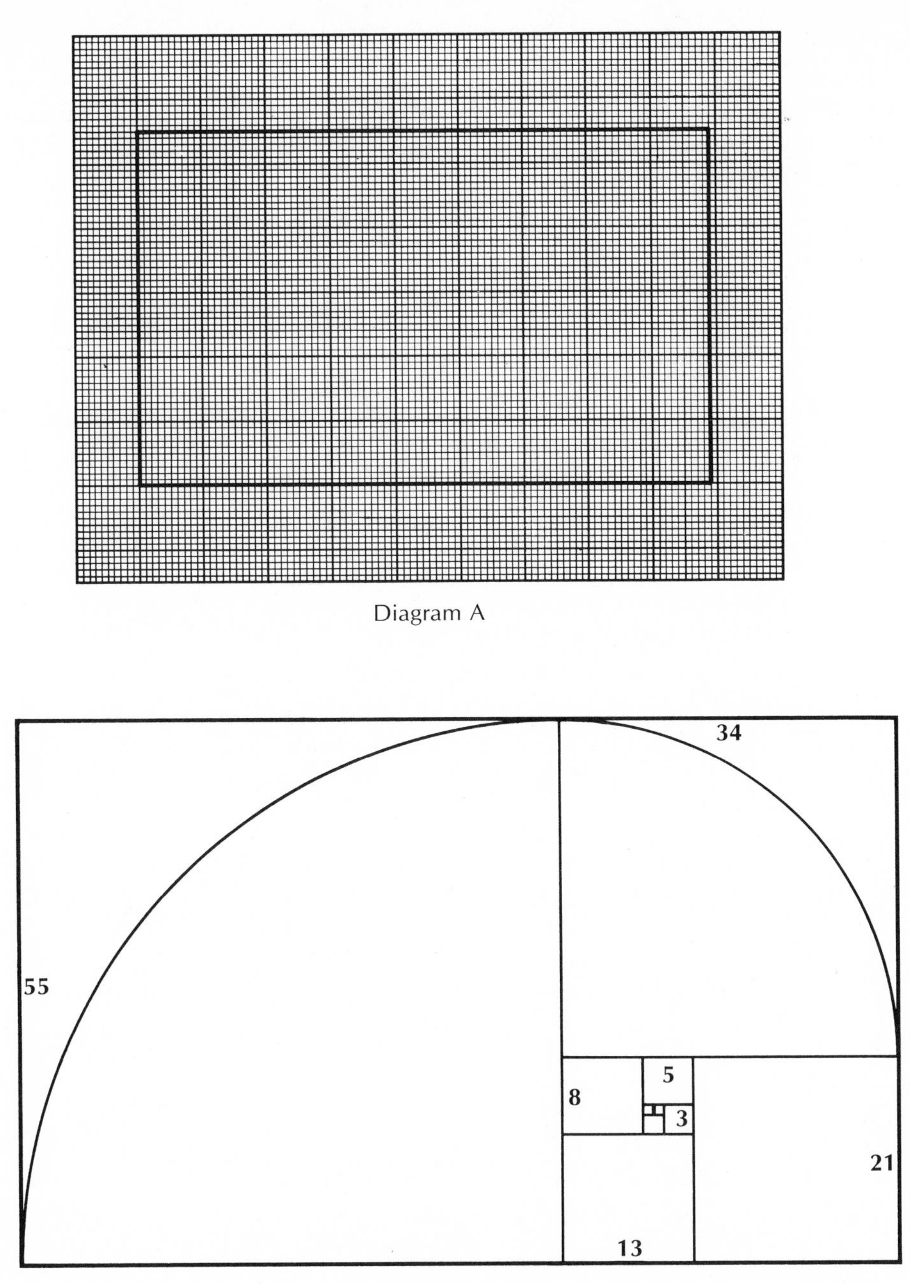

Diagram A

Diagram B

Now draw an arc that is a quarter circle from a corner of each square, starting with the largest square, to form a continuous curve. The diagram shows the first two of these arcs drawn in.

The resulting curve, based on the Fibonacci sequence, is close to the shape of the chambered nautilus!

INTERESTING READING

Mathematics, by David Bergamini, a book in Time-Life's "Life Science Library" series, 1963: "The Mathematics of Beauty in Nature and Art," pp. 88–102.

The Universe, by David Bergamini, a book in Time-Life's "Life Nature Library" series, 1962: "Other Island Universes," pp. 155–165.

Lesson 6

The Cycloid

AS a wheel rolls along a straight line, a point on its rim follows a path called a **cycloid**. The diagram shows part of this mathematical curve. It might seem as if a point on the rim would move

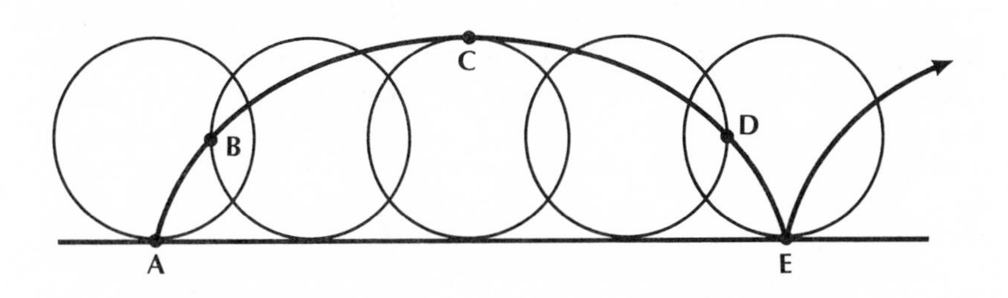

at a constant speed as the wheel rolls steadily forward, yet this is not the case. Five positions of the wheel are shown in the diagram; the first and last show the wheel with the point on the rim touching the ground; the other three show intermediate positions in one revolution: after turning 90°, 180°, and 270°.

Notice that as the wheel rolls equal distances along the ground, the point on the rim does not travel equal distances along its cycloid path. The distance from B to C is longer than the distance from A to B, so that the point on the rim must move faster as it approaches C. Its greatest speed is at the moment it has reached the top, and as the point moves downward to E, it slows down until it stops for an instant at the bottom of the curve before beginning another arc. Points A and E of the cycloid, where the curve suddenly changes direction, are called "cusps."

Although the early Greek mathematicians knew a lot about the conic sections, the ellipse, parabola, and hyperbola, and about the spiral of Archimedes as well, they apparently never thought of the cycloid. In fact, no one knows who discovered this remarkable curve. Galileo, in the seventeenth century, suggested that it would be a good shape for an arch of a bridge. A great French mathematician named Pascal later studied the cycloid as a result of suffering from a toothache! He decided that he needed something interesting to think about to take his mind off the pain and, as a result, made many discoveries about this "curve of a rolling wheel."

EXERCISES

Set I

Experiment. The Path of a Moving Point

Part 1. The Cycloid To draw a cycloid, we will actually roll a "wheel" along a straight line and mark different positions taken by a point on its rim. With a compass, draw a circle the size of the one below on a 3 × 5 card. The radius should be $1\frac{1}{4}$ inches.

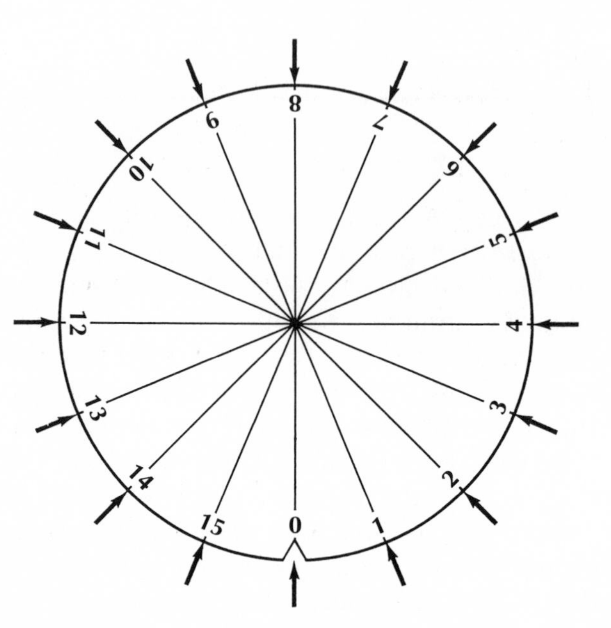

Cut the circle out, place it on the circle on the page, and mark 16 points on the rim as indicated by the arrows. Take your circle away from the page and, with a ruler, draw a set of diameters to join the points you have marked. You now have a wheel with "spokes." Number the spokes 0 through 15 as shown and cut a small notch in the rim of the wheel at the spoke numbered 0.

Take a sheet of graph paper ruled 4 units per inch and turn it sideways. Then draw a line across it about 3 inches from the top. Mark the line with a series of 21 points that are 2 units apart. Number the points from the left 0 through 15 and 0 through 4 as shown in this diagram.

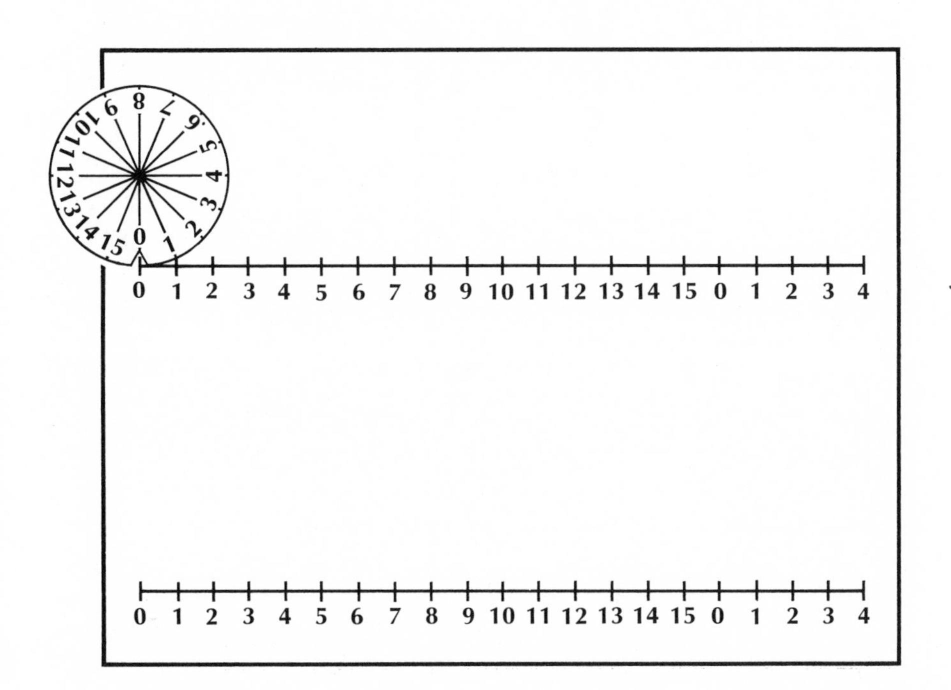

Place your wheel on the line so that the spoke numbered 0 points downward and the notch is at the point on the line numbered 0. Mark a heavy dot on the graph paper at the notch. Then rotate the wheel to the right so that the spoke numbered 1 points downward and just touches the point on the line numbered 1. Mark another heavy dot on the graph paper through the notch in the wheel.

Repeat this all the way across the paper so that each spoke points downward to the point on the line with its corresponding number; mark a dot in the notch each time.

Now join the dots with a smooth curve; you will get one full arc and part of a second arc of a cycloid.

Part 2. A Prolate Cycloid What does the path of a point on the *inside* of a wheel that rolls along a straight line look like? It follows a curve called a "prolate cycloid." To draw this curve, we will use the same method that we used in drawing the cycloid.

Use a paper punch to punch a hole in the center of the spoke numbered 0 on your wheel.

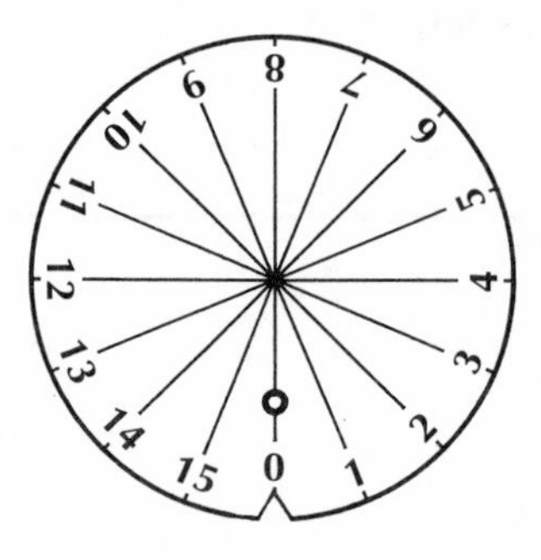

Draw another line on the same sheet of graph paper about 1 inch from the bottom. Mark the line with a series of points and number them from 0 to 15 and 0 to 4, as you did with the other line. Place the wheel in the appropriate position on the line at each point and mark a dot on the graph paper through the center of the punched hole each time. When the dots are joined with a smooth curve, a prolate cycloid is the result.

Part 3. A Curtate Cycloid In this part, we will find out what the path of a point *outside* a wheel that rolls along a straight line looks like. The curve of this path is called a "curtate cycloid."

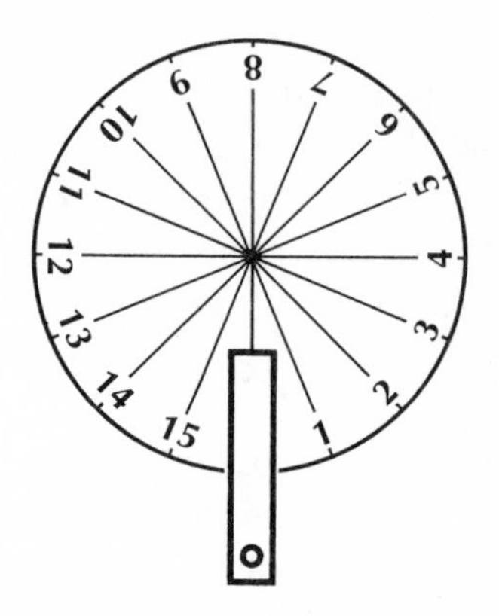

First, cut a small strip from the card out of which you cut the circle, punch a hole in the end of this strip, and tape it along the spoke numbered 0, as shown in the diagram at the bottom of page 296.

Place the wheel at each point on the same line as the previous curve and mark a series of dots on the graph paper to show each position of the hole in the strip. A curtate cycloid results when the dots are joined in order with a smooth curve.

Set II

As a wheel rolls steadily forward, each point *on its rim* comes to a *stop* for an instant when it touches the ground.

1. Where is each point on the rim when it is moving the *fastest?*
2. Look at your drawing of the path of a point *inside* the wheel. Do you think each point inside the wheel also comes to a stop at certain instants?
3. Look at your drawing of the path of a point *outside* the wheel. The motion of such a point with respect to the wheel at certain times is very strange. Do you see why?
4. Are there any points on a wheel that move forward at a *steady rate* as the wheel rolls along? If so, where?

This diagram shows the relationship between the length of a circle (its circumference) and the circle's diameter.

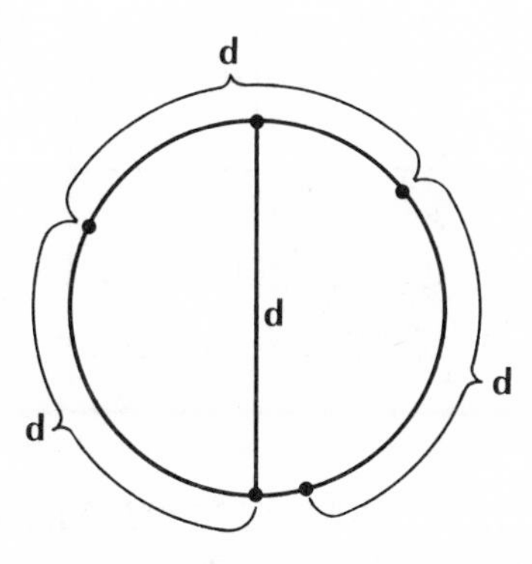

Do you see that the circumference is a little more than 3 times as long as the diameter? In fact, it is about 3.14 times as long and this number is commonly called "pi." *

5. The diameter of the wheel you have been using is 10 units. What is its approximate circumference?

It is impossible to determine the *exact* length of a circle from its diameter because the exact value of pi cannot be found. The cycloid is certainly a more complicated curve than the circle, and so it would seem that the exact length of one arc of a cycloid could not be found either.

Sir Christopher Wren, a great English architect, made the surprising discovery that the length of an arc of a cycloid is exactly equal to a certain small whole number multiplied by the length of the diameter of the wheel that generates the cycloid.

6. Look at this drawing of a cycloid and its wheel to find out what this number is.

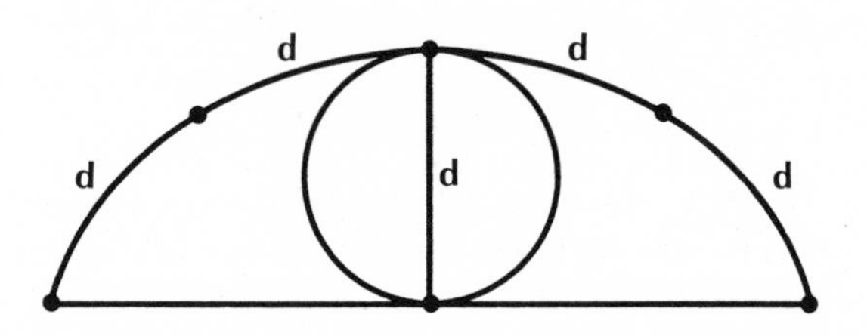

This diagram shows the relationship between the area of a circle (the amount of surface inside it) and the square of the circle's radius.

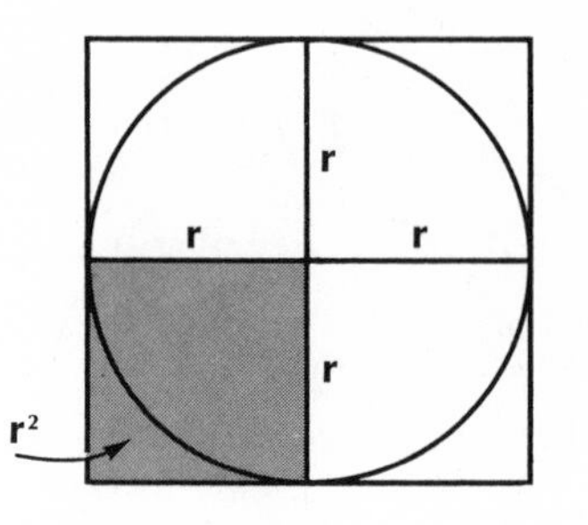

* More precisely, pi is 3.14159265358979323846264338327950288 41971693993751058209749445923078164062862089986280348253 421170679, but even this is not exact! Pi as a decimal number never ends nor repeats in any pattern. It has been computed to more than 500,000 decimal places!

Notice that the area of the large square surrounding the circle is exactly 4 times the square of its radius. The area of the circle, then, is somewhat less than this; it is about 3.14 (or pi) times the square of the radius.

7. What is the radius of the wheel you have been using?

8. Find the wheel's approximate area.

Galileo guessed that the area under one arc of a cycloid is equal to pi times the area of the wheel that produces it. He was wrong, for it was discovered later that the area of each of the 3 regions shown in this figure is exactly the same.

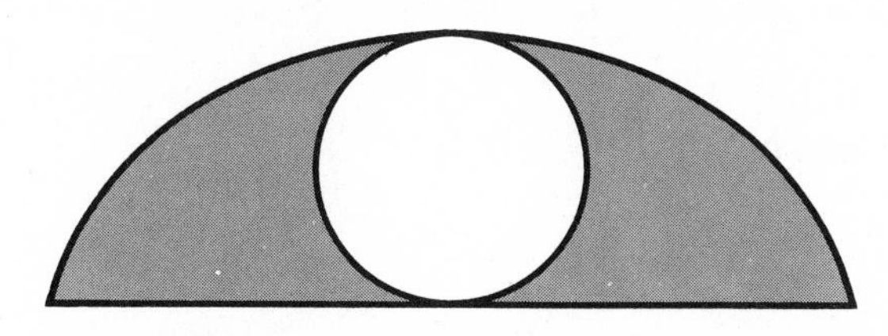

9. This means that the area under the cycloid is not pi times larger, but how many times larger than the area of the wheel that produced it? (Notice that the area of the wheel in this position is also part of the area under the cycloid.)

10. Use your answers to the previous two questions to find the approximate area under the cycloid you drew.

Set III

An especially surprising property of the cycloid can be discovered by turning it upside-down. If a curved ramp is built in the shape of a cycloid and a couple of marbles are released from any two points on the ramp, they will always reach the bottom at the same time!

Can you explain how this could be possible, since one marble will usually have farther to roll than the other?

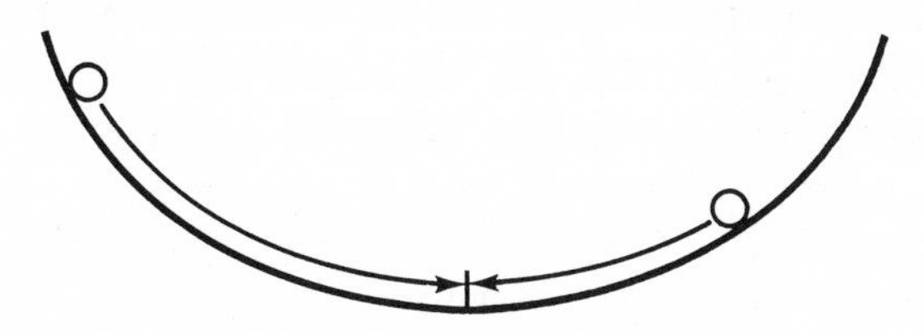

Chapter 6 / Summary and Review

IN this chapter we have become acquainted with:

The conic sections *(Lessons 1, 2, & 3)* The *circle:* the set of all points in a plane that are at the same distance from a fixed point called the center.

The *ellipse:* the set of all points in a plane such that the sum of the distances of each point from two fixed points, called the foci, is the same.

The *parabola:* the set of all points in a plane such that the distance of each point from a fixed point, called the focus, is the same as its distance from a fixed line.

The *hyperbola:* the set of all points in a plane such that the difference between the distances of each point from two fixed points, called the foci, is the same.

Circle

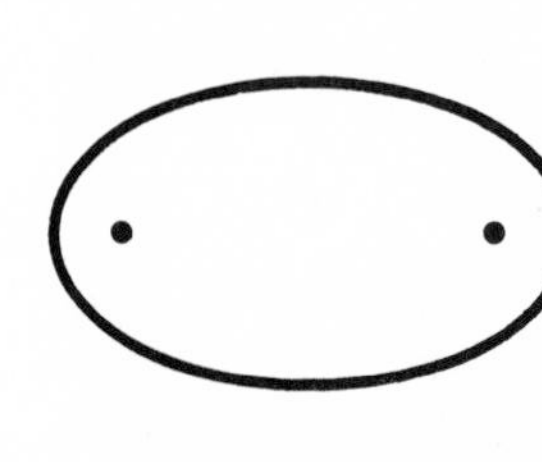

Ellipse

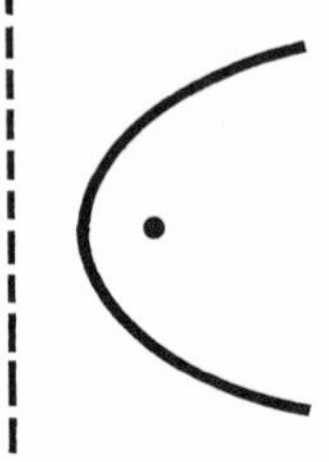

Parabola

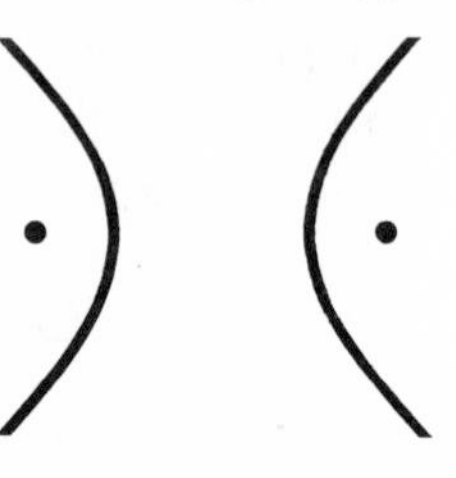

Hyperbola

The sine curve *(Lesson 4)* The *amplitude* of a sine curve is its maximum distance from the x-axis. The *period* of a sine curve is the distance along the x-axis of one wavelength.

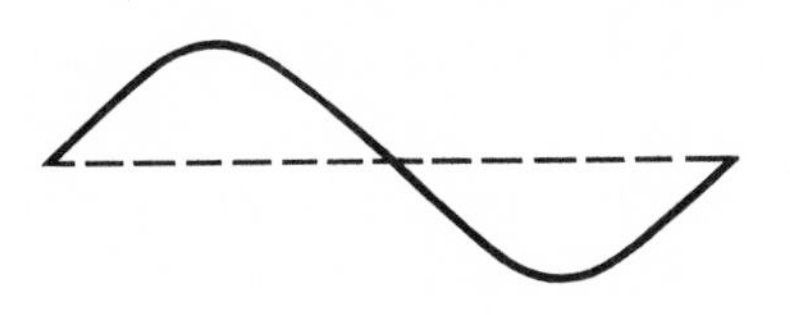

Spirals *(Lesson 5)* A spiral is a curve traced by a point that moves around a fixed point, called the pole, from which the point continually moves away.

The distances of an *Archimedean spiral* from its pole increase in arithmetic sequence.

The distances of a *logarithmic spiral* from its pole increase in geometric sequence.

The cycloid *(Lesson 6)* A cycloid is the curve traced by a point on the rim of a wheel rolling along a straight line.

EXERCISES

Set I

1. Here is a photograph of a fountain in which the jets of water are pointed in many directions. What curve is the shape of the path of each jet?

2. There is a whispering gallery in the Capitol building in Washington, D.C. The ceiling of this room is a curved surface, formed by rotating an ellipse about one of its axes. If you stand at one spot in this large room and whisper very softly, someone standing at a certain spot many yards away can hear you clearly.

   What do you think these two spots are called with respect to the ellipse? (Hint: See the drawing of the elliptical pool table on page 254.)

3. The graph of an alternating electric current is a sine curve.

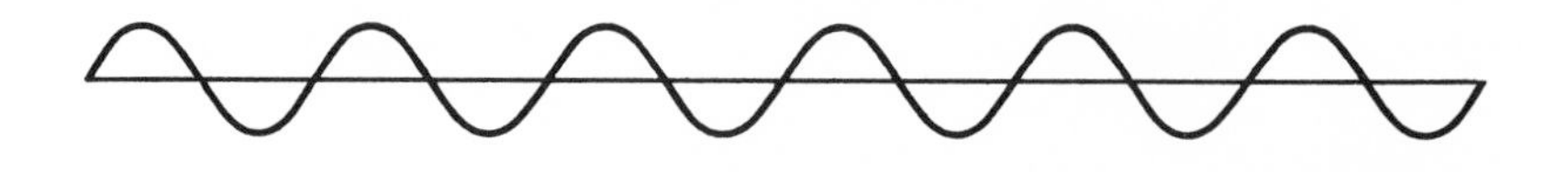

a) How many wavelengths of this curve are shown?

b) The graph of an ordinary alternating current has 60 wavelengths per second. What fraction of a second is shown above?

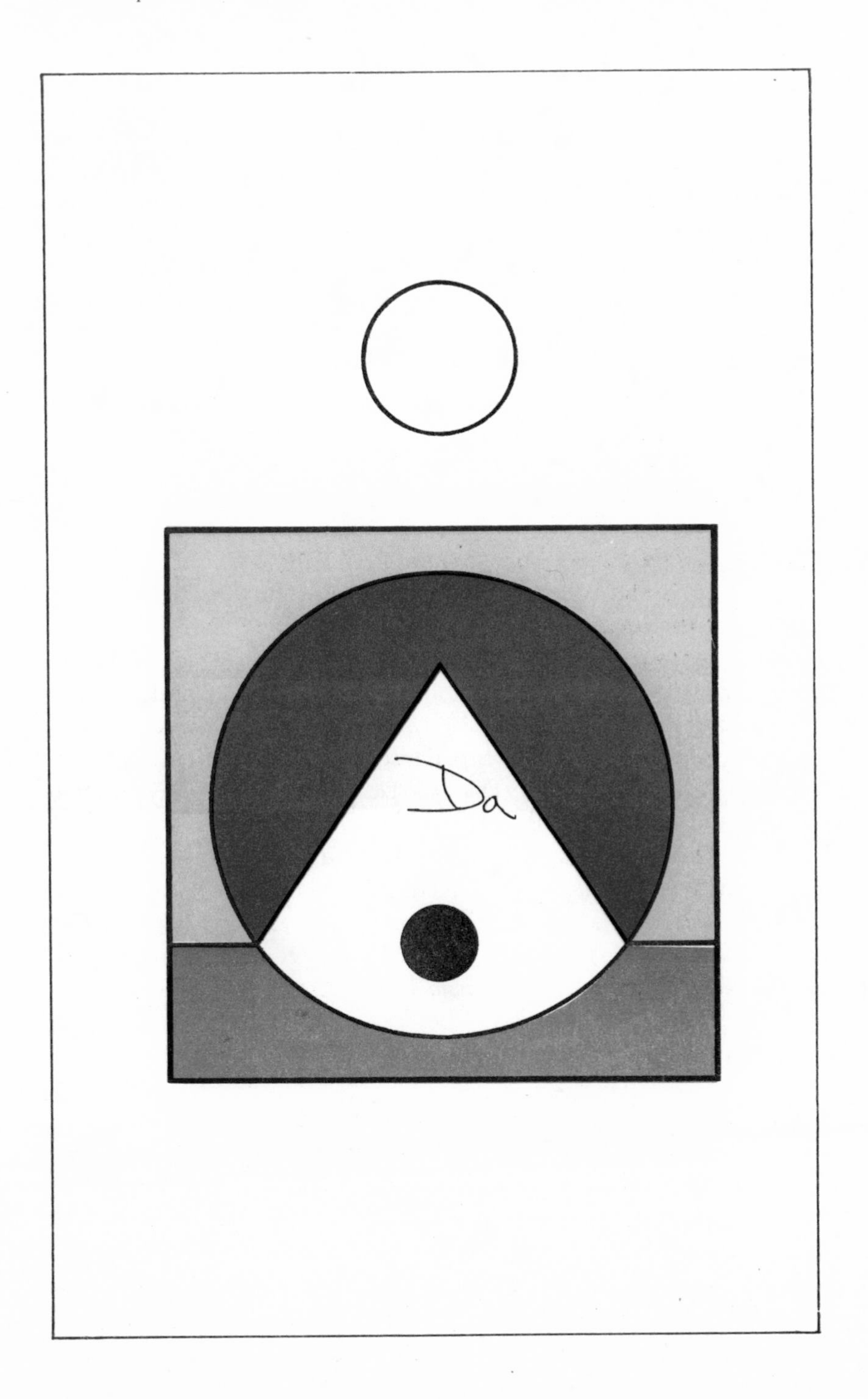
Da

"Logo" is a Greek word meaning "word" or "thought." The Logo of The Free Communion Church is a Sacred Idea, Communication, Word, Symbol, or Image. It is an esoteric ("secret" or "holy") visual symbol or description of the whole and Awakened body-mind of Man. The entire Logo, and not any single part, is an archetype of the body-mind and Transcendental Condition of the Enlightened Spiritual Master and the Enlightened Devotee. Through the geometry of its colors and shapes, it points to the Transcendental Reality that pervades and animates the body-mind as its true Condition, Source, Nature, Help, and Destiny.

The elements of the Logo are the three primary geometrical forms (the circle, the square, and the triangle), the three primary colors (red, yellow, and blue), and the principal achromatic colors (black and white).

Red and Yellow: Extroversion and the Verbal Mind

Red is the color of earth, of vitality. In the Logo the red portion of the square symbolizes the lower body or "navel" of Man. Red also represents the sympathetic nervous system, which primarily conducts Life-Energy and attention downward and outward through the motor activities of the vital-physical functions. This extroverted and active function of the sympathetic nervous system is related to the right side of the body, characterized by heat, expansiveness of energy, and objectivity.

Yellow is the color of the verbal mind, the analytical, linear, discriminative faculties of the mind of Man. The yellow portion of the square represents the left hemisphere of the brain, which is the root of the sympathetic nervous system and which controls the right side of the body. Thus, yellow is considered higher or senior to red. Red and yellow together represent the "rajasic" or active aspect of the body-mind.

Black and Blue: Introversion and the Intuitive Mind

The black circle represents the passive, introverted, and subtle structures and functions of the lower body-mind. It symbolizes the parasympathetic division of the autonomic nervous system, which primarily conducts energy and attention inward and upward via the sensory currents to the root of the senses in the brain. This introverted and passive function is related to the left side of the body, characterized by coolness, contraction of energy, and subjectivity.

4. When an alpha particle is shot toward the nucleus of an atom, the particle is repelled and changes direction. A great British scientist named Ernest Rutherford showed that the particle's path is along one of the two branches of a mathematical curve we have studied. It is one of the conic sections. Which one?

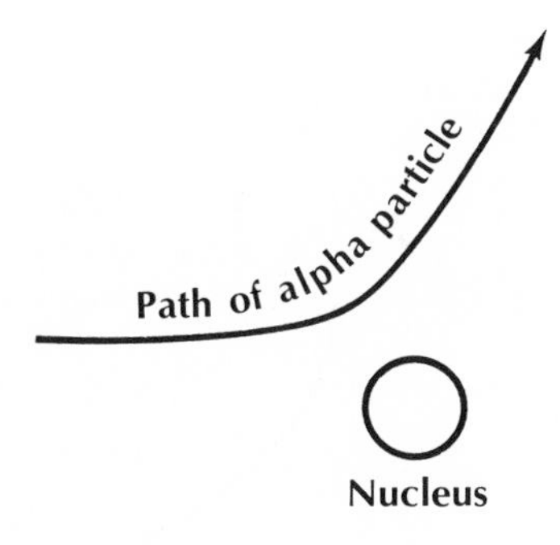

5. Four bugs are at the corners of a square. At the same time, each bug begins crawling directly toward the bug at the

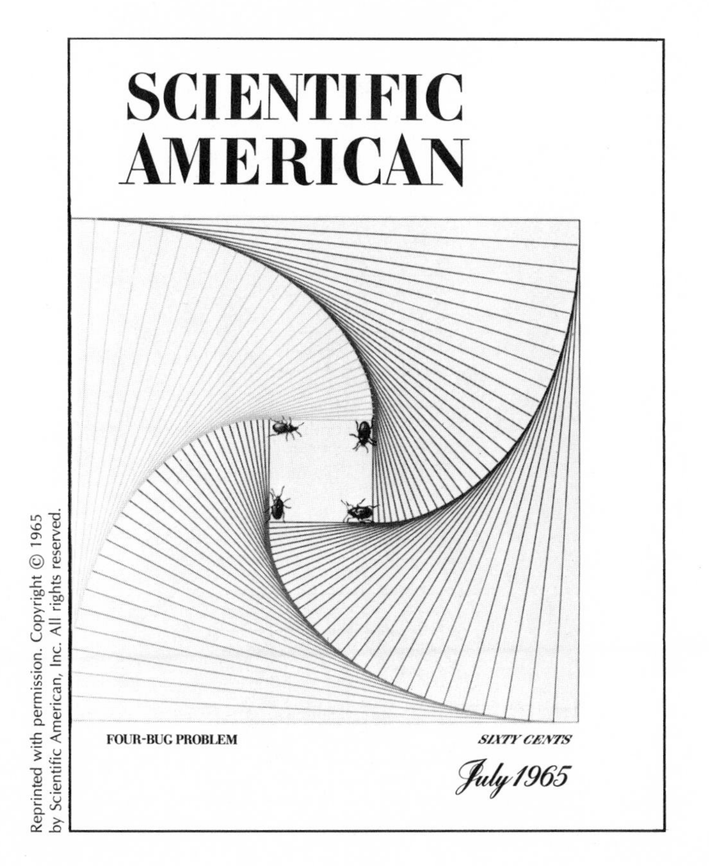

next corner. If the bugs all move at the same speed, they will follow the paths shown in the diagram below and will meet in the center of the square. What kind of curve is each bug's path?

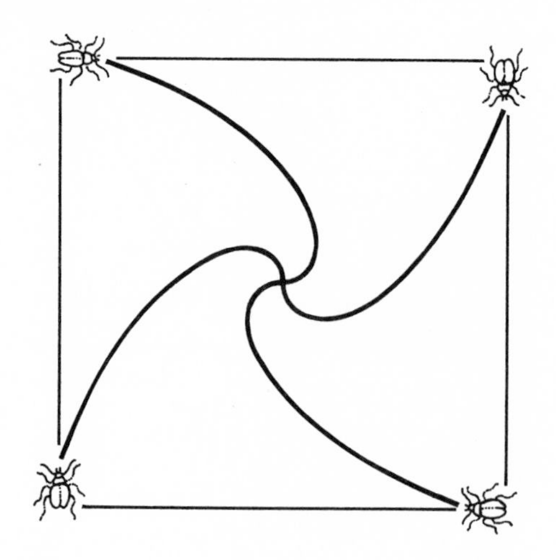

Set II

1. Find the lengths of the major and minor axes of the following ellipses A, B, C, and D. (The major axis of the first ellipse is 6 units long and its minor axis is 4 units.)

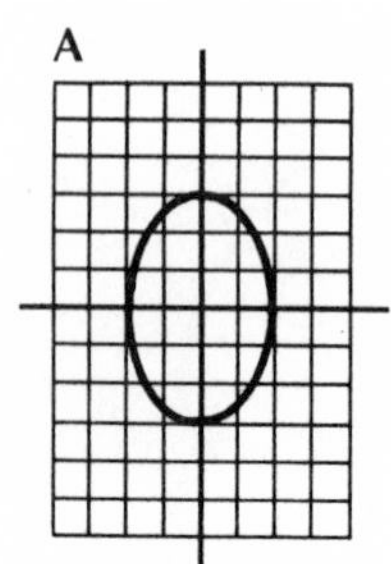

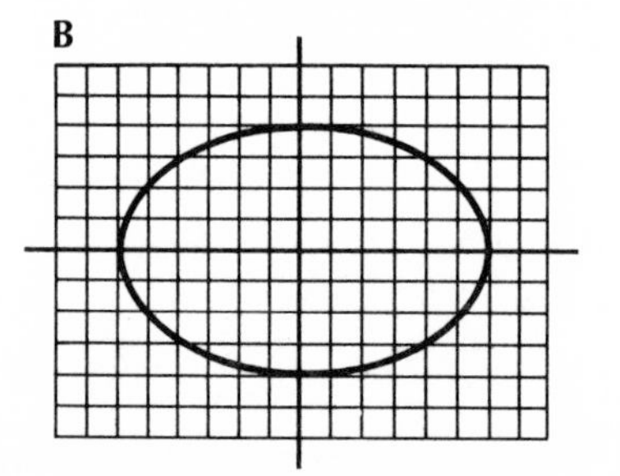

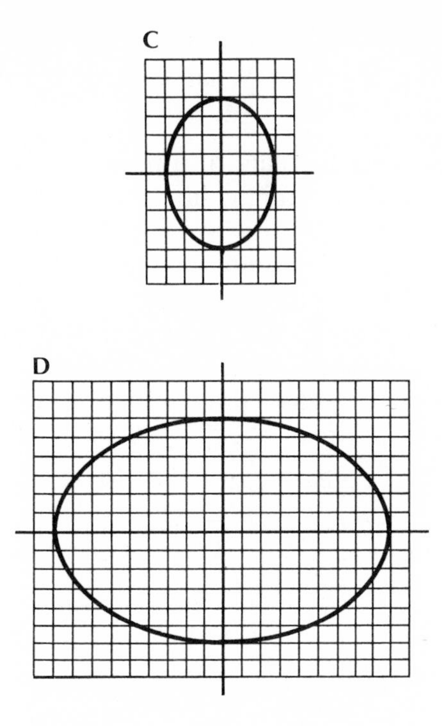

2. Find the ratio of the lengths of the major and minor axes for each ellipse. The ratio for ellipse A is 6/4 = 3/2.

3. Which of these ellipses do you think have the same shape?

4. Draw an angle of about 45° so that each side extends at least 3 inches. Mark points on each side of the angle at $\frac{1}{4}$ inch intervals from the vertex and number them as shown in this diagram.

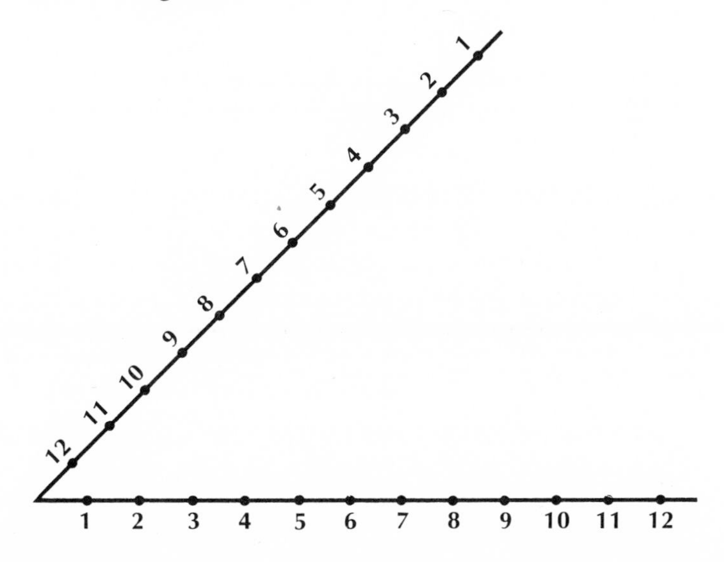

Join each pair of points having the same number with a straight line segment. What kind of curve do you think appears as a result?

5. You know what the path of a point on the rim of a wheel that rolls along a straight line looks like: it is a cycloid.

   If the wheel rolls around the inside of a circle, the shape of its path depends upon the relative sizes of the wheel and circle. For example, if the diameter of the circle is 3 times the diameter of the wheel, the path of a point on the rim of the wheel is a "hypocycloid" with 3 cusps as shown in the figure below.

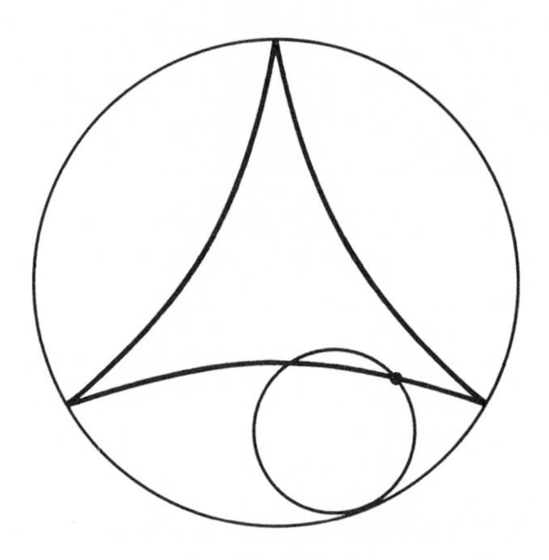

   Make a sketch showing what you think the path would look like if the diameter of the circle were 4 times the diameter of the wheel.

6. What do you think the path would look like if the diameter of the circle were twice the diameter of the wheel?

7. If $100 is invested at 5% interest, compounded annually, its growth rate can be shown by a logarithmic spiral. In the graph at the top of the facing page, the lines represent the years, and the increasing distances of the points from the pole represent the increasing values of the money invested. The spiral begins here with a point representing the original $100 and shows the rate of growth to $300.

   About how many years does it take for the $100 to increase to $150?

8. How long does it take for the $100 to double?

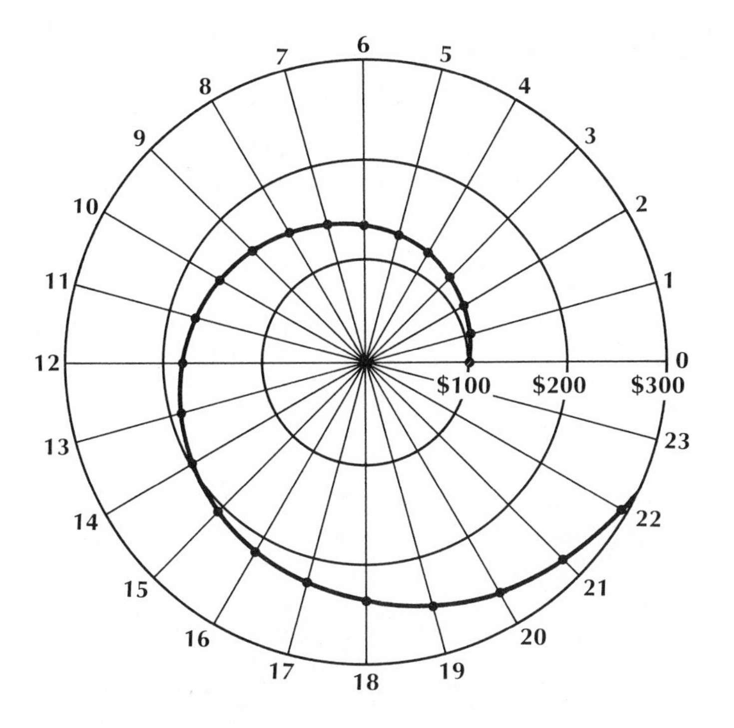

Set III

Experiment 1. Another Curve by Folding Paper

Use a compass to draw a circle with a radius of about 3 inches on a sheet of unlined paper. Mark a point approximately 1 inch inside the circle. Then cut the circle out and fold it so that its edge falls on the point you marked; make a sharp crease.

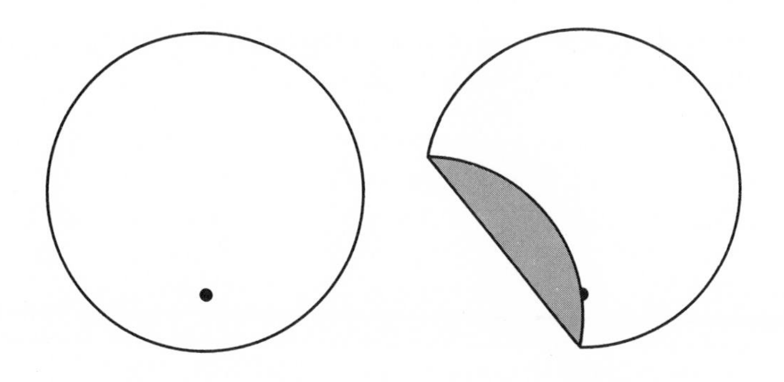

Unfold the paper and fold it again in a different direction so that the edge again falls on the point. Repeat this many times so that points all around the edge have been folded onto the point.

1. What kind of curve is the result?
2. Where are its foci?
3. What curve do you think would have appeared if the center of the circle had been chosen as the original point?

Experiment 2. Curves from Cutting a Candle

Cut out a strip of paper about 4 inches long and 1 inch wide. Wind it tightly around a small "birthday cake" candle and, with a razor blade, cut through the paper and candle at a slant, as shown in this photograph.

1. Look at the cross-section of the candle you have cut. What curve do you see?
2. Unwind the paper. What curve results?

INTERESTING READING

The Attractive Universe, by E. G. Valens, World, 1969: Chapter 1, "Falling," and Chapter 2, "Curving."

SOME METHODS OF COUNTING

Lesson 1

A Fundamental Counting Principle

LINUS is taking a true or false test and seems to be guessing every answer. If there are 20 questions, how many different "patterns" of answers are possible?

In order to figure this out, we will list all of the possible answer patterns for some very short true-false tests. A test with only 2 questions has 4 possible sets of answers: TT, TF, FT, FF. With 3 questions, the number of possible sets of answers increases to 8: TTT, TTF, TFT, TFF, FTT, FTF, FFT, FFF, and with 4 questions, there are 16 possibilities: TTTT, TTTF, TTFT, TTFF, TFTT, TFTF, TFFT, TFFF, FTTT, FTTF, FTFT, FTFF, FFTT, FFTF, FFFT, FFFF!

The number of sets of answers grows rapidly as the number of questions increases. Another way to find the number of answers is to use a diagram to represent all sets of possible answers with a box for each answer. The answers for a test with 2 questions can be represented by $\boxed{}\boxed{}$. Since there are 2 ways to answer each question (true or false), we write a 2 in each box, $\boxed{2}\boxed{2}$. The number of possible sets of answers, 4, can be found by multiplying these numbers. Notice that with this method we did not have to list all of the possibilities and then count them.

In the same way, a test with 3 questions, $\boxed{2}\boxed{2}\boxed{2}$, has $2 \times 2 \times 2 = 8$ different sets of answers, and a test with 4 questions, $\boxed{2}\boxed{2}\boxed{2}\boxed{2}$, has $2 \times 2 \times 2 \times 2 = 16$ sets of answers.

How many different sets of answers are possible for a test

having 20 questions?

$$2 \times 2 \times 2 \times 2 \times 2 \times 2 \times 2 \times 2 \times 2 \times 2 \times 2 \times 2 \\ \times 2 \times 2 \times 2 \times 2 \times 2 \times 2 \times 2 \times 2 =$$

1,048,576 different sets!

Someone would really be lucky if he guessed the correct set of answers out of all these possibilities!

We will call the method we have used in figuring this out the **fundamental counting principle.** It says:

► To find the number of ways of making several decisions in succession, multiply the numbers of choices that can be made in each decision.

Here is another example of how this principle works. Suppose you are taking a multiple-choice quiz in which there are 5 questions, each with 4 choices. If you guess, how many ways of answering the questions are possible? We can represent the answers with 5 boxes: $\boxed{}\boxed{}\boxed{}\boxed{}\boxed{}$. Since you can choose from 4 answers for each question, we can write: $\boxed{4}\boxed{4}\boxed{4}\boxed{4}\boxed{4}$. By the fundamental counting principle, there are

$$4 \times 4 \times 4 \times 4 \times 4 = 1{,}024$$

ways of guessing the answers!

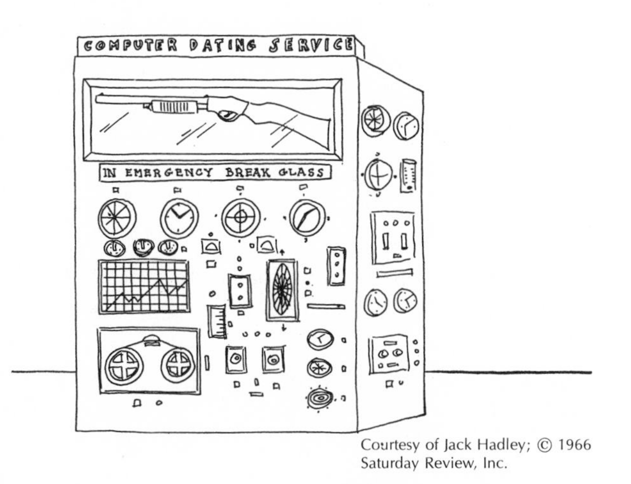

Courtesy of Jack Hadley; © 1966 Saturday Review, Inc.

EXERCISES

Set I

If a Computer Dating Service has cards for 230 men and 480 women, how many different dates can it arrange?

Each date involves 2 decisions, which we represent by: $\boxed{}\boxed{}$.

There are 230 men to choose from and 480 women, so we can write: | 230 | 480 | . Now, using the fundamental counting principle,

$$230 \times 480 = 110{,}400$$

different dates can be arranged.

1. If the Dating Service gets 1 additional card for a man and 1 additional card for a woman, how many different dates can it arrange?

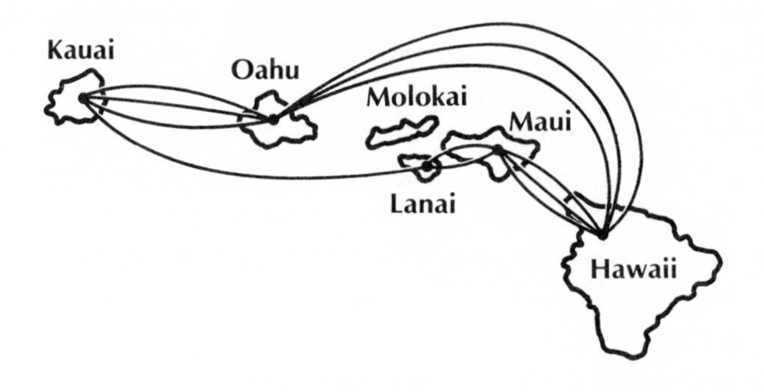

On a tour of the Hawaiian Islands, Captain and Mrs. Cook plan to travel from the island of Oahu to the island of Kauai, and then to Lanai, Maui, and Hawaii in that order, before returning to Oahu. The number of ways they can travel between each island (including both planes and boats) is shown on the map. For example, they can travel from Lanai to Hawaii in $2 \times 3 = 6$ different ways.

2. In how many different ways can the Cooks plan their entire tour?

3. If they learn later that they can travel in only one way from Lanai to Maui, in how many ways can their tour be made?

Suppose you are at Bob's Restaurant and like the following items on the menu:

Dinners		Beverages	
Big Boy Combination	1.05	Coffee	.15
Fried Chicken	1.30	Coca-Cola	.15
Shrimp	1.55	Root Beer	.15
Fish and Fries	1.05	Milk	.15
Bob's Special Steak	1.95	Orange Drink	.15

Desserts	
Hot Fudge Sundae	.50
Apple Pie	.30
Banana Cream Pie	.35
Strawberry Pie	.40
Cheese Cake	.30
Fudge Brownies	.20

4. If you order a dinner and a beverage, without dessert, how many different combinations can you choose from?
5. If you also order a dessert, how many combinations are possible?
6. How many choices do you have if you order a dinner, beverage, and dessert, but decide that you don't want sea food?

Some people, especially politicians and businessmen, like to throw big words around when they want to sound important without really saying anything. A member of the U.S. Public Health Service has come up with a list of 30 words from which impressive-sounding 3-word phrases can be chosen at random. Here is his list.*

1	2	3
integrated	management	options
total	organizational	flexibility
systematized	monitored	capability
parallel	reciprocal	mobility
functional	digital	programming
responsive	logistical	concept
optional	transitional	time-phase
synchronized	incremental	projection
compatible	third-generation	hardware
balanced	policy	contingency

To make up a phrase, choose a word from the first column, followed by a word from the second column, followed by a word from the third. Some examples are: "integrated logistical mobility" and "compatible management programming."

* *Time,* September 13, 1968, p. 22.

7. How many different 3-word phrases can be made up from this list in the same way?

8. If you are the "chairman of the budget" of a big company, and you want a phrase that begins with the word "balanced," how many phrases are possible?

Biologists are trying to figure out the "genetic code," that is, the way in which the genes of a living organism determine the structure of the organism's proteins.

The genes contain long molecules, which are chains of small molecules. There are 4 kinds of small molecules, which are represented by the initials of their names: A, C, G, and T.* Groups of these small molecules along the chain make up the code.

9. Make a list of all possible groups of 2 molecules (AA, AC, AG, etc.). We will consider groups such as AC and CA to be different.

10. How many different groups of 2 molecules are possible?

11. How many different groups containing 3 molecules are possible? (Don't make a list; use the fundamental counting principle.)

Set II

A standard slot machine has 3 dials, each having 20 symbols as shown in this table:

Symbol	Dial 1	Dial 2	Dial 3
Bar	1	3	1
Bell	1	3	3
Plum	5	1	5
Orange	3	6	7
Cherry	7	7	0
Lemon	3	0	4
	20	20	20

Each dial can stop on any one of its 20 symbols, so there are

$$20 \times 20 \times 20 = 8{,}000$$

*The molecules are called "adenine," "cytosine," "guanine," and "thymine."

combinations possible. (Many of these look the same since most of the symbols appear several times on each dial.)

1. The biggest payoff is for 3 bars. How many ways are there of getting 3 bars?

2. It is obviously impossible to get 3 cherries, since there are no cherries on the third dial. How many ways are there of getting 3 of each of the other symbols?

3. How many combinations that *look* different can turn up on the machine? (Hint: 6 different symbols can appear on the first dial; how many can appear on the second and third?)

4. A mouse is in a maze in which there are 8 places where it can go to the left or right. In how many different ways can the mouse run through the maze? (Hint: The mouse has 8 decisions to make, and in each decision there are 2 possible choices.)

5. Keys of different shapes are designed by choosing from several patterns for each of their parts. The keys of General Motors cars have 6 parts for which there used to be 2 patterns each. How many different key designs were possible?

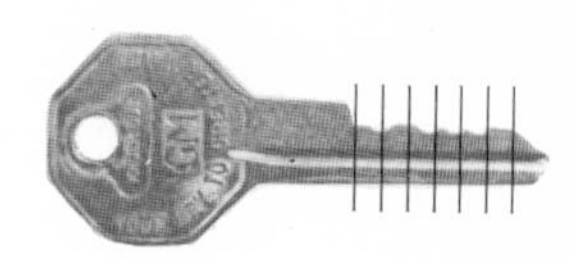

6. Recently, General Motors increased the number of patterns for each part to 3. How many different key designs do they have now?

7. Suppose the number of patterns for each part were increased again to 4. How many key designs would be possible?

8. You know that we use 10 digits, 0, 1, 2, 3, 4, 5, 6, 7, 8, and 9, to write numbers. How many different 3-digit numbers can be written whose first digit is *not* 0, whose second digit *is* 0, and whose third digit can be anything?

9. The call letters of radio and television stations in the United States begin with either K or W. How many sets of call letters having 3 letters each are possible? (Hint: There are 3 decisions to make, with 2 choices in the case of the first; how many choices are there for the second and third?)

10. How many sets of call letters having 4 letters each are possible?

Set III

A popular puzzle called "Instant Insanity" consists of 4 blocks, whose faces have different colors: some red, some green, some blue, and some white. None of the blocks has the same arrangement of colored faces. The object is to stack them in such a way that all 4 colors show on each side of the stack. Although this may seem to be very easy, anyone who has tried it will agree that the puzzle is a difficult one to solve.

1. How many faces does a single cube have?

A cube can be placed with any one of its faces on top, and with any of 4 other faces in front.

2. Altogether, in how many different-looking positions can one cube be placed?
3. In how many different-looking ways can a stack of 4 cubes be made? The answer to this question has something to do with how the puzzle got its name!

Lesson 2

Permutations

Drawing by Bill Hoest; © 1967 Look Magazine.

"Next time, let me handle the seating arrangements."

THE host of this dinner party hasn't done a very good job of planning the seating arrangements. It looks like everyone would rather be somewhere else. Suppose the 14 people at this table had decided to exchange places before they sat down. In how many different ways could they arrange themselves?

The answer to this question is a very large number:

87,178,291,200 different ways!

In fact, this number is so large that if the 14 people were able to trade places every *second* without stopping, it would take more than *2,700 years* for them to arrange themselves in every possible order!

Mathematicians call these arrangements *permutations.*

► A **permutation** is an *arrangement* of things in a *definite order.*

Suppose there are only 2 people seated at the table and that we call them A and B. Clearly there are only 2 orders in which they can sit:

AB or BA.

Three people can be seated at the table in 6 different orders. If we represent the third person by the letter C, they are:

ABC	BAC	CAB
ACB	BCA	CBA.

With 4 people at the table, the number of orders jumps to 24!

ABCD	BACD	CABD	DABC
ABDC	BADC	CADB	DACB
ACBD	BCAD	CBAD	DBAC
ACDB	BCDA	CBDA	DBCA
ADBC	BDAC	CDAB	DCAB
ADCB	BDCA	CDBA	DCBA

How do you suppose the number of arrangements of 14 people at the table was figured out, since it would take a lot of time and space to list every order and count them? In other words, how can the number of permutations of a group of things be determined more easily?

We can use the *fundamental counting principle*. If we apply it to the case of 4 people, each position at the table can be represented by a box: [ ][ ][ ][ ] .

There are 4 ways of filling the first position, since any one of the 4 people can be seated there. Then there are 3 ways left to fill the second position, 2 ways to fill the third, and only 1 way left for the fourth: [4][3][2][1] . The number of arrangements for 4 people is $4 \times 3 \times 2 \times 1 = 24$.

In the same way, the number of arrangements for 2 people is $2 \times 1 = 2$, and the number of arrangements for 3 people is $3 \times 2 \times 1 = 6$.

A short way to write each of these products is to use a mathematical symbol that looks like an exclamation mark but is called a **factorial** symbol. $4 \times 3 \times 2 \times 1$ can be written as 4! (read as "four factorial"), $3 \times 2 \times 1$ as 3!, and 2×1 as 2!.

► In general, $n!$ means to multiply the consecutive numbers from n all the way down to 1, and the number of permutations of n different things is $n!$.

So 14 people can be seated at a table in

$$14! = 14 \times 13 \times 12 \times 11 \times 10 \times 9 \times 8 \times 7 \times 6 \times 5 \times 4 \times 3 \times 2 \times 1 = 87{,}178{,}291{,}200$$

different orders. It would be impossible for someone to make a list of them and then count them one by one. Do you know why?

EXERCISES

Set I

1. Copy and complete the following table of factorials by finding the values of 5! through 12!. You can check your work by finding the value of 13! and comparing it with the number given.

$$1! = 1$$
$$2! = 2$$
$$3! = 6$$
$$4! = 24$$
$$\vdots$$
$$13! = 6{,}227{,}020{,}800$$

2. Do you think the last digit of *every* factorial number larger than 4! is 0?

Use your table to help decide whether each of the following is true or false.

3. $7! = 7 \times 6!$
4. $2! + 2! = 4!$
5. $\frac{8!}{4} = 2!$
6. $6! \times 7! = 10!$

Use your table to answer each of the following questions.

7. In how many ways can 10 people line up at a theater box office?
8. In how many ways can the names of 5 candidates for the same office be listed on a ballot?
9. In how many ways can a hand of 13 different cards be arranged?
10. How many ways are there of scrambling the letters of the word SCRAMBLE?
11. In how many ways can 7 different digits of a telephone number be arranged?

"Blue Boy seems to be holding back a bit."

Courtesy of Joseph Zeis; © 1960 The Saturday Evening Post.

Set II

If 8 horses are entered in a race, in how many ways can the "win," "place," and "show" (the first 3 places) be taken?

The answer is not 8!, because we are not concerned with the number of orders in which all 8 horses can come in, but only the first 3. We can represent the first 3 places with boxes: $\boxed{\ }\boxed{\ }\boxed{\ }$. Any one of the 8 horses can come in first (assuming that none of them behave like Blue Boy), leaving 7 horses that can come in second, and 6 horses that can come in third: $\boxed{8}\boxed{7}\boxed{6}$. Using the fundamental counting principle, we determine that there are $8 \times 7 \times 6 = 336$ ways in which the "win," "place," and "show" can be taken.

1. If there are 60 floats in Pasadena's Rose Parade on New Year's Day, in how many different ways can the first and second floats be chosen?
2. In how many different ways can the first 3 moves in a game of tic-tac-toe be made? (There are 9 squares in which the first move can be made; how many places are left for the second move? How many places for the third move?)
3. A three-dimensional version of tic-tac-toe is played on a board with 4 levels, each level having 16 squares. In how many different squares can the first move be made?
4. In how many ways can the first *2 moves* be made?

5. In how many ways can the first *3 moves* be made?

6. In how many ways can 5 boys and 6 girls be arranged for a school club photograph, if the girls are seated in a row with the boys standing in a row behind them? (Hint: First find the number of ways of arranging the boys in one row and the number of ways of arranging the girls in the other; then multiply the numbers of ways.)

A television network is planning its program schedule for the new season. During "prime time" on Monday evenings (7 P.M. to 10 P.M.), the network plans to schedule 1 western series (60 minutes), 3 comedy series (30 minutes each), and 1 quiz show (30 minutes). It wants one comedy series to begin at 7:00 and another to begin at 9:30.

7. In how many ways may these 2 time spots be filled?

8. In how many ways can the remaining time between 7:30 and 9:30 be filled? (Hint: There are 3 shows left; in how many ways can they be arranged?)

9. In how many ways can the complete Monday evening "prime time" schedule be arranged? (Hint: Multiply the numbers of ways you found in problems 7 and 8.)

10. In how many ways can 2 couples be seated in a row in a theater if each boy sits beside his own date? (To answer this, make a list of all of the possibilities, letting A and B represent one couple, and X and Y the other couple. Two possibilities are: ABXY and XYBA.)

Set III

"There was a table set out under a tree in front of the house, and the March Hare and the Hatter were having tea at it: a Dormouse was sitting between them, fast asleep, and the other two were using it as a cushion, resting their elbows on it, and talking over its head. . . .

"The table was a large one, but the three were all crowded together at one corner of it. 'No room! No room!' they cried out when they saw Alice coming. 'There's *plenty* or room!' said Alice indignantly, and she sat down in a large arm-chair at one end of the table."

—*Alice's Adventures in Wonderland,* by Lewis Carroll.

1. If the table had 12 places around it, in how many different ways could Alice, the March Hare, the Hatter, and the Dormouse be seated at it?

2. Out of all these different ways, in how many ways can the 4 sit next to each other with no empty places in between? (First consider how many different sets of 4 consecutive places there are at the table. Then consider how many ways 4 people can be seated in a row.)

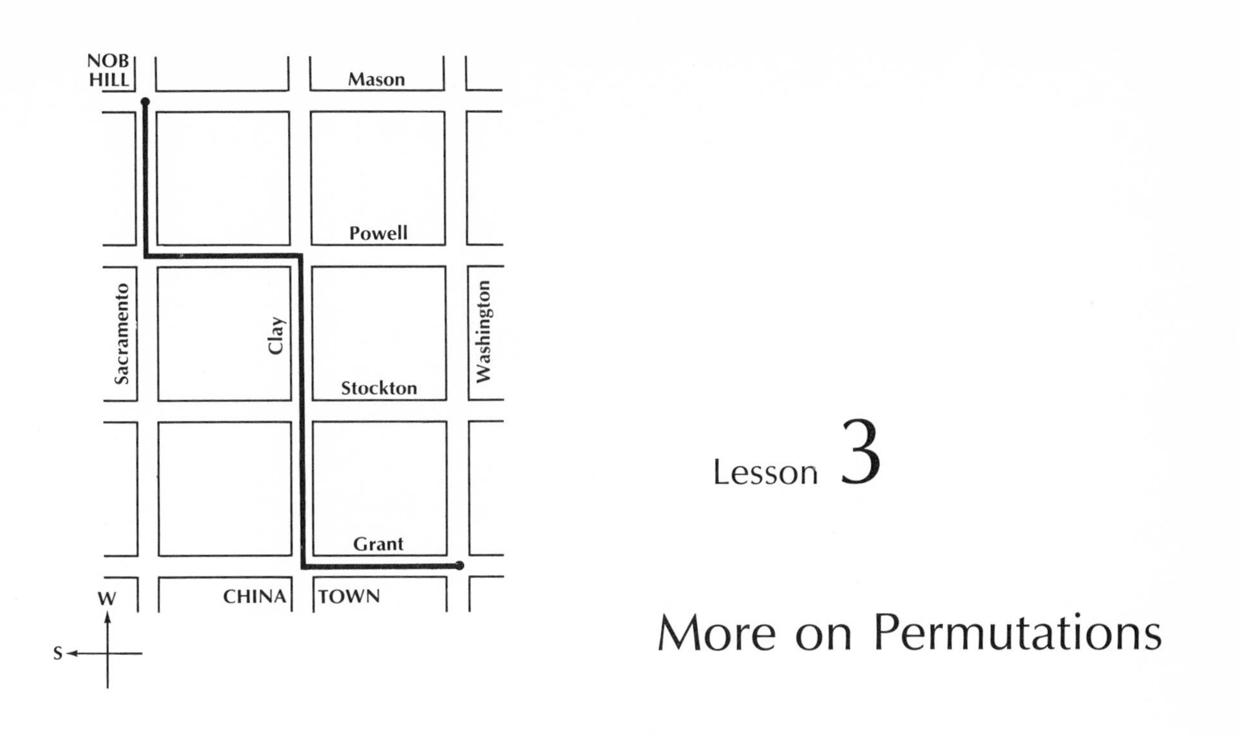

Lesson 3

More on Permutations

MR. and Mrs. Fairmont have been walking through Chinatown in San Francisco and are standing at the corner of Grant Avenue and Washington Street. They plan to return to their hotel at the corner of Mason and Sacramento streets on Nob Hill, 2 blocks south and 3 blocks west of where they are. If the Fairmonts do not want to walk more than 5 blocks, how many different paths do they have to choose from?

To figure this out without drawing maps of all the possible paths, we might represent each block the Fairmonts walk south with an S and each block they walk west with a W. The path shown on the map would be shown as

SWWSW

since they walk 1 block south, 2 blocks west, 1 block south, and 1 block west.

The problem of the number of paths the Fairmonts can choose from is equivalent to the problem: in how many ways can 2 S's and 3 W's be arranged in a row?

We have already solved many problems similar to this one. The number of permutations of 5 different things is 5!.

$$5! = 5 \times 4 \times 3 \times 2 \times 1 = 120$$

But there are surely not 120 different paths to choose from!

Look again at the path shown as

SWWSW.

Suppose we tag the 2 S's with numbers. The *2 different* arrangements

$$S_1WWS_2W \text{ and } S_2WWS_1W$$

would be *identical* without the numbers. This means that when we assumed that the S's were different and found the number of paths, we counted twice as many as there are if the S's are the same. If we divide 5! by 2! (the number of ways of arranging the 2 S's in a particular row), we "make up" for this.

In the same way, suppose we tag the 3 W's with numbers. The *6 different* arrangements

$$\begin{array}{lll} SW_1W_2SW_3 & SW_2W_1SW_3 & SW_3W_1SW_2 \\ SW_1W_3SW_2 & SW_2W_3SW_1 & SW_3W_2SW_1 \end{array}$$

would be *identical* without the numbers. This means that when we assumed that the W's were different and found the number of paths, we counted 6 times as many as there are if the W's are the same. Dividing 5! by 3! (the number of ways of arranging the 3 W's in a particular row) takes care of this.

In other words, the number of different permutations of 5 things, 2 of which are alike and 3 of which are alike, is

$$\frac{5!}{2! \times 3!}.$$

It is easy to simplify this fraction, by first writing out each factorial and then reducing:

$$\frac{5 \times 4 \times 3 \times 2 \times 1}{2 \times 1 \times 3 \times 2 \times 1} = \frac{5 \times \overset{2}{\cancel{4}} \times \cancel{3} \times \cancel{2} \times \cancel{1}}{\cancel{2} \times 1 \times \cancel{3} \times \cancel{2} \times \cancel{1}} = \frac{10}{1} = 10.$$

So there are actually 10 different paths from Chinatown to the hotel.

► In general, the number of permutations of n things, of which a things are alike, another b things are alike, another c things are alike, and so forth, is

$$\frac{n!}{a! \times b! \times c! \ldots}.$$

EXERCISES

Set I

1. Show that 10 is the correct number of different paths from Chinatown to the Fairmonts' hotel, by making a list of all the ways in which 2 S's and 3 W's can be arranged together in a row. Then draw a set of diagrams like this one to show all 10 paths.

The name of the magazine published by the San Diego Zoo is ZOONOOZ. This name is unusual because spelled backwards it is the same word. In how many *different* ways can the letters of ZOONOOZ be arranged? Since the word has 7 letters of which 2 are alike and 4 are alike, the number of ways is

$$\frac{7!}{2! \times 4!} = \frac{7 \times 6 \times 5 \times 4 \times 3 \times 2 \times 1}{2 \times 1 \times 4 \times 3 \times 2 \times 1} = 105.$$

2. There was once a British race horse named POTOOOOOOOO. This strange name was pronounced like the name of a common vegetable. Do you see why? In how many different ways can the letters of POTOOOOOOOO be arranged?

3. In how many different ways can the letters of the word ANTEATER-EATER be arranged?

By permission of Johnny Hart and Field Enterprises, Inc.

Here is a problem that includes a riddle.

A customer walks into a hardware store to buy something and asks the clerk how much 1 would cost. The clerk tells him 25¢. He then asks how much 10 would cost, the clerk says 50¢. The customer says, "I'll buy 4515," and pays the clerk $1.

4. How many different numbers can be made by arranging the digits 4515 in all possible ways?
5. Show that your answer is correct by making a list of all of these arrangements.
6. Now the riddle: What was the customer buying?
7. If you have a set of 10 poker chips in which 5 are white, 3 are red, and 2 are blue, into how many different looking stacks can you pile them?
8. Mrs. Maxwell, who drinks a lot of coffee and likes it with cream, claims that by tasting a cup she can tell whether the cream was poured into the cup before or after the coffee was. To find out whether she is telling the truth, we decide to make 8 cups of coffee, 4 in which the cream is added first and 4 in which the coffee is added first. The cups are given to her one at a time for her to tell how they were made.

   In how many different orders can the cups of coffee be given to Mrs. Maxwell?
9. How many different tunes can be made by rearranging these notes in the title song of the popular Broadway musical and film, "Hello, Dolly!"?

Courtesy of Jerry Herman; © 1963.

   (Assume that the time values do not matter, so that the second note will be considered the same as the sixth note.)
10. In studying the genetic code,* a biologist wonders about the number of possible arrangements of 12 molecules in a chain, if the chain contains 4 different kinds of molecules and 3 molecules of each kind. One arrangement can be represented as:

    AAACCCGGGTTT.

    How many different arrangements are possible in all?

* See page 314.

Set II

If 4 identical coins are arranged in a row, each one can show either heads or tails, so that there are $2 \times 2 \times 2 \times 2 = 2^4 = 16$ ways that they can appear.

There is only 1 way in which there can be 4 heads. How many ways are there in which 3 heads can appear? (In other words, in how many ways can 3 heads and 1 tail be arranged?)

The number of ways is

$$\frac{4!}{3!} = \frac{4 \times 3 \times 2 \times 1}{3 \times 2 \times 1} = 4,$$

and they are: THHH, HTHH, HHTH, and HHHT.

1. How many ways are there in which 2 heads can appear? (In other words, in how many ways can 2 heads and 2 tails be arranged?) Check your answer by making a list of them.
2. How many ways in which 1 head appears?
3. How many ways in which 0 heads appear?
4. Copy and complete the following table:

Ways of Arranging 4 Coins

No. of heads	4	3	2	1	0
No. of ways	1	▒▒▒	▒▒▒	▒▒▒	▒▒▒

The numbers on the second line should add up to 16.

5. Make a table to show the numbers of ways in which 5 identical coins can be arranged in a row. Notice that the number of ways for 5 heads can be written as

$$\frac{5!}{5!},$$

the number of ways for 4 heads (and 1 tail) as

$$\frac{5!}{4! \times 1!},$$

the number of ways for 3 heads (and 2 tails) as

$$\frac{5!}{3! \times 2!},$$

and so on.

Ways of Arranging 5 Coins

No. of heads	5	4	3	2	1	0
No. of ways	▒▒▒	▒▒▒	▒▒▒	▒▒▒	▒▒▒	▒▒▒

6. The sum of the numbers on the second line should be equal to 2^5. What is it?

7. Use the same method to make a table for 6 coins.

8. In what way are the second lines of all three of your tables alike?

Set III

The World Series is held early in October each year between the winners of the American and National League pennants. The two teams play until one team has won 4 games.

Suppose the New York Yankees are playing the Los Angeles Dodgers. How many different orders of winners in the games played are possible?

To figure this out, let's consider the number of orders possible in which the Yankees win the series. The series may last 4, 5, 6, or 7 games, depending upon whether the Dodgers win 0, 1, 2, or 3 games.

If it lasts only 4 games, the Yankees win all 4, and there is only 1 possible order of winners: YYYY.

If the series lasts 5 games, and the Yankees win it, there are several possible orders of winners: DYYYY, YDYYY, YYDYY, and YYYDY. Since the winner of the series always wins the last game, the last letter in the row is Y. Then the problem is one of finding the number of orders of the other 4 letters in the row: 3 Y's and 1 D.

$$\frac{4!}{3!} = \frac{4 \times 3 \times 2 \times 1}{3 \times 2 \times 1} = 4.$$

1. How many orders of winners are possible if the series lasts 6 games? (The Yankees win the last game, so the problem

is: In how many ways can 5 letters, 3 Y's and 2 D's, be arranged?)

2. How many orders of winners are possible if the series lasts 7 games?

3. Copy and complete the following table:

Ways in Which the Yankees Win the Series

No. of games in the series	4	5	6	7
No. of orders of winners	1	4	▒	▒

4. The sum of the numbers on the second line of the table is the number of orders of winners possible if the Yankees win. How many is that?

5. How many different orders of winners do you think are possible if the Dodgers win the series instead?

6. How many different orders of winners in a World Series are possible altogether?

Lesson 4

Combinations

GAMES played with dominoes have long been popular all over the world. A domino is a rectangular tile divided into two squares. The squares in a standard set are either blank or marked with from 1 to 6 spots. If every domino is different, how many are there in a complete set?

First, there are 7 dominoes called "doubles" on which both squares are alike.

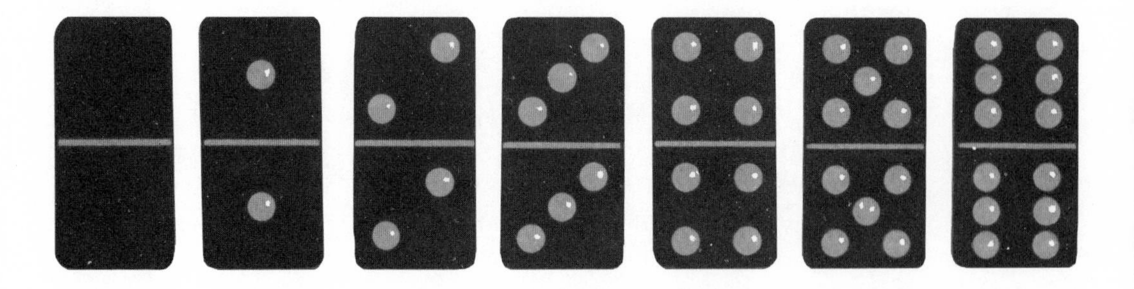

The rest of the dominoes are called "singles" and have different numbers of spots on each square. Let's figure out how many "singles" there are.

There are 7 choices for one square of each domino (blank or from 1 to 6 spots), and 6 choices left for the other square. Using the fundamental counting principle, we multiply the numbers of choices and get

$$7 \times 6 = 42 \text{ dominoes.}$$

Let's look at a couple of "singles" dominoes.

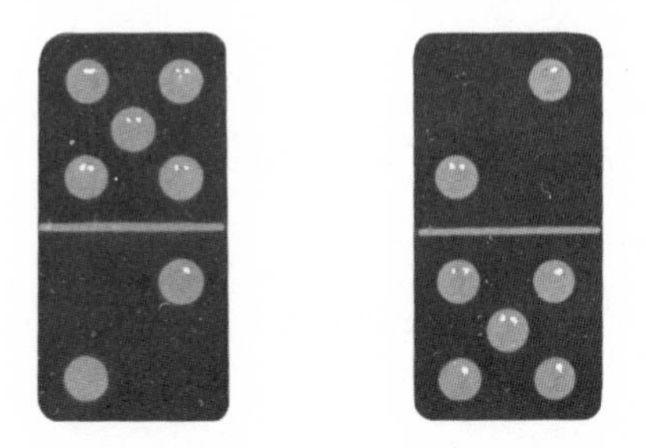

It is easy to see that these dominoes are not really different. They both contain 2 spots and 5 spots, and if one were turned around, they would look alike. In other words, the *order* in which the spots appear on the domino is *not important.*

We have counted each "singles" domino twice, and if we divide by 2, we find that there are really only

$$\frac{7 \times 6}{2} = 21 \text{ different "singles" dominoes.}$$

Adding the 7 "doubles," we find that there are 28 dominoes in a complete set.

The problem we have just solved is called a *combination* problem.

► A **combination** is a *selection* of things in which the *order doesn't matter.*

► To find the number of combinations possible, first *use the fundamental counting principle* and then *divide by the number of ways in which the things can be arranged.*

Here is another example of how this works. An offer of the Book-of-the-Month Club to new members is a choice of 3 books for $1 from a list of 37 books. How many combinations of 3 different books can you choose from this list?

By the fundamental counting principle, there would be $37 \times 36 \times 35$ ways to choose the 3 books in order. The number of ways in which 3 books can be arranged is $3! = 3 \times 2 \times 1$. Therefore, the number of choices you can make is

$$\frac{37 \times 36 \times 35}{3 \times 2 \times 1} = \frac{37 \times \overset{6}{\cancel{36}} \times 35}{\cancel{3 \times 2 \times 1}} = 7{,}770.$$

EXERCISES

Set I

The standard set of dominoes described in this lesson is sometimes called a "double-six" set. A larger set of dominoes is a "double-nine" set. The squares on these dominoes are either blank or marked with from 1 to 9 spots.

1. How many "doubles" dominoes are in a "double-nine" set?
2. How many "singles" dominoes are in this set?
3. How many dominoes are there altogether?
4. How many dominoes are there (both "doubles" and "singles") in a "double-twelve" set?
5. How many lines can be drawn through 5 points on a circle? Since exactly one line can be drawn through 2 points, this question is the same as: How many different combinations of 2 points can be chosen from a set of 5 points?

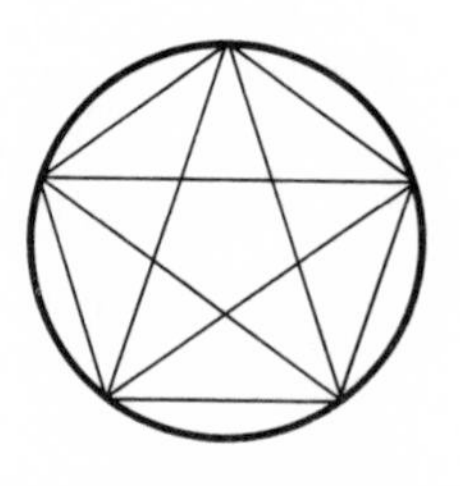

Use the method suggested in this lesson to figure this out, and check your answer by counting the number of lines in this figure.

6. How many lines are there in the figure on page 136?
7. At the beginning of every meeting of a very friendly club, each of the 20 members shakes hands with each of the others exactly once. How many handshakes are there in all, when every member is present?
8. How many fewer handshakes would there be if 3 members are absent? (Hint: Find the number of handshakes between 17 members and then subtract.)
9. There are 6 "stand-bys" waiting for seats on the midnight flight of the "Fly-By-Night" Airlines. If there are 3 seats

available on the plane, how many different combinations of "stand-bys" can be chosen to fill them?

Courtesy of Jack Hadley; © 1966 Saturday Review, Inc.

10. How many different combinations of coins have a sum of exactly 25¢? To find out, copy and complete the following table:

Kind of coin	25¢	10¢	5¢	1¢
Combination 1	1	0	0	0
Combination 2	0	2	1	0
Combination 3	0	2	0	5
and so on.				

Set II

1. A new Senate Committee to be in charge of "Trivial Affairs" is to have 3 members. If all 100 Senators are available to serve on it, how many different committees can be chosen?

2. It is possible to color a map of the United States with only 4 colors so that no 2 states with a common border have the same color. (But the problem of proving that 4 colors is enough to color any map so that no bordering states or

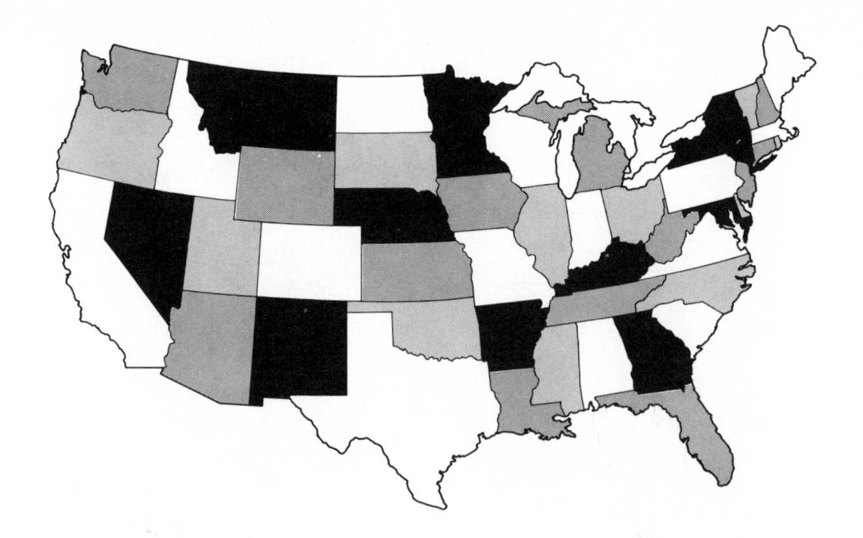

countries have the same color is one whose solution still eludes mathematicians.)

If there are 10 colors available, how many different sets of 4 colors may be chosen?

Auditions are being held for a new musical group to be called the "Pandemonians." The group will consist of 5 guitar players and 2 drummers, but 12 guitar players and 5 drummers try out.

3. How many combinations of guitar players can be chosen?
4. How many combinations of drummers can be chosen?
5. How many different sets of musicians can be chosen for the group?

The "Hungry Puppy" Hot Dog Stand sells hot dogs for 20¢ and for an extra 5¢ each you can choose from the following to go on them: mustard, ketchup, pickle relish, cheese, onions, chili, and sauerkraut.

6. If you pay 40¢ for a hot dog, how many different combinations can you order to have put on it?
7. If you pay 50¢ for a hot dog, how many different combinations do you have to choose from?
8. A hand in the game of Poker consists of 5 cards. How many different Poker hands can be drawn from 52 cards? (The arithmetic for this problem requires some patience.)

Set III

According to the ancient Hindu astrologers, the end of the world was supposed to take place when the sun, moon, and the 5 planets then known, Mercury, Venus, Mars, Jupiter, and Saturn, appeared to all come together in the same place in the sky. In fact, they thought, as do modern astrologers, that the coming together of just 2 of these heavenly bodies influenced events on the earth.

A mathematician in the twelfth century figured out the number of combinations of 2 or more heavenly bodies that might come together.

1. Find the number of combinations of the 7 heavenly bodies that contain:

   a) 2 of them,

   b) 3 of them,

   c) 4 of them,

   d) 5 of them,

   e) 6 of them,

   f) all 7 of them.

2. Add these numbers to find out what number this 12th century mathematician got.

Chapter 7/Summary and Review

In this chapter we have become acquainted with:

A fundamental counting principle *(Lesson 1)* To find the number of ways of making several decisions in succession, multiply the numbers of choices that can be made in each decision.

Permutations *(Lessons 2 & 3)* A permutation is an arrangement of things in a definite order.

$n!$ means to multiply the consecutive numbers from n all the way down to 1.

The number of permutations of n different things is $n!$.

The number of permutations of n things, of which a things are alike, another b things are alike, another c things are alike, and so forth, is

$$\frac{n!}{a! \times b! \times c! \ldots}.$$

Combinations *(Lesson 4)* A combination is a selection of things in which the order doesn't matter.

To count the number of combinations possible, first use the fundamental counting principle and then divide by the number of ways in which the things can be arranged.

EXERCISES

Set I

Do you know what the signal "... --- ..." means? Translated from Morse Code into the English alphabet, it reads "S O S."

The letter E is represented by a dot in Morse Code and the letter T is represented by a dash. All of the other letters are represented by 2 or more symbols. Four letters can be represented by 2 symbols, since the different possibilities are: .., .-, -., and --.

1. If 3 symbols (dots or dashes) are used, 8 more letters can be represented. Two of the possibilities are: ... and .--. Make a list showing all 8 possibilities.
2. Show how the number of possibilities can be determined by using the fundamental counting principle instead of a list.
3. Use the fundamental counting principle to find how many additional letters can be represented by 4 symbols, such as -.--.

The number of letters that can be represented by 3 symbols or less is 14, since 2 letters can be represented by 1 symbol each, 4 letters by 2 symbols, 8 letters by 3 symbols, and

$$2 + 4 + 8 = 14.$$

So every letter of the alphabet cannot be represented by 3 symbols or less.

4. Is it possible to represent every letter of the alphabet with 4 symbols or less?

A conference is being held between representatives of the following Central American countries: Guatemala, El Salvador, Honduras, Nicaragua, and Costa Rica. The flags of the 5 countries are all flown on one pole, and each day they are arranged in a different order.

5. Are the arrangements of the flags on the pole called *permutations* or *combinations?*
6. Show that the number of orders in which the flags can be flown on the pole is 120.

7. Suppose that a representative from Panama joins the conference, so that there are 6 flags instead of 5. How many days would the conference have to last if the flags were flown in all possible arrangements?

The set of 2 initials that a person might have ranges all the way from Archibald Aardvark's A.A. to Zelda Zilch's Z.Z.

8. Can you show that there are 676 different sets of 2 initials possible?

9. Most people have a middle name. How many different sets of 3 initials are possible?

"Well, finally! I thought this thing would never end."

Suppose that the train just going by in this cartoon has 100 cars, while the train coming in the other direction has 102 cars. The number of different ways of arranging the cars on the first train is 100! while the number of ways for the second train is 102!.

10. What is the symbol ! called in mathematics and what does it mean?

11. The numbers 100! and 102! are so large that it would be difficult to determine their exact values without a computer. Nevertheless, it is not very hard to figure out how many times larger 102! is than 100!. Can you do it? (Hint: 102! is 102 times larger than 101!.)

The Pepsi-Cola Company had a "Matching Picture" Contest in which the object was to match 5 pictures of Miss Americas with 5 baby pictures. One pair of pictures was already matched, leaving 4 pairs for the contestants to figure out. Contestants were allowed to enter as many times as they wished and the first prize was $10,000!

12. How many different entries would you have to send in to be sure of having all the pictures matched correctly? (The winners were selected at random drawings from correct entries.)

Set II

The Hawaiian language has only 12 letters: the vowels a, e, i, o, u, and the consonants h, k, l, m, n, p, and w. How many different 3-letter Hawaiian "words" are possible if they must each contain 2 different consonants with a vowel between them?

There are 7 choices for the first consonant, 5 choices for the vowel, and 6 choices left for the second consonant. By the fundamental counting principle, the number of possible words is

$$7 \times 5 \times 6 = 210.$$

1. How many different 3-letter Hawaiian "words" are possible if they must each contain two different vowels with a consonant between them?
2. How many different 3-letter "words" are possible if they must start with a consonant followed by 2 vowels which *may be the same?*

Most California automobile license plates have 3 letters followed by 3 digits, and each symbol can be used more than once.

3. By the fundamental counting principle, the number of different license plates possible is

$$26 \times 26 \times 26 \times 10 \times 10 \times 10.$$

   How many is this?
4. How many of these license plates end with a zero?
5. How many of them contain the word CAR?
6. A package of assorted Life Savers contains 11 in all, of which there are 3 cherry and 2 each of orange, pineapple,

lemon, and lime. In how many different ways can these be arranged in the package?

Ordinarily all 12 members of a jury must agree before a case can be decided. There are 12 different ways in which a jury can be deadlocked where 1 person disagrees with the rest, since any one of the jurors can be that person.

7. In how many ways can a jury be deadlocked in which 2 members disagree with the rest? (Hint: How many combinations of 2 people can be chosen from 12?)
8. How many ways are there in which 3 members of the jury disagree with the rest?
9. Twenty players are entered in a tennis tournament. How many singles matches are necessary in order for each person to play each of the others exactly once?
10. The Baskin-Robbins Ice Cream Stores have 31 flavors of ice cream. How many different 3-scoop ice cream cones can you choose from if each scoop is a different flavor and the order of the scoops on the cone doesn't matter?

Set III

Buying a car can be a confusing matter, because there are so many models and features from which to choose. For example, Volkswagen sedans come in 7 colors, with a choice of either cloth or leatherette upholstery; Volkswagen convertibles come in 6 colors with no choice in upholstery.

1. How many different choices of cars can you make?

Los Angeles Times Photo.

Volkswagens photographed from the Goodyear blimp after being unloaded from a ship.

Among the options available for each car are:

automatic stick shift transmission,

whitewall tires,

hinged rear side windows,

sliding steel sunroof,

radio and antenna, and

air conditioner.

2. How many combinations of choices are available for each car? (Hint: Notice that in the case of each option, you have 2 choices: you can take it or leave it. You have 6 decisions to make, with 2 choices in each case.)

3. If a Volkswagen dealer wanted to have one car with each possible combination of choices on his lot, how many cars would he have to have?

   Even though Volkswagens are small, he would need a very big lot!

THE MATHEMATICS OF CHANCE

Lesson 1

Probability—The Measure of Chance

EVERY time Lucy promises to hold the football for Charlie Brown, she whisks it away at the last moment. This time she claims that she had an involuntary muscle spasm, even though the probability of this is very small. Where did Lucy find those "odds" and how long have such numbers been calculated?

The branch of mathematics called probability theory had its origin in the 16th century, when an Italian physician and mathematician named Jerome Cardan wrote the first book on the subject, *The Book on Games of Chance*. For many years "the mathematics of chance" was used primarily to solve problems dealing with gambling. It has come a long way since then. Today the theory of probability is, according to one 20th century mathematician, "a cornerstone of all the sciences." *

Some of the first questions about probability had to do with games played with dice. An ordinary die is a cube whose faces are marked with from 1 to 6 spots. When it is rolled, it is equally likely that the die will land with any one of its 6 faces pointing upward. Since there is 1 way out of 6 for a die to turn up any one of its numbers, we say that the probability of rolling that number —for example, a 3—is 1/6. In symbols, we can write:

$$P(3) = \frac{1}{6}.$$

* Mark Kac, in his article titled "Probability," *Scientific American*, September 1964, pp. 92–108.

This suggests the following as a definition of probability:

► **Probability of an event** $= \dfrac{\text{number of favorable ways}}{\text{total number of ways}}$

Let's use this definition to figure out some more probabilities. What is the probability of rolling an even number with a die? Since 3 faces of a die have even numbers (2, 4, and 6),

$$\text{P(even number)} = \frac{3}{6}.$$

Since we have defined probability as an ordinary fraction, and fractions can often be reduced, we can also write:

$$\text{P(even number)} = \frac{1}{2}.$$

What is the probability of rolling a 7 with a die? Since no face is marked with 7 spots,

$$\text{P}(7) = \frac{0}{6} = 0.$$

The *probability of an impossible event* is *0*.

What is the probability of rolling a number less than 10 with a die? Since every face is marked with less than 10 spots,

$$\text{P(number less than 10)} = \frac{6}{6} = 1.$$

The *probability of an event that is certain* to happen is *1*. No event can have a probability of more than 1 when it is expressed as a fraction, since the number of favorable ways cannot be larger than the total number of ways.

EXERCISES

Set I

If you roll a die, what is the probability of getting:

1. A 4?
2. An odd number?
3. A number less than 3?
4. A number more than 6?

If a die is rolled many times in succession, the best guess for the number of times that a 3 will appear can be found by multiplying 1/6, the probability of getting a 3 in one roll, by the number of rolls. For example, if you roll a die 30 times, the best guess is that you will get a 3:

$$30 \times \frac{1}{6} = \frac{30}{6} = 5 \text{ times.}$$

Use the same method to determine the best guess for the number of times each of the following would appear in the number of rolls given:

5. A 4, in 60 rolls.
6. An odd number, in 100 rolls.
7. A number less than 3, in 300 rolls.
8. A number more than 6, in 1,000 rolls.

Suppose the 8 letters of the words "HOT COCOA" are written

on cards and the cards are then shuffled. If a card is chosen at random, the probability that it will contain the letter O is 3/8. What is the probability that it will contain:

9. The letter C?
10. A vowel?
11. A letter in the word "CHOCOLATE"?
12. A letter in the word "CREAM"?

A big piggy bank contains 80 pennies, 25 nickels, 10 dimes, and 5 quarters. If it is equally likely that any one of the coins will fall out when the bank is turned upside-down and shaken, what is the probability that the coin:

13. Will be a penny?
14. Will be either a nickel or a quarter?
15. Will not be a quarter?
16. Would be at least enough to pay for a phone call?

Set II

An American roulette wheel has 38 compartments around its rim. Two of these are numbered 0 and 00 and are colored green; the others are numbered from 1 to 36, of which half are colored red and the other half are colored black.

While the wheel is spun in one direction, a small ivory ball is rolled in the opposite direction along its rim. If the wheel is a fair one, the chances of the ball falling into any one of the 38 compartments as it slows down are equally likely.

The probability of the ball landing at 7 is 1/38. To predict the number of times in 100 spins that the number 7 will win, it is convenient to express its probability as a percent* — 3 percent (%), rounded off. Since 3% = 3/100, this means that in the long run, the number 7 will win approximately 3 times out of every 100 spins.

1. Show why the percent probability that a black number will win is 47%.

Find the percent probability of each of the following coming up:

2. Any number from 1 through 12.
3. Either 0 or 00.
4. A red number.
5. A red number, if the 26 numbers which had come up previously were all black.**
6. A number that is not red. (Remember that there are more than 2 colors.)

Probabilities are often expressed in terms of "odds."

► **Odds in favor of an event** $= \dfrac{\text{number of favorable ways}}{\text{number of unfavorable ways}}.$

What are the odds in favor of getting a 3 in one roll of a die? Since there is 1 favorable way and 5 unfavorable ways, the odds are 1/5, or 1 to 5.

On one roll of a die, what are the odds in favor of getting:

7. A number less than 3?
8. An even number?

---

*See page 501 if you don't remember how to figure out percents.
**This actually happened on August 18, 1913 at the Casino in Monte Carlo.

9. How do you think the odds *against* an event should be defined?

On one roll of a die, what are the odds against getting:

10. A 2?
11. An odd number?
12. Lucy says that the odds against her having an involuntary muscle spasm at the moment Charlie Brown tries to kick the ball are 10,000,000,000 to 1. What are the odds *in favor* of this happening?

Set III

An absent-minded lady wrote 3 letters and addressed 3 envelopes. Then she put the letters into the envelopes without paying attention to which belonged in which.

Can you find the probability that:

1. One or more letters gets mailed to the right person?
2. Exactly one letter goes to the right person?
3. Exactly one letter goes to the wrong person?

   (Hint: Suppose the envelopes are represented by the letters A, B, and C, and the letters are represented by the letters a, b, and c. Then a list showing all of the possible ways in which the envelopes can be paired with the letters looks like this:

	A	B	C
1.	a	b	c
2.	a	c	b
3.	b	a	c
4.	b	c	a
5.	c	a	b
6.	c	b	a

Notice that in the first of these cases all 3 letters get mailed to the right persons, while in each of the other cases at least one letter goes to the wrong person.)

Lesson 2

Some Dice Probabilities

THOR and B.C. are playing a dice game, and Thor wins on the first roll if the dice come up either 7 or 11. Are the chances of getting each of these numbers equally likely? What is the probability of rolling "snake eyes" (getting a 1 on each die) instead?

In order to answer these questions, suppose we have a pair of dice, one of which is red and the other white. Since there are 6 ways of each die turning up, there are

$$6 \times 6 = 36 \text{ ways}$$

in which they can turn up together. We can list all of the possible combinations and their sums in a table (see facing page).

Since each of the combinations shown in the table is equally likely, the probability of getting any sum can be found by counting the number of favorable ways and dividing it by 36. There are 6 ways of rolling a 7, so the probability of the dice coming up 7 is

$$\frac{6}{36} = \frac{1}{6}.$$

There are only 2 ways of rolling an 11, so

$$P(11) = \frac{2}{36} \equiv \frac{1}{18}.$$

Since there are 8 ways in all of getting either a 7 or 11, the prob-

By permission of Johnny Hart and Field Enterprises, Inc.

ability that Thor will win the game on the first roll is

$$\frac{8}{36} = \frac{2}{9}.$$

The chance of rolling "snake eyes" is rather slim, since there is only 1 way of getting a 2.

$$P(2) = \frac{1}{36}.$$

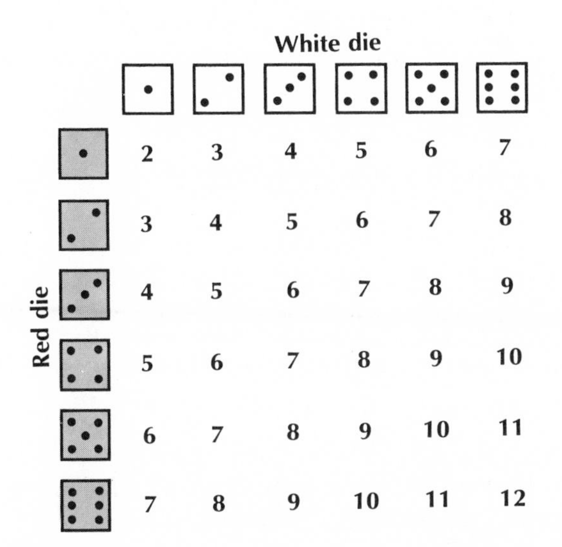

White die (columns); **Red die** (rows)

Red die \ White die	⚀	⚁	⚂	⚃	⚄	⚅
⚀	**2**	**3**	**4**	**5**	**6**	**7**
⚁	**3**	**4**	**5**	**6**	**7**	**8**
⚂	**4**	**5**	**6**	**7**	**8**	**9**
⚃	**5**	**6**	**7**	**8**	**9**	**10**
⚄	**6**	**7**	**8**	**9**	**10**	**11**
⚅	**7**	**8**	**9**	**10**	**11**	**12**

It is easier to compare probabilities when they are expressed as percents.

$$P(7) = \frac{1}{6} \qquad \frac{1}{6} \times 100 = \frac{100}{6} = 17\%$$

$$P(11) = \frac{1}{18} \qquad \frac{1}{18} \times 100 = \frac{100}{18} = 6\%$$

$$P(2) = \frac{1}{36} \qquad \frac{1}{36} \times 100 = \frac{100}{36} = 3\%$$

Each of these numbers is rounded to the nearest percent.

If Thor rolls the dice 100 times, 7 will come up about 17 times, 11 about 6 times, and "snake eyes" only 3 times. That is, if the dice don't keep rolling into that cave!

EXERCISES

Set I

1. Copy the following table, and then use the information in the table on the previous page to complete it.

Sum of 2 dice	2	3	4	5	6	7	8	9	10	11	12
Number of ways	1					6				2	
Probability	$\frac{1}{36}$					$\frac{1}{6}$				$\frac{1}{18}$	

2. Add a fourth line to the table, showing the probabilities as percents. Round off each percent to the nearest whole number.

Percent probability	3					17				6	

3. In the game that Thor and B.C. are playing, Thor loses on the first throw if a sum of 2, 3, or 12 comes up. How many ways are there altogether of getting any of these sums?

4. What is the probability that Thor will lose on the first throw?

5. If Thor rolls a 4, 5, 6, 8, 9, or 10, then he neither wins nor loses immediately. How many ways are there altogether of getting any of these numbers?

6. What is the probability that Thor will neither win nor lose on the first throw?
7. If Thor gets one of these numbers (4, 5, 6, 8, 9, or 10), he must roll the same number again *before* rolling a 7 in order to win. Which of these numbers are least likely to come up?
8. Which are most likely to come up?
9. Suppose Thor gets a 6 on his first roll. Which is more likely: that he will roll another 6 or a 7 first?

Look closely at the pair of "doctored dice" shown in this illus-

tration from an article in *Mad* magazine. Assuming that the faces you can't see are like the faces you can see:

10. What is the probability of rolling a 7 with these dice?
11. What is the probability of rolling "snake eyes" with them?

Set II

The great scientist Galileo became interested in probability when he was asked by some gamblers about the chances in a game played with 3 dice. He figured out the probabilities by making a table.

1. Since each of the 3 dice can turn up in 6 ways, in how many ways can they turn up together? (Use the fundamental counting principle.)

You have already made a table showing the results when 2 dice are thrown, which looks like this:

Sum of 2 dice	2	3	4	5	6	7	8	9	10	11	12
Number of ways	1	2	3	4	5	6	5	4	3	2	1

If the third die turns up 1 with each of these combinations, we have:

Sum of 3 dice	3	4	5	6	7	8	9	10	11	12	13
Number of ways	1	2	3	4	5	6	5	4	3	2	1

If the third die turns up 2 with each of the same combinations, we have:

Sum of 3 dice	4	5	6	7	8	9	10	11	12	13	14
Number of ways	1	2	3	4	5	6	5	4	3	2	1

There are 4 more sets of combinations to be considered: those in which the third die turns up 3, 4, 5, or 6. We can write all 6 sets of combinations in one table.

2. Copy and complete the table below:

Number of ways each possible sum may be obtained with 3 dice

Number on 3rd die	Sum of 3 dice															
	3	4	5	6	7	8	9	10	11	12	13	14	15	16	17	18
1	1	2	3	4	5	6	5	4	3	2	1					
2		1	2	3	4	5	6	5	4	3	2	1				
3			1	2	3	4	5	6	5	4	3	2	1			
4																
5																
6																
Total number of ways	1	3	6													

3. Check to see that the numbers on the last line of your table add up to the number you figured out in problem 1.

4. Plot the results in your table on a graph, representing the sums of the 3 dice on the x-axis and the numbers of ways of getting them on the y-axis. Convenient scales for the axes are shown below. After you have plotted the points, use a ruler to join them in order with straight line segments. The first part of the graph is shown here.

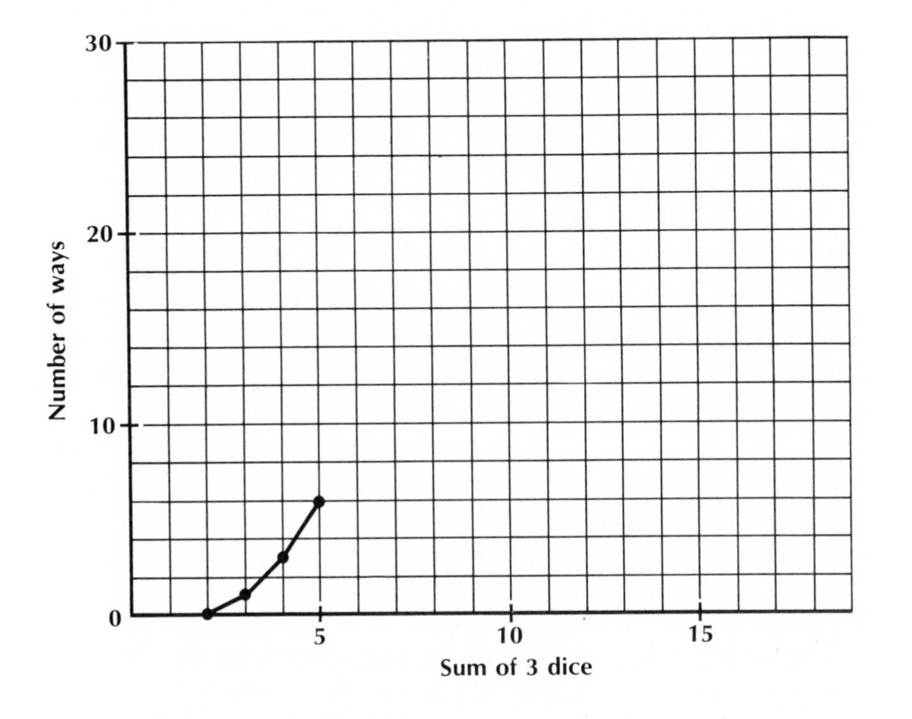

A point has been included at (2, 0) since the number of ways of getting a sum of 2 when 3 dice are thrown is 0.

5. How do the numbers of ways of getting sums of 7 and 17 compare?

6. What sums are most likely to turn up when 3 dice are thrown?

7. What is the probability of getting a sum of 11? Express the probability both as a fraction and as a percent.

8. Which is more probable: that the sum that turns up on the 3 dice will be an *odd* number or that it will be an *even* number?

9. What is the probability that all 3 dice will show a 1?
10. What is the probability that all 3 dice will show the same number? (In other words, all 1's or all 2's or all 3's, etc.)

Set III

Experiment. Rolling Two Dice

You know that 2 dice can come up in any one of 36 equally likely ways. You made a table of these ways at the beginning of the exercises in Set I, and that table is repeated again here.

Sum of 2 dice	2	3	4	5	6	7	8	9	10	11	12
Number of ways	1	2	3	4	5	6	5	4	3	2	1

This table does not mean that if you roll a pair of dice 36 times, a sum of 7 will turn up exactly 6 times and that each of the other sums will appear exactly the number of times shown. What it does claim is that if an experiment of rolling a pair of dice 36 times is repeated over and over again, the averages of the results will come closer and closer to these numbers.

To try this out, take a pair of dice and carry out the experiment 3 times. This means rolling the dice $3 \times 36 = 108$ times in all; in order to keep from losing count, number 18 lines on a sheet of paper and draw 6 columns to the right.

Then roll the dice and record each sum that turns up, until you have a number on each line of each of the 6 columns.

Count the number of times each of the numbers from 2 through 12 appears in your list, divide each one by 3 to find the averages, and record them in a table like this:

Sum of 2 dice	2	3	4	5	6	7	8	9	10	11	12
Average number of times											

For example, if the number 7 appears 20 times in your list,

$$\frac{20}{3} = 6.7$$

and the average 6.7 should be recorded in the table.

On a graph like the one on the facing page, compare your results (average numbers of times) with the predictions by plotting

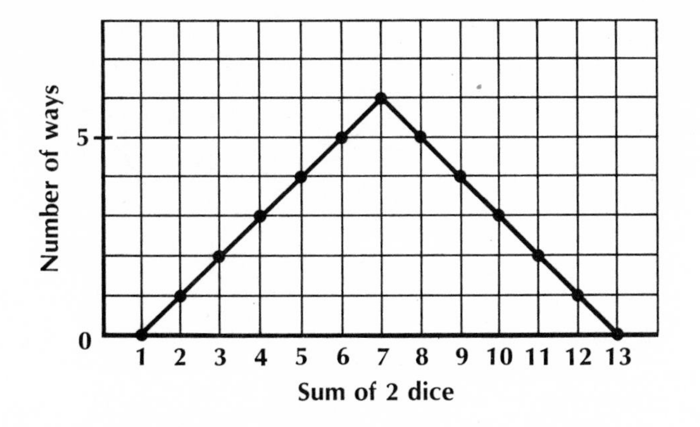

both sets of points on the same pair of axes. Connect each set of points in order with straight line segments. Use black for the predictions and red for your results. (The points for the predictions are shown on the graph.)

Lesson 3

Binomial Probability

IF a coin is tossed in the air, the probability that it will come up heads is the same as that of coming up tails: each is $\frac{1}{2}$. In claiming this we are assuming several things: that the coin is symmetrical so that there is no reason for it to come down more often on one side than the other, that it is not tossed in some special way, and that it will never land on its edge.

Suppose a coin is tossed 10 times. What do you think the chances are that it might come up heads every time? Suppose 10 coins are tossed at once. How likely do you think it is that exactly half of them will come up heads and the other half tails?

Before trying to answer these questions, let's consider some probabilities with just 2 or 3 coins. If 2 coins are tossed at the same time, there are 4 equally likely outcomes, since each coin can come up in 2 different ways. One way of showing these outcomes is with a **tree diagram.**

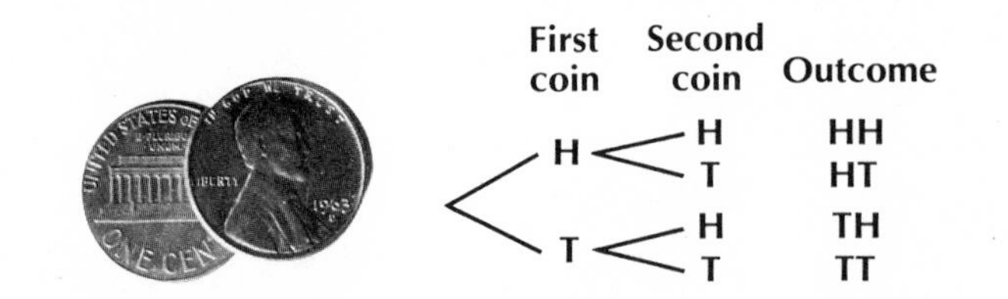

The probability of getting 2 heads is $\frac{1}{4}$, or 25%, since there is only 1 way out of 4 for this to happen; the probability of getting

By permission of Johnny Hart and Field Enterprises, Inc.

2 tails is the same. Since there are 2 ways out of 4 of getting a head and a tail, the probability of this is $\frac{2}{4} = \frac{1}{2}$, or 50%. We can summarize these results in a table:

Number of heads	2	1	0
Number of ways	1	2	1
Probability	$\frac{1}{4}$	$\frac{2}{4}$	$\frac{1}{4}$

When 3 coins are tossed, there are 8 equally likely outcomes. Notice how the tree diagram is drawn and how the outcomes can be read by reading along its branches.

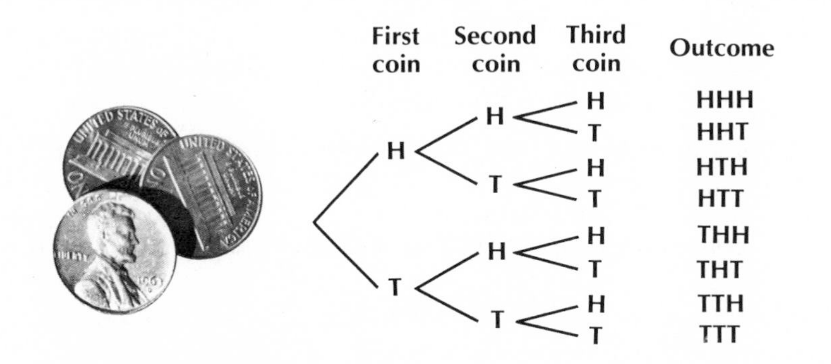

Counting the numbers of ways of getting different numbers of heads, we have the following table:

Number of heads	3	2	1	0
Number of ways	1	3	3	1
Probability	$\frac{1}{8}$	$\frac{3}{8}$	$\frac{3}{8}$	$\frac{1}{8}$

The chances of getting heads or tails when several coins are tossed is one example of *binomial* probability.

► The probabilities in any situation in which there are *2 possible outcomes* for each part are called **binomial.**

Many problems, including such widely different ones as predicting the chances of a pair of newlyweds having a certain number of boys in a row, to figuring out, on the basis of his batting average, how likely it is that a baseball player will get at least one hit in a certain number of times at bat, involve binomial probability.

EXERCISES

Set I

1. Draw a tree diagram to show the possibilities of boys and girls in a family with 2 children.

We will assume that the probability of a child being a boy is $\frac{1}{2}$, even though it actually is slightly more than $\frac{1}{2}$.*

2. In a family with 2 children, what is the probability that both are boys?
3. What is the probability that one is a boy and the other a girl?
4. Draw a tree diagram to show the possibilities of boys and girls in a family with 3 children.
5. In a family with 3 children, what is the probability that all 3 are girls?
6. What is the probability that the family has at least one boy?

At the top of the facing page is a tree diagram showing the 16 equally likely outcomes when 4 coins are tossed.

7. Copy and complete the following table, showing the numbers of ways of getting different numbers of heads.

Number of heads	4	3	2	1	0
Number of ways	▯	▯	▯	▯	▯

* In the United States P(boy) = 0.515.

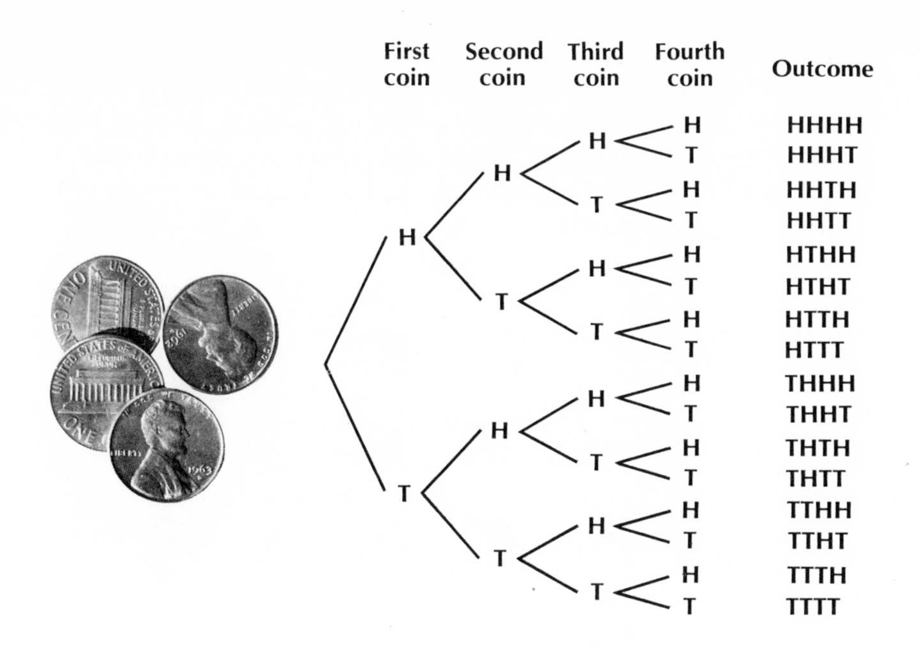

8. The sum of the numbers on the second line of your table should be the number of equally likely outcomes. What is it?
9. When 4 coins are tossed, what is the probability of getting 2 heads and 2 tails?
10. Which is more likely: getting 1 head or getting 3 heads?

A tree diagram showing the possibilities of boys and girls in a family with 4 children would look very much like the diagram for tossing 4 coins shown above.

11. In a family with 4 children, what is the probability that 2 are boys and 2 are girls?
12. What is the probability that all 4 children are the same sex?

Set II

An Austrian monk named Gregor Mendel founded the science of genetics in the 19th century with his experiments with the cross-breeding of plants. His discoveries, written up in a report titled *The Mathematics of Heredity,* were perhaps the first in which the theory of probability was applied to science.

Here is an example similar to Mendel's original experiments.

If some plants having red flowers are cross-bred with some plants having white flowers, the seeds produced grow into plants having pink flowers. When these plants with pink flowers are bred with each other, the new seeds produce 3 kinds of plants: approximately 1 plant with red flowers and 1 plant with white flowers for every 2 plants with pink flowers.

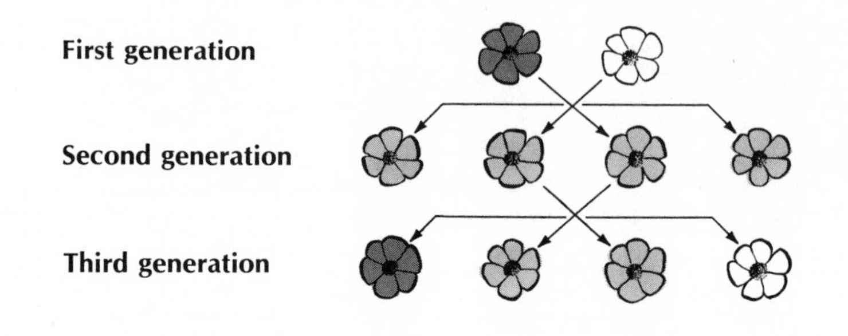

We can explain the appearances of plants with flowers of different colors in the following way. Suppose we represent with a couple of letters the genes of each plant that determine the color of its flowers.

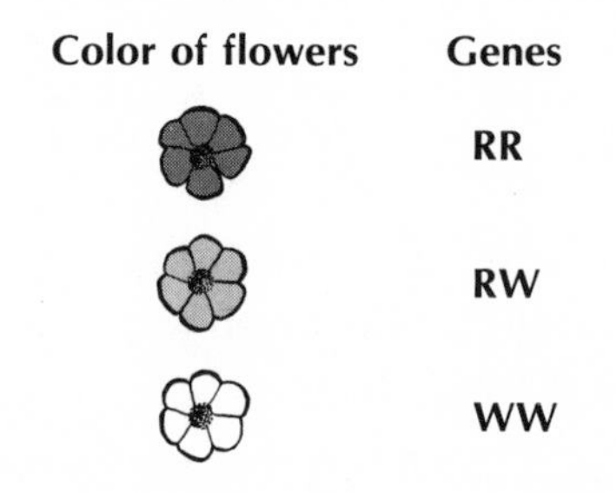

Each plant contributes one of its color genes to each of its offspring. The cross-breeding of the first generation to produce the second might be shown in a diagram like this:

Plant with white flowers

Plant with red flowers		W	W
	R	RW	RW
	R	RW	RW

1. What is the probability of getting a plant with pink flowers in the second generation?

2. What is the probability of getting a plant with red or white flowers in the second generation?

3. Copy and complete this diagram to show the 4 equally likely outcomes of cross-breeding of the second generation to produce the third.

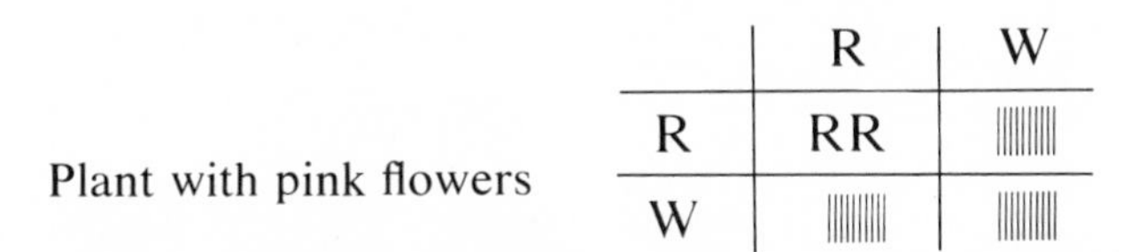

Plant with pink flowers

Plant with pink flowers		R	W
	R	RR	
	W		

4. What is the probability of getting a plant with pink flowers in the third generation?

5. What is the probability of getting a plant with red flowers in the third generation? A plant with white flowers?

Set III

Here is another heredity experiment that can be explained with probability. Several blue parakeets are mated with some yellow ones and every one of the offspring is green. When the all-green generation is mated, the colors of its offspring are green, blue, yellow, and white.

To explain the appearance of parakeets with 4 different colors, it is necessary to suppose that 2 pairs of genes determine the color instead of 1 pair. Again we will represent each gene with a letter. Suppose the genes of the first generation are:

Color of parakeet	Genes
Blue	BW-BW
Yellow	YW-YW

Then the breeding of this first generation will result in a second generation of all green parakeets, like this.

Yellow parakeet

Blue parakeet		YW	YW
	BW	BWYW green	BWYW green
	BW	BWYW green	BWYW green

We will assume that if the genes of a parakeet include both B and Y, it is green, and that W does not affect the color unless it appears without the other letters.

The figure below shows the relative numbers of green, blue, yellow, and white parakeets in the third generation, which resulted from the breeding of an all-green second generation.

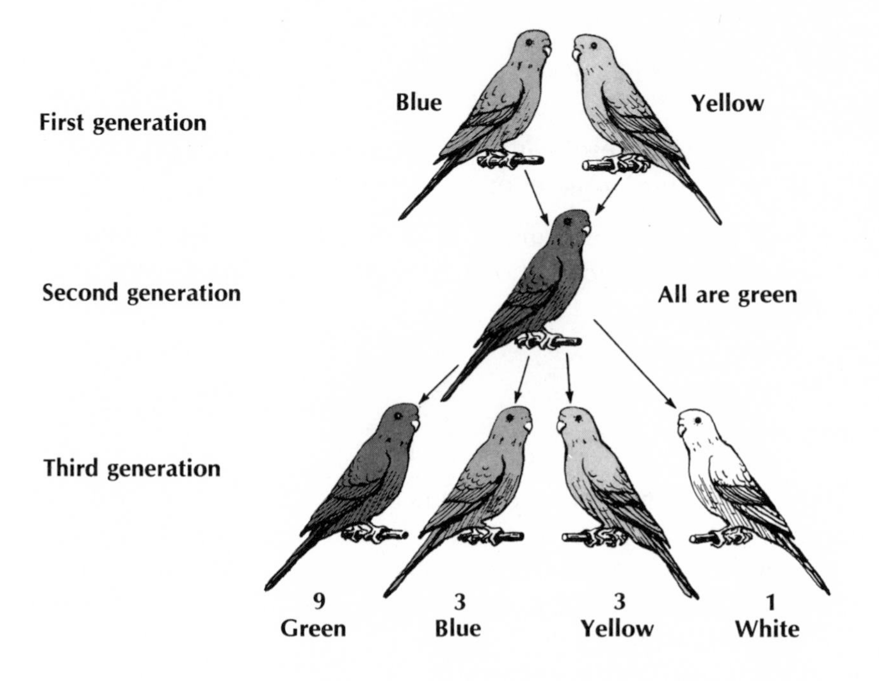

1. Copy and complete the following diagram showing the 16 equally likely outcomes of breeding the second generation to produce the third.

		Green parakeet			
		BY	BW	YW	WW
	BY	BYBY green	BYBW green		
Green parakeet	BW		BWBW blue		
	YW				
	WW				

2. Compare the results to the diagram on the previous page showing the relative numbers in the third generation. Then copy and complete the table below:

Color of parakeet in third generation	Green	Blue	Yellow	White
Probability	$\frac{9}{16}$			

Lesson 4

Wide World Photos.

Pascal's Triangle

ALL 8 children in this unusual family are girls. Since the probabilities of a child being a boy or girl are about the same, it would seem as if, in a family of this size, 4 girls and 4 boys would be much more likely. Exactly how do the probabilities of a family having 8 girls and a family having 4 girls and 4 boys compare?

We could figure this out by drawing a tree diagram, but there is an easier way. It involves a pattern of numbers known for so many centuries that no one knows who first discovered it. The pattern, called **Pascal's triangle,** is named after a great French mathematician who lived in the 17th century. One of the pioneers in the theory of probability, Pascal wrote a book about the triangle and its properties. Pascal's triangle looks like this:

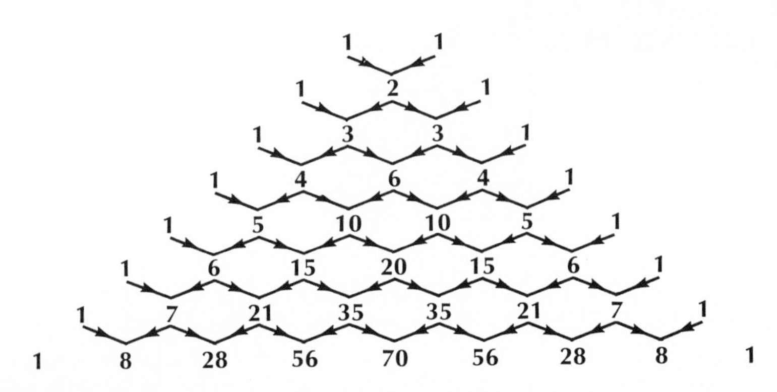

Each number within the triangle is found by adding the pair of numbers directly above it at the left and right. For example, look at the numbers in the fourth row: 4 = 1 + 3, 6 = 3 + 3, and 4 = 3 + 1. More rows of numbers can be included in the triangle by first writing a 1 at each end of the row and then adding the pairs of numbers in the previous row.

Do you see that the second row of the triangle,

1 2 1,

represents the numbers of ways in which 2 coins can turn up when they are tossed?

Number of heads	2	1	0
Number of ways	1	2	1
Probability	$\frac{1}{4}$	$\frac{2}{4}$	$\frac{1}{4}$

The third row of the triangle,

1 3 3 1,

represents the numbers of ways in which 3 coins can turn up.

Number of heads	3	2	1	0
Number of ways	1	3	3	1
Probability	$\frac{1}{8}$	$\frac{3}{8}$	$\frac{3}{8}$	$\frac{1}{8}$

The first row shows the numbers of ways in which 1 coin can turn up, and you probably recognize the fourth row as showing the numbers of ways of getting different numbers of heads with 4 coins.

Now look at the eighth row of the triangle. This row shows the numbers of ways in which 8 coins can turn up.

Number of heads	8	7	6	5	4	3	2	1	0
Number of ways	1	8	28	56	70	56	28	8	1

Adding the numbers of ways, we find that there are

$$1 + 8 + 28 + 56 + 70 + 56 + 28 + 8 + 1 = 256$$

equally likely outcomes in all. This means that the probability of getting 8 heads when 8 coins are tossed is 1/256 or less than 1%, and that the probability of getting 4 heads (and 4 tails) is 70/256, or about 27%.

The probabilities in a family of 8 children are the same:

$$P(8 \text{ girls}) = \frac{1}{256}, \text{ and}$$

$$P(4 \text{ girls \& } 4 \text{ boys}) = \frac{70}{256}.$$

So the chances of a family having 4 girls and 4 boys are 70 times greater than the chances of a family having 8 girls.

EXERCISES

Set I

1. Copy the 8 rows of Pascal's triangle shown in this lesson and then add 2 more rows.
2. At the left side of your triangle, write the number of each row in a column. At the right side, write the sum of the numbers in each row in another column:

Number of row		Sum of numbers in row
1	1 1	2
2	1 2 1	4
3	1 3 3 1	8
	etc.	

3. The numbers in the sum column are part of a simple number sequence. What kind of sequence is it?
4. Which row shows the numbers of ways in which 6 coins can turn up when they are tossed?
5. Copy the table below showing the numbers of ways in which 6 coins can turn up, using Pascal's triangle to fill in the missing numbers.

Number of heads	6	5	4	3	2	1	0
Number of ways	▥	▥	▥	▥	▥	▥	▥

The probability that all 6 coins will show heads when they are tossed is 1/64, or approximately 2%.

6. Find the percent probabilities for each of the other possibilities when 6 coins are tossed and write them on a third line in your table:

Percent probability	2						

7. Notice that the probability of getting exactly 4 heads when 6 coins are tossed is the same as the probability of getting another number of heads. What is that number?

8. The probability of getting exactly 3 heads when 10 coins are tossed is the same as getting another number of heads. What is that number?

9. The probability of getting all heads when 7 coins are tossed is 1/128. What is the probability of getting all heads when 8 coins are tossed?

10. How does the probability of getting all heads when a set of coins is tossed change if the number of coins is increased by 1?

11. Copy and fill in the missing numbers in the following table, showing the probabilities of getting equal numbers of heads and tails when a set of coins is tossed.

Number of coins	2	4	6	8
Probability of equal numbers of heads and tails	$\frac{2}{4}$	$\frac{6}{16}$	$\frac{20}{64}$	
Percent probability	50	38		

12. What happens to the probability of getting equal numbers of heads and tails as the number of coins is increased?

At the beginning of Lesson 3, a couple of questions were raised that we haven't answered yet. They are repeated here. Use Pascal's triangle to figure out the answers.

13. If a coin is tossed 10 times, what is the probability that it will come up heads every time? (This question is equivalent to that of finding the probability of getting 10 heads when 10 coins are tossed.)

14. If 10 coins are tossed at once, what is the probability that exactly half of them will come up heads and the other half tails?

Set II

Experiment. Tossing Six Coins

Using Pascal's triangle, you have figured out the percent probabilities for the different numbers of heads that can appear when 6 coins are tossed (Set I, exercise 6). On the basis of this table, we might predict that if they are tossed 100 times, all 6 coins will come up heads about 2 times, 5 heads will come up about 9 times, and so forth.

Number 25 lines on a sheet of paper and draw 4 columns to the right.

Now take 6 pennies and toss them at the same time. Count the number of heads that come up and record it on the first line of the first column. Toss the 6 pennies again, record the result, and continue in the same way until you have done this 100 times.

Count the number of times each of the numbers from 6 through 0 appears in your list, and write the results in a table like this:

Number of heads	6	5	4	3	2	1	0
Number of times							

Check to see that the sum of the numbers on the second line is 100.

Compare the results of this experiment with the percent probabilities you had determined previously.

Set III

Pascal's triangle has other interesting properties in addition to those related to probability. In fact, it has so many that Pascal said that more properties had been omitted from his book on the triangle than had been included.

1. By multiplying, find the values of 11^2, 11^3, and 11^4.
2. What do these numbers have to do with Pascal's triangle?
3. One sloping row of numbers has been marked in the Pascal's triangle at the top of the facing page.

   Starting at the top of the marked row of numbers, what is the sum of:

   Its first and second numbers?
   Its second and third numbers?

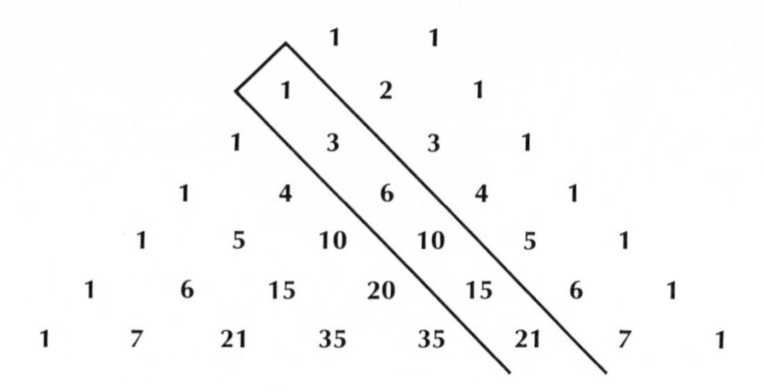

Its third and fourth numbers?
Its fourth and fifth numbers?
Its fifth and sixth numbers?

4. These sums are part of a certain number sequence. What sequence is it?

5. The numbers in the Pascal's triangle shown below have been separated into a series of rows tilted in another direction.

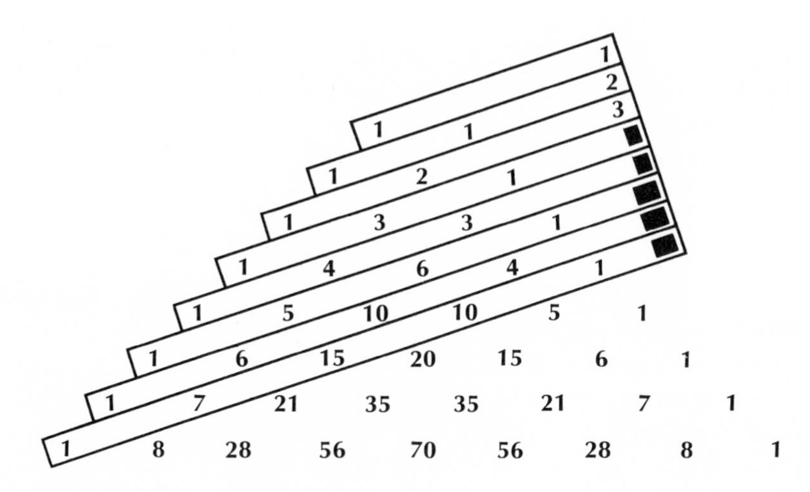

The sums of the numbers in the first three rows are 1, 2, and 3. What are the sums of the numbers in the next five rows?

6. These 8 sums are part of a famous number sequence. What sequence is it?

Lesson 5

Independent and Dependent Events

A standard deck of playing cards consists of 52 cards which are divided into 4 suits of 13 cards each: spades, hearts, diamonds, and clubs. The cards in each suit are: ace, king, queen, jack, 10, 9, 8, 7, 6, 5, 4, 3, and 2.

Suppose a deck of cards is shuffled and then one card dealt from it. What is the probability that the card will be a heart? Since there are 13 favorable ways of drawing a heart compared to 52 ways of drawing any card,

$$\text{P(heart)} = \frac{13}{52} = \frac{1}{4}.$$

There is 1 chance in 4 that a heart will be dealt.

Now suppose that the card is put back into the deck where it is shuffled with the others. If another card is drawn, what is the probability that it will also be a heart? It is the same as before: 1 chance in 4, or $\frac{1}{4}$.

If the dealer claims at the beginning that both cards drawn will be hearts, what is the probability of this happening? To find out, we can multiply the probabilities that each card will be a heart, in the same way that we have used the fundamental counting principle in many problems.

► To find the probability of several things happening in succession, multiply the probabilities of each thing happening.

Since the probability of drawing a heart from a full deck is $\frac{1}{4}$,

"All right, Cavendish—stop dealing off the bottom of the deck."

the probability of drawing 2 hearts in a row, the second one after the first had been put back, is

$$\frac{1}{4} \times \frac{1}{4} = \frac{1}{16}.$$

Since the first card was put back into the deck before the second card was drawn, what happened in the first draw has no effect on what happens in the second.

► Events such as these which have *no influence* on each other are said to be **independent.**

Now suppose 2 cards are dealt from the deck in succession, the second one *without* putting the first one back. What is the probability that they will both be hearts? This is a different situation. What happens in the first draw: whether the card is a heart or not, *does* affect what happens on the second draw: the probability that the second card will be a heart.

► Events that *are influenced* by other events are said to be **dependent.**

We can find the probability of drawing 2 hearts in succession without putting the first card back by using the same multiplication principle. It works like this. Although the probability that the first card will be a heart is 13/52, the probability that the second card will be a heart is slightly different. Assuming that the first card is a heart, there are 12 hearts left in the 51 cards remaining in the deck, so the probability of drawing a second heart is 12/51. The probability, then, of drawing 2 hearts in a row without putting the first one back is

$$\frac{13}{52} \times \frac{12}{51} = \frac{1}{4} \times \frac{4}{17} = \frac{1}{17}.$$

EXERCISES

Set I

Here again is a table showing the numbers of symbols on each of the 3 dials of a standard slot machine:

Symbol	Dial 1	Dial 2	Dial 3
Bar	1	3	1
Bell	1	3	3
Plum	5	1	5
Orange	3	6	7
Cherry	7	7	0
Lemon	3	0	4
	20	20	20

Each dial moves independently of the others. To find the probability of getting 3 bars, we can multiply the probabilities of getting a bar on each dial:

$$\frac{1}{20} \times \frac{3}{20} \times \frac{1}{20} = \frac{3}{8{,}000}.$$

Find the probability of:

1. Getting 3 bells.
2. Getting 3 oranges.
3. Getting 3 lemons.
4. *Not* getting any cherries. (Hint: The probability of not getting a cherry on the first dial is 13/20.)

You know that if a pair of dice are thrown once, the probability that they will come up "snake eyes" (a couple of 1's) is 1/36.*

---

* Look again at the diagram on page 351.

5. Suppose a pair of dice are thrown twice. Does the result of the first throw affect the result of the second? In other words, are the 2 events dependent?

If a pair of dice are thrown twice, the probability of getting "snake eyes" both times is

$$\frac{1}{36} \times \frac{1}{36} = \frac{1}{1,296}.$$

What is the probability in throwing a pair of dice twice, that you will get:

6. "Box-cars" (a couple of 6's) each time?
7. A total of 7 each time?
8. A total of 7 the first time and a total of 11 the second time?

A little old lady keeps her money in a sugar bowl. The bowl contains two \$100 bills, four \$50 bills, five \$20 bills, eight \$10 bills, and three \$5 bills. If the bills are all mixed up and the little old lady reaches into the bowl and grabs one bill and then another, what is the probability that they will both be \$20 bills?

Since the bowl contains five \$20 bills and 22 bills in all,

$$\text{P(first is a \$20 bill)} = \frac{5}{22}.$$

The events are dependent, since the removal of one \$20 bill leaves four \$20 bills in a bowl containing 21 bills.

$$\text{P(second is a \$20 bill)} = \frac{4}{21}.$$

The probability that both are \$20 bills is

$$\frac{5}{22} \times \frac{4}{21} = \frac{20}{462} = \frac{10}{231}.$$

In each of the following problems, assume that the little old lady starts with all 22 bills in the bowl. What is the probability that:

9. If she pulls out 2 bills, she will get both \$100 bills?
10. If she pulls out 3 bills, all of them will be \$50 bills?
11. If she pulls out 2 bills, they will together be worth only \$10?

Set II

What is the probability of drawing 3 aces in succession from a deck of 52 cards, assuming that no cards are put back?

Since there are 4 aces in the deck, the probability is

$$\frac{4}{52} \times \frac{3}{51} \times \frac{2}{50} = \frac{1}{13} \times \frac{1}{17} \times \frac{1}{25} = \frac{1}{5{,}525}.$$

Assume in each of the following problems that you start with a complete deck of 52 cards. Express each probability as a fraction in reduced form.

1. What is the probability of drawing 2 kings?
2. What is the probability of drawing the jack of clubs, followed by the queen of diamonds?
3. How many red cards are in the deck? (The hearts and diamonds are red; spades and clubs black.)
4. What is the probability of drawing 3 red cards?
5. Change this probability to a percent.
6. If you drew 3 cards from a complete deck 100 times in succession, about how many times would you expect all 3 to be red?
7. How many face cards are in a deck? (The kings, queens, and jacks are face cards.)
8. What is the probability of drawing 3 face cards?
9. Change this probability to a percent.
10. If you drew 3 cards from a complete deck 100 times in succession, about how many times would you expect all 3 to be face cards?

A flush is a poker hand of 5 cards all of the same suit. When you are dealt a flush, it doesn't matter what card you get first. That means the probability of getting the first card is 52/52. After the first card has been dealt, however, there are only 12 cards left in the deck that are of the same suit.

11. Show how to find the probability of being dealt a flush. Do not reduce any of the fractions and do not multiply the answer out.

"Then again, he could be bluffing."

Drawing by Herb Green; © 1961 The Saturday Evening Post.

12. The product of the fractions you have written is approximately equal to 1/500. How frequently are you likely to be dealt a flush?

A "Yarborough" is a hand of 13 cards that contains no aces, kings, queens, or jacks.

13. Show how to find the probability of being dealt a "Yarborough" by writing it as a product. Do not multiply the answer out. (Hint: Since 16 cards cannot appear in the hand, the first card dealt can be any one of 36 cards.)
14. The product of the fractions you have written is approximately equal to 1/270. Which is more probable: a flush in a game of poker, or a "Yarborough" in a game of bridge?

Set III

Could a person be convicted of a crime on the basis of probability? Suppose an old man is robbed of his wallet while he is walking in the park. A witness sees a blond girl, her hair in a ponytail, run out of the park with a bulldog and jump into a yellow car driven by a young man with a beard and mustache.

A couple matching this description is found and charged with the crime. The prosecutor uses probability to convince the jury that this couple is the guilty pair. He estimates the probabilities of 6 events in the case as follows:

$$P(\text{a girl has blond hair}) = \frac{1}{4}$$

$$P(\text{a girl has a ponytail}) = \frac{1}{10}$$

$$P(\text{a yellow car}) = \frac{1}{10}$$

$$P(\text{a man has a beard}) = \frac{1}{10}$$

$$P(\text{a man has a mustache}) = \frac{1}{3}$$

$$P(\text{a couple with a bulldog}) = \frac{1}{1{,}000}$$

1. What does the prosecutor mean when he says that the probability of seeing a yellow car is 1/10?
2. What do you suppose the prosecutor would claim to be the probability of a girl having blond hair and a ponytail?
3. What would he calculate for the probability that a couple would match the complete description given?
4. What do you think he told the jury this number meant?

In using these probabilities in this way, it is very important that the events involved are all independent. To see why, suppose that the probability of seeing identical twins is 1/1,000 and the probability of seeing two children dressed exactly alike is 1/1,000. The probability of seeing two identical twin children dressed exactly alike is *not*

$$\frac{1}{1{,}000} \times \frac{1}{1{,}000} = \frac{1}{1{,}000{,}000}$$

because these events are not independent. (Identical twins are more frequently dressed exactly alike than other children.)

5. Do you think the probability would be *more* or *less* than

$$\frac{1}{1{,}000{,}000}?$$

6. Two of the six events listed in the prosecutor's case don't seem to be really independent of each other. Which ones are they?

In an actual case* very similiar to the one described here, the couple involved were convicted and sent to prison. Later, however, the conviction was reversed by a higher court. Among the reasons given were that the prosecutor had failed to prove that the probabilities he had used were even roughly accurate, and that the six events were not completely independent.

7. Do you think that, on the basis of probability, someone could be convicted of a crime with absolute certainty?

---

* *Time* magazine, January 8, 1965 and April 25, 1968.

Lesson 6

Complementary Probabilities

AN amazing problem in probability is that of the "Coinciding Birthdays." If 30 people are chosen at random, what do you think the probability is that at least 2 of them have their birthday on the same day? Since there are 365 days in a year (ignoring leap years), the chances of this would seem rather small. However, this is not the case. In fact, the probability that at least 2 people in a group of 30 share the same birthday is about 70%!

In order to understand how this was figured out, you need to know something about *complementary probabilities.* The idea is that since an event either happens or doesn't happen,

$$\text{P(it happens)} + \text{P(it doesn't happen)} = 1.$$

► The probabilities that an event will happen and that it will not happen are called **complementary.**

Since the sum of 2 complementary probabilities is always 1, we can find either one by subtracting the other from 1.

Here is an example of how this works. If a pair of dice are rolled, what is the probability that they will not come up 7? You may recall* that the probability of rolling a 7 is

$$\frac{6}{36} = \frac{1}{6}.$$

*See page 350 if you don't.

By permission of Johnny Hart and Field Enterprises, Inc.

$$\text{P(not rolling a 7)} = 1 - \text{P(rolling a 7)}$$
$$= 1 - \frac{1}{6} = \frac{5}{6}.$$

Now let's see how this applies to the "Coinciding Birthdays" problem. What is the probability that in a group of just 2 people, both will share the same birthday? The first person's birthday can be any day. The probability that the second person's birthday is *not* the same day is 364/365, since there are 364 days in the year left to choose from (ignoring leap year).

$$\text{P(2 share birthday)} = 1 - \text{P(2 do not share birthday)}$$
$$= 1 - \frac{364}{365} = 1 - 0.997 = 0.003 \text{ (approximately)}$$

This probability is a very small number as we would expect.

What is the probability that in a group of 3 people, 2 will have the same birthday? The probability that the third person's birthday is *not* the same as either of the other two is 363/365. Using the multiplication principle for probabilities, the probability that both the second and third persons' birthdays will be different from the first's is

$$\frac{364}{365} \times \frac{363}{365}.$$

Finally,

$$P(2 \text{ share birthday}) = 1 - P(2 \text{ do not share birthday})$$
$$= 1 - \frac{364}{365} \times \frac{363}{365} = 1 - 0.992 = 0.008.$$

This probability, although still very small, is larger than that for only 2 people.

Reasoning in the same way, the probability that 2 people in a group of 30 will share the same birthday is

$$P(2 \text{ share birthday}) = 1 - P(2 \text{ do not share birthday})$$
$$= 1 - \frac{364}{365} \times \frac{363}{365} \times \frac{362}{365} \times \ . \ . \ .^{*} \times \frac{336}{365}$$
$$= 1 - 0.294 = 0.706.$$

Changing this to a percent, we get

$$0.706 \times 100 = 70.6\%.$$

So, contrary to our intuition, it is rather likely that 2 people in a group of 30 will have their birthday on the same day!

EXERCISES

Set I

1. In the same way as shown in this lesson, show how the probability that 2 people in a group of 5 have the same birthday can be figured out. (Don't work out the answer.)
2. The probability that 2 people in a group of 5 do *not* have the same birthday is about 0.973. What is the probability that 2 of them *do* share the same birthday?
3. Change this probability to the nearest percent.

The graph on the facing page shows how the probability of 2 coinciding birthdays increases as the number of people in the group increases.

Use this graph to answer the following questions.

---

*The three dots represent the fractions 361/365, 360/365, and so forth down to 337/365.

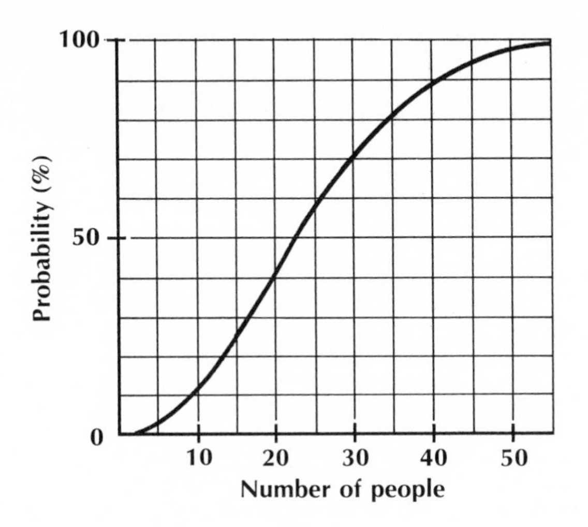

4. Does the probability that you gave in problem 3 agree with the one shown in the graph?

5. What is the probability that 2 people in a group of 40 people have the same birthday?

6. How many people must be in a group in order for the probability that 2 share the same birthday is 50%?

As you can see, the curve comes closer and closer to the line representing 100% as the number of people increases. In a group of 50 people, it is extremely likely that 2 birthdays will fall on the same day, yet such a coincidence is not certain.

7. How many people do you think would have to be in a group to make it *absolutely certain* (a 100% probability) that 2 have the same birthday? (Hint: Use your common sense and don't ignore leap year.)

If a pair of dice are rolled, the probability that they will come up 11 is

$$\frac{2}{36} = \frac{1}{18}.$$

8. What is the probability that they will *not* come up 11?

9. If a die is rolled once, what is the probability that it will not come up 6?

10. If a die is rolled twice, what is the probability that it will not come up 6 either time?

11. If a die is rolled twice, what is the probability that it will come up 6 at least once? (Hint: This probability is complementary to the probability in the previous problem.)
12. If 5 coins are tossed, what is the probability that they will all come up tails?
13. What is the probability that at least one of them will come up heads? (In other words, what is the probability that they will *not* come up tails?)

Set II

The Arthur Murray Studios once offered $25 worth of dancing lessons to anyone who had a "lucky dollar." A "lucky dollar" was a bill whose serial number included a 2, 5, or 7. If you have any dollar bills with you, check to see whether or not you would have been a winner.

Since our number system uses 10 digits, the probability that any one of the digits *will* be one of the lucky 3 is 3/10.

1. What is the probability that any one of the digits *will not* be a lucky one?
2. There are 8 digits in the serial number of a dollar bill. What is the probability that all 8 digits will not be lucky ones?
3. Change your answer to decimal form and round it off to the nearest hundredth.
4. You have just found the probability that a dollar bill *will not* be a lucky one. What, then, is the probability that someone with a dollar bill has a "lucky dollar"?
5. Out of 100 people with dollar bills, how many do you think will be winners?

Set III

Rube Goldberg is a 20th century artist who is so well-known for his cleverly ridiculous inventions that, according to one dictionary,* his name has come to mean, "having a fantastically com-

---

**The Random House Dictionary* of the English Language, unabridged edition, 1966.

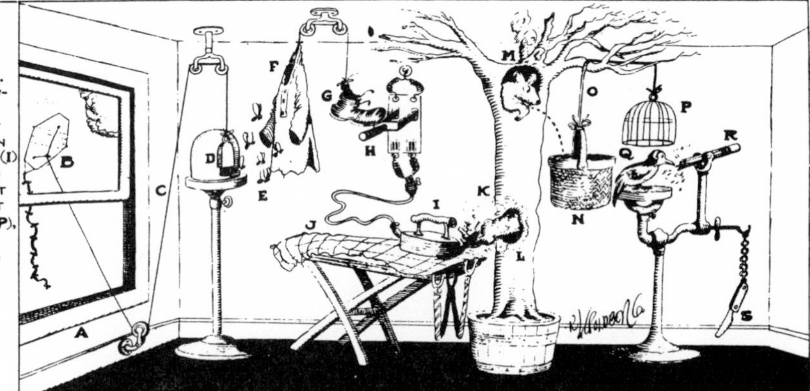

plicated, improvised appearance" and "deviously complex and impractical."

The invention for sharpening pencils shown here makes use of an opossum and a woodpecker and includes an emergency knife in case either animal gets sick and can't work. Suppose the probability that the opossum will get sick is 3/10 and the probability that the woodpecker will get sick is 1/10. Can you figure out the probability that *neither one* will get sick so that the knife isn't needed?

Chapter 8 / Summary and Review

In this chapter we have become acquainted with:

Probability—The measure of chance *(Lessons 1 & 2)* A definition of probability:

$$\text{Probability of an event} = \frac{\text{number of favorable ways}}{\text{total number of ways}}$$

The probability of an impossible event is 0; the probability of an event that is certain to happen is 1.

Probabilities can be expressed either as fractions or as percents (%); they may also be expressed in terms of odds.

$$\text{Odds in favor of an event} = \frac{\text{number of favorable ways}}{\text{number of unfavorable ways}}$$

Binomial probability and Pascal's triangle *(Lessons 3 & 4)* Tree diagrams are sometimes useful for showing all of the possibilities in a situation.

The probabilities in a situation in which there are 2 possible outcomes for each part are called binomial.

Pascal's triangle is a pattern of numbers from which binomial probabilities can be easily determined. Each number within the triangle is found by adding the two numbers above it at the left and right. The triangle begins like this:

1 1
1 2 1
1 3 3 1
1 4 6 4 1

Independent and dependent events *(Lesson 5)* To find the probability of several things happening in succession, multiply the probabilities of each thing happening.

Events that have no influence on each other are said to be independent; events that are influenced by other events are called dependent.

Complementary probabilities *(Lesson 6)* The probabilities that an event will happen and that it will not happen are called complementary.

P(an event happens) + P(an event doesn't happen) = 1.

EXERCISES

Set I

A box contains 90 disks for a bingo game. The disks are numbered from 1 through 90. If one disk is drawn from the box, find the probability that it contains:

1. A number more than 60.
2. An even number.
3. A two-digit number.
4. A square number.
5. A number that is divisible by 5.

A popular song of the 1920's began like this:

"Keep your sunny side up, up!
Hide the side that gets blue.
If you have 9 sons in a row
Baseball teams make money, you know!"*

---

*"Sunny Side Up," copyright © 1929 by DeSylva, Brown & Henderson, Inc. Copyright renewed. Used by permission of Chappell & Co., Inc.

6. What is the probability of a child being a boy?
7. What is the probability of a family having 9 sons in a row?
8. Here is a diagram from a Chinese book written in 1303. What is it?

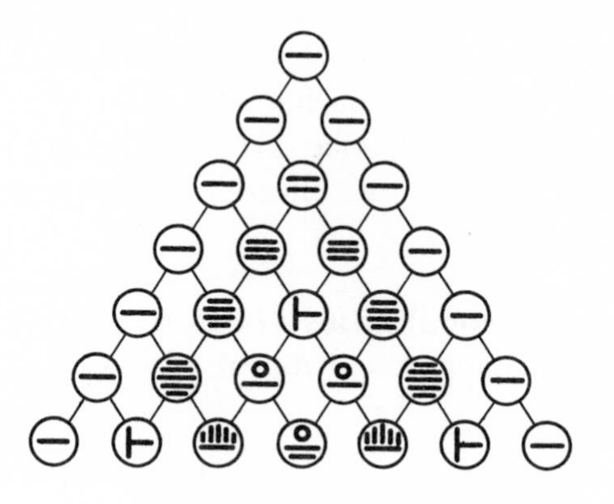

9. What is the Chinese symbol for 20?
10. Four ladies starting to play bridge each draw a card from the deck to decide who will deal first. What is the probability that each one will turn up an ace?*
11. Suppose the ladies had first divided the deck into 4 separate piles, each pile containing all of the cards of one suit. If each lady took one pile and drew a card from it, what is the probability now that each one will turn up an ace?

Set II

To test for extrasensory perception (ESP), a special deck of 25 cards is sometimes used in which there are 5 cards marked like each of those shown on the facing page.

1. If the deck is shuffled and a card drawn at random, what is the probability that someone could guess the symbol on it simply by luck?

---

*This incident was reported in the *Los Angeles Times*. In addition to giving the probability, the article pointed out that if these women played 5,000 bridge hands a year, which is an awful lot of bridge, this would happen only once every 54 years.

Cards used in extrasensory perception test.

2. What is the probability that someone could correctly guess 4 cards in succession, if each card is put back and the deck reshuffled before the next card is drawn?

3. One card marked with each of the symbols is taken from the deck and the 5 cards are arranged face down in random order. What is the probability of guessing the correct order by luck? (Hint: In this case, the events of guessing the cards are dependent. The probabilities for the first 2 cards are $\frac{1}{5}$ and $\frac{1}{4}$.)

4. About how many times would you expect this to happen if the experiment were repeated 500 times?

The Counterfeiter's Club is having a raffle and each of the 25 members buys a ticket. The 25 stubs are put into a box from which one will be drawn to determine which member will win the prize.

5. What is the probability that Fraudulent Fred, one of the members, will win the prize?

6. Fred decides to improve his chances by forging 7 copies of his stub and sneaking them into the box. Now what is the probability that he will win?

7. Now what is the probability that any one of the other members will win?

8. Just before the drawing, Fred finds out that after the first stub is drawn to determine the winner, a second stub will be drawn to decide who will get a booby prize. What is the probability that Fred will win both prizes, so that his forgery is discovered?

9. The first stub is drawn and it is not one of Fred's. What is the probability that he will still get the booby prize?

Set III

It has been speculated that if a group of monkeys were given typewriters and the monkeys hit the keys at random, over a long period of time they would eventually write every book that has been written, including this one!

Suppose that a typewriter has 40 keys and that the chance of a monkey hitting any one of them is the same.

What is the probability that:

1. A monkey will type the letter M?
2. An ape will type the word APE? (Assume that the events of the monkey hitting each key are independent.)
3. A gibbon will type the word GIBBERISH?

INTERESTING READING

How to Take a Chance, by Darrell Huff, Norton, 1959.

Mathematics by David Bergamini, a book in Time-Life's "Life Science Library" series, 1963: "Figuring the Odds in an Uncertain World" and "The Fascinating Game of Probability and Chance," pp. 126–147.

AN INTRODUCTION TO STATISTICS

Lesson 1

Frequency Distributions

MANY people think of the word "statistics" in the sense that Schroeder and Charlie Brown are using it here: as meaning numbers or tables of numbers. Mathematicians, however, use the word in a more general sense. The subject of statistics is a branch of mathematics that deals with the collection, organization, and interpretation of numerical facts. It has become an important tool in many fields, including such widely different ones as atomic physics, medicine, the advertising industry, and even the study of history! H. G. Wells, the English author and historian who wrote such books as *The War of the Worlds* and *The Outline of History*, once said: "Statistical thinking will one day be as necessary for efficient citizenship as the ability to read and write."

Here is a problem in statistics of interest to the manager of a baseball team, who might want to compile some figures on how well his team is playing. Suppose a list of the team's scores for 30 games looks like this:

7	5	9	3	5
4	2	3	0	7
10	9	2	4	2
0	4	7	12	3
3	2	4	1	8
6	5	4	3	4

It is easy to see from this list that the team scored 12 runs in one game, while in a couple of others they didn't score at all. However, since the numbers are not arranged in order, it is not so easy to conclude much else from them.

A **frequency distribution** is a convenient way of organizing data to reveal what patterns it may have. First, we list the numbers 0 through 12 in a column to show the scores, and then we go through the original list in the order in which it is written, putting tally marks on the appropriate lines of our table. In a third column, we can list the *frequency* of the score, which is the number of times that it appears.

Score	Tally marks	Frequency
0	II	2
1	I	1
2	IIII	4
3	~~IIII~~	5
4	~~IIII~~ I	6
5	III	3
6	I	1
7	III	3
8	I	1
9	II	2
10	I	1
11		0
12	I	1
	Total =	30

Now it is easier to see that the team scored 4 runs more times than any other number: more specifically, in

$$\frac{6}{30} = \frac{1}{5} = 20\% \text{ of the games.}$$

The team's score in 12 of the 30 games, or 40% of the games, was less than 4, and it was more than 4 in the same number of games.

The frequency distribution we have made could be made shorter by grouping together the scores in intervals. If we choose an interval of 3 runs, the distribution looks like this:

Score	Frequency
0-3	12
4-6	10
7-9	6
10-12	2
Total =	30

We can get a general idea of the team's scores from a quick glance at this table, but notice that some of the information from the previous table is lost in this one. A frequency distribution arranged in group intervals is generally necessary whenever there are many different numbers involved.

EXERCISES

Set I

Here is a list of the Presidents of the United States and the number of children of each.

1. Washington, 0
2. J. Adams, 5
3. Jefferson, 6
4. Madison, 0
5. Monroe, 2
6. J. Q. Adams, 4
7. Jackson, 0
8. Van Buren, 4
9. W. H. Harrison, 10
10. Tyler, 14
11. Polk, 0
12. Taylor, 6
13. Fillmore, 2
14. Pierce, 3
15. Buchanan, 0
16. Lincoln, 4
17. A. Johnson, 5
18. Grant, 4
19. Hayes, 8
20. Garfield, 7
21. Arthur, 3
22. Cleveland, 5
23. B. Harrison, 3
24. McKinley, 2
25. T. Roosevelt, 6
26. Taft, 3
27. Wilson, 3
28. Harding, 0
29. Coolidge, 2
30. Hoover, 2
31. F. D. Roosevelt, 6
32. Truman, 1
33. Eisenhower, 2
34. Kennedy, 3
35. L. B. Johnson, 2
36. Nixon, 2

1. Make a frequency distribution of the numbers of children per president. Label the 3 columns: "Number of children," "Tally marks," and "Frequency."

Use your frequency distribution to answer these questions.

2. What is the most frequent number of children per president?
3. How many presidents had 3 children or less?

The percent frequency of presidents who had no children is

$$\frac{6}{36} \times 100 = \frac{1}{6} \times 100 = \frac{100}{6} = 17\%.$$

What is the percent frequency of presidents who had:

4. Two children?
5. More than 7 children?

The melting points in degrees Fahrenheit of 45 common chemical elements are given in the list below.

1. Aluminum, 1220
2. Antimony, 1167
3. Argon, −309
4. Arsenic, 1502
5. Barium, 1337
6. Bismuth, 520
7. Boron, 4170
8. Bromine, 19
9. Cadmium, 610
10. Calcium, 1550
11. Carbon, 6420
12. Chlorine, −150
13. Chromium, 3430
14. Cobalt, 2725
15. Copper, 1981
16. Fluorine, −363
17. Gold, 1945
18. Helium, −457
19. Hydrogen, −434
20. Iodine, 204
21. Iron, 2795
22. Lead, 622
23. Lithium, 354
24. Magnesium, 1172
25. Manganese, 2271
26. Mercury, −38
27. Neon, −416
28. Nickel, 2647
29. Nitrogen, −346
30. Oxygen, −361
31. Phosphorus, 81
32. Platinum, 3216
33. Potassium, 147
34. Radium, 1290
35. Silicon, 2570
36. Silver, 1761
37. Sodium, 208
38. Strontium, 1416
39. Sulfur, 240
40. Tin, 450
41. Titanium, 3045
42. Tungsten, 6170
43. Uranium, 2070
44. Xenon, −170
45. Zinc, 787

6. Make a frequency distribution of these melting points by grouping the temperatures together in intervals of 500 degrees. Number the first column in your table like this:

Melting point, °F
−500-0
1-500
501-1,000
1,001-1,500
1,501-2,000
2,001-2,500
2,501-3,000
3,001 & higher

What percent of these elements have melting points that are:

7. Less than 500°F?
8. More than 2,000°F?
9. Approximately half of the elements in this table have melting points that are less than a certain temperature. What is it?

Set II

The rapid development of computers in the middle of the 20th century has been astonishing. Problems that formerly would have taken years to figure out can now be solved in a matter of

"Now, here's an earlier model I can give you a real buy on."

seconds. A good example is the calculation of the decimal value of pi.

A very popular book on mathematics written in 1940 and still available in most bookstores today says about the number pi: "Even today it would require 10 years of calculation to determine pi to 1,000 places."* Yet a little more than 20 years later, the

---

* *Mathematics and the Imagination,* by Edward Kasner and James Newman, Simon and Schuster, 1940, pp. 77–78.

value of pi was determined to more than *100,000 places* in less than 9 hours by a computer at the IBM Data Center in New York City!

One question that mathematicians have about pi concerns the frequencies of the digits in its decimal form. Do the digits occur with the same frequency or do certain digits appear much more frequently than others?

Here is the value of pi to 200 decimal places. The digits are separated into groups of 5 to make them easier to work with. The vertical rule divides the digits into 2 sets of 100 digits each.

3.14159 26535 89793 23846 | 26433 83279 50288 41971
69399 37510 58209 74944 | 59230 78164 06286 20899
86280 34825 34211 70679 | 82148 08651 32823 06647
09384 46095 50582 23172 | 53594 08128 48111 74502
84102 70193 85211 05559 | 64462 29489 54930 38196

1. If each of the 10 digits has the same frequency, how many times would you expect to see each one appear in a series of 100 digits?
2. Make a frequency distribution of the first 100 digits after the decimal point. List the digits in the first column in order from 0 through 9.
3. Now make a frequency distribution of the second 100 digits, by adding 2 more columns to your previous table: one for the new tally marks and one for the new frequencies.
4. Add a sixth column to your table in which you show the frequency total for each digit. The top of your completed table should look like this:

Digit	Tally marks	Frequency	Tally marks	Frequency	Total frequency
0	卌 III	8	卌 卌 I	11	19

5. Examine your frequency totals. Does any digit stand out in your table as occurring much more frequently in the decimal value of pi than the others?
6. Does any digit seem to be falling behind the rest in the number of times it appears as the value of pi is read to more and more decimal places?

Set III

Have you heard of the language "Esperanto"? It was created in the late nineteenth century as an international language. Several million people in different countries now speak Esperanto and thousands of books have been written in it.

Here is a sample of the language:

"En la komenco Dio kreis la cielon kaj la teron. Kaj la tero estis senforma kaj dezerta, kaj mallumo estis super la abismo; kaj la spirito de Dio svebis super la akvo. Kaj Dio diris: Estu lumo; kaj farigis lumo."

1. This passage is from the beginning of a world famous book. Do you have any idea what it says?
2. The passage contains 164 letters; print the alphabet in a column on your paper and make a frequency distribution of them.
3. Although Esperanto uses the letter "h," which is not in the quotation, it does not use certain other letters of the English alphabet. What do you think they are?
4. What letter occurs most frequently in this passage?
5. In general, what kind of letters seem to have the highest frequency in this language?

Lesson 2

An Application of Statistics— The Breaking of Ciphers and Codes

IN one of the Sherlock Holmes stories, called "The Adventure of the Dancing Men," the methods of statistics are applied to deciphering a series of mysterious messages. The messages were written as rows of little stick figures in which each figure stands for a letter of the alphabet. The first message looked like this:

As you can see, some of the figures are holding flags, and Sherlock Holmes concluded that they were used to indicate the ends of words. This message, then, consists of 4 words. Of the 15 figures it contains, one appears 4 times:

Mr. Holmes, knowing that the letter E is the most frequently occurring letter in written English, decided that this symbol represented E. This seems especially likely since two of these men are holding flags and E is the last letter of many words.

Later more messages appeared, and Holmes, by putting a number of clues together, was able to decipher what they said. For instance, the first word of the longest message,

is 5 letters long and begins and ends with the letter E. The name of the woman to whom it was sent was Elsie, so it seemed probable that this word was ELSIE, so that the symbols

stand for L, S, and I respectively.

Continuing to reason in the same way with some other messages, Holmes decided that the first one said:

AM HERE ABE SLANEY,

and that the other message shown here said:

ELSIE, PREPARE TO MEET THY GOD.

Holmes trapped the villain, named Abe Slaney, by writing a message to him with the stick figure symbols. Slaney, who assumed that only Elsie could read the cipher, thought that the message was from her and so came to her home, where he was arrested.*

Many detective stories have been written about secret messages. Did you know that the writing and breaking of ciphers and codes is of great importance in national security and has actually influenced the course of history? Messages on the "hot line" between Washington and Moscow are kept secret by means of code. The U.S. Department of State receives and sends several million coded words every week. The National Security

* If you are interested in reading the entire story, "The Adventure of the Dancing Men" is included in the series of adventures titled *The Return of Sherlock Holmes*, by Sir Arthur Conan Doyle.

Agency, which develops and breaks codes, has more than ten thousand employees, and, since modern codes are so complex, is thought to have more computers than any other organization in the world. Today, cryptology, the science of code writing and breaking, makes use of many mathematical ideas, especially in the fields of advanced algebra and statistics.

EXERCISES

Set I

The messages in "The Adventure of the Dancing Men" are written in a "simple substitution cipher," a cipher in which each letter is always represented by the same symbol. Here is a list of the letters and symbols used in the 2 messages that are given in this lesson.

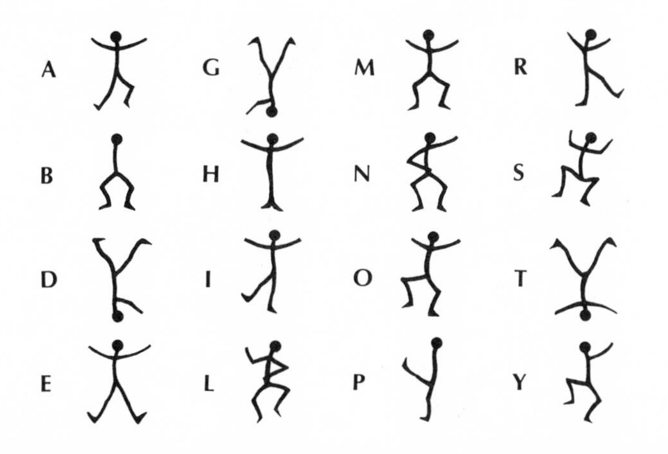

1. Most substitution ciphers do not have any pattern in the order of symbols used for the various letters of the alphabet. Does there seem to be any pattern in the symbols in this "dancing men" cipher, or do they appear to be chosen at random?

2. What does the following message say?

3. Here is the message that Sherlock Holmes sent to Abe

Slaney. Although it uses a symbol that does not appear in the list, can you figure out what it says?

Determining the frequency of each letter in a cipher is usually the first step in solving it. This is useful since the frequencies of the letters in any written language are always about the same in every large sample of it. Here is a list of the percent frequencies of the letters in ordinary English, in decreasing order.

Letters	Percent frequency
E	13
T	9
A, O	8
N	7
I, R	6.5
S, H	6
D	4
L	3.5
C, U, M	3
F, P, Y	2
W, G, B	1.5
V	1
K, X, J	0.5
Q, Z	0.2

Do you see that, according to this list, the letter T occurs about 9 times in an ordinary sample of 100 letters, while the letter P occurs about twice?

Use the list to estimate the number of letters in each of the following:

4. E's in 200 letters.

5. W's in 1,000 letters.

6. Z's in 2,000 letters.*

An extraordinary sentence often used as a typing exercise is:

"Pack my box with five dozen liquor jugs."

7. Write the alphabet in a column and make a frequency distribution of the letters in this sentence.
8. Find the percent frequency of the letter E in this sentence.
9. Is it greater or less than in ordinary English?
10. What is the percent frequency of the letter Q in this sentence?
11. How does it compare to the frequency of Q in ordinary English?
12. If the "liquor jugs" sentence were written in code, do you think that it would be relatively easy or relatively difficult to decipher?

Set II

Ciphers and codes are of especially great importance during times of warfare. This telegram was sent to Tokyo from the

RCA
RADIOGRAM
R.C.A. COMMUNICATIONS, INC.
A RADIO CORPORATION OF AMERICA SERVICE
TO ALL THE WORLD — BETWEEN IMPORTANT U S CITIES

Send the following Radiogram "Via RCA" subject to terms on back hereof, which are hereby agreed to

CDE

GAIMUDAIJIN TOKIO

1941 DEC 6 PM 6 01

SIKYU 02540 GUGUG OAHEL FCOIC JUUQH COGAA LMOPE LENUR WPJFI
EUENE QILDO TAOEO DJAEO PPOCF DBIQO IOGLP UIBOY LUEJA ODGIJ
EEDON NREEE IEBJO GDYFR DOOBW OEIIA AAUAQ HOYAH IIEDO RAAEJ
BDUIV BMNLU FQFEE OENON ADODM IJBVN AEGRE UQEEU NOIKE BOKAG
YHUOG DGYQH UEVTA TOREH

KITA

Department of the Navy.

* These estimates are only approximate and are for ordinary English. What effect would knowing that the 2,000 letters were in a magazine article about zebras in Zanzibar have on this estimate?

Japanese consulate in Honolulu on the evening before the attack on Pearl Harbor.

It reported on the ships of the United States fleet that were in port. The telegram is written in a cipher in which the words are run together and the letters grouped in blocks of 5.

Here is another cipher in which the words are not separated so that there are no clues about their lengths. For convenience, the message has been written in groups of 10 letters each; it contains 140 letters in all.

```
V Z B M X C V B T A   V N V B A A H D K B
T V Z B G N C X L H   T B H J V Z B F N Y
N X B A B X N K S C   X V Z B A B L H X T
P H E D T P N E V Z   C A Y D N S B T N X
C G Y H E V N X V E   H D B C X V Z B N G
B E C L N X K C L V   H E S N V V Z B T B
L C A C K B W N V V   D B H J G C T P N S
```

1. Copy the cipher on your paper, leaving 2 blank lines between each line of letters. Then check to see that every letter has been copied correctly.
2. To get some clues about what letter each letter represents, write the alphabet in a column and make a frequency distribution of the letters in the cipher.
3. The most frequent letter is probably E, so write an E underneath each place this letter appears in the cipher.
4. The second most frequent letter is probably T. Write a T underneath each place this letter appears.
5. A very common word in English is THE. If this word appears several times in the cipher, you should be able to guess what letter represents H.
6. A strong clue in solving a cipher is some knowledge of certain words that are likely to appear in it. For instance, this cipher mentions the United States. If you can find where the words UNITED STATES appear, you will know what several more letters represent.
7. The most frequent letters in written English are:

   E, T, A, O and N.

   So far, you know what letters represent all of these except O. Look at the letters with the highest frequencies in your frequency distribution and you should be able to guess what letter stands for O.

8. Solving for the remaining letters in the cipher is left to you. What does the cipher say?

Set III

"This is a highly unusual paragraph. Do you know why? If you try to find out what is odd about it too quickly, it probably won't occur to you. Study it without hurrying and you may think of what it is. Good luck."

(Hint: If this paragraph were written in code, something about it would make it difficult to decipher.)

INTERESTING READING

The Codebreakers, by David Kahn, Macmillan, 1967.

Lesson 3

Measures of Central Tendency

THOR and B.C. are playing golf. Although they are about the same at driving the ball onto the green, Thor is better at putting. On the ninth hole, however, B.C. makes a 5 and Thor gets stuck in a sand trap. By the time he finally gets out of it and makes it to the hole, Thor has a 15 and is so mad that he quits.

Here are the scores on the 9 holes:

Hole	1	2	3	4	5	6	7	8	9	Total
Thor	4	3	6	4	3	3	4	3	15	45
B.C.	6	4	7	5	5	3	5	4	5	44

Who is the better player on the average?

B.C. claims that he is, since his total score is 1 less than Thor's. He has figured out their averages by dividing each total score by the number of holes, getting:

B.C.'s average	Thor's average
$\frac{44}{9} = 4.9$	$\frac{45}{9} = 5$

Thor looks this over and wonders how B.C.'s average can be less than his, since he either beat or tied B.C. on every hole except one. Thor doesn't think that B.C.'s method of working out averages gives a true picture of the situation, so he tries a different method. He starts by making a frequency distribution of

By permission of Johnny Hart and Field Enterprises, Inc.

the scores that he and B.C. got on each hole:

B.C.		Thor	
Score	Tally	Score	Tally
3	I	3	IIII
4	II	4	III
5	IIII	6	I
6	I	15	I
7	I		

After studying these lists for awhile, Thor tells B.C.: "Your average score is 5, since you made a 5 on more holes than any other number. My average score is 3, as you can easily see." Who do you think is right: Thor or B.C.?

The average B.C. figured out is called the *mean,* or arithmetic average. It is the one most people think of when they hear the word "average."

► The **mean** of a set of numbers is found by adding them and dividing the result by the number of numbers added.

Thor, on the other hand, is using another kind of average, called the *mode.*

► The **mode** of a set of numbers is the number that occurs most frequently. If no number occurs more than once, then there is no mode. It is also possible, however, for a set of numbers to have several modes.

A third kind of average is called the *median.*

► The **median** of a set of numbers is the middle number when the numbers are arranged in order of size.

B.C.'s scores on the 9 holes arranged in order are:

3 4 4 5 5 5 5 6 7
↑ median (the fifth score, 5)

and Thor's scores are:

3 3 3 3 4 4 4 6 15.
↑ median (the fifth score, 4)

B.C.'s median score is 5 and Thor's median score is 4.

In general, an average is one number used to represent a set of numbers. When the individual numbers in a set are not known, the average should give us an idea of a typical number in it. Because of this, averages are called *measures of central tendency.*

Let's compare the three averages as measures of the ability of B.C. and Thor to play golf.

Average	B.C.	Thor
Mean	4.9	5
Mode	5	3
Median	5	4

If we look back at the original list of scores, it seems that the mean is not very appropriate here. Although Thor is a quitter, he does play better than B.C. most of the time, and this is clearly shown by the median and mode.

EXERCISES

Set I

1. Are the 5 numbers below part of an arithmetic sequence or a geometric sequence?

   2 5 8 11 14

2. Find the mean average of these 5 numbers.
3. From what kind of sequence are the 5 numbers below?

   2 4 8 16 32

4. Find their mean average.
5. Here is part of the Fibonacci sequence.

   3 5 8 13 21

   Find the mean average of these 5 numbers.

6. Which average is the same for all 3 sets of numbers above: the mean, median, or mode?
7. Which average does not exist for these sets of numbers?

The sequence of squares of the numbers 1 through 8 is:

1 4 9 16 25 36 49 64.

What is the median average of these squares, since there is no single middle number in an even number of numbers? In such a case, the median is the number halfway between the middle two numbers. It is:

$$\frac{16 + 25}{2} = \frac{41}{2} = 20.5.$$

8. How does this number compare to the mean average of these squares?

The sequence of cubes of the numbers 1 through 6 is:

1 8 27 64 125 216.

9. Which average of this set of numbers do you think is easier to determine: the mean or the median? What is it?

Words in the English language vary in length from "a" to "supercalifragilisticexpialidocious" and even longer. What is the average length, measured in number of letters, of words used in ordinary English?

10. To get an idea, make a frequency distribution of the numbers of letters in the words of this paragraph, a translation of the original written in Italian by Galileo. Notice that the longest word in the paragraph, "mathematical," is 12 letters long.

    "That great book which stands forever open before our eyes, the universe, cannot be read until we have learned the language and become familiar with the symbols in which it is written. It is written in mathematical language, without which means it is humanly impossible to comprehend a single word."

11. What is the mode average of the length in letters of the words used?

12. Use your frequency distribution to find the median average of the word length.

13. Find the mean average. (The paragraph contains 243 letters in all.)

14. Which average do you think is the *least* appropriate for indicating the length of a typical word in this paragraph?

An average can sometimes give a very misleading idea of the set of numbers it is supposed to represent. Here is a good example of how this can happen.

The Mighty Mousetrap Company is owned by two partners and has 13 employees. The partners pay themselves salaries of $15,000 each, and of the 13 employees, 3 earn $6,000 each, 4 earn $4,500, and 6 earn $4,000.

15. Make a list of the 15 salaries in order from largest to smallest.

16. In reporting the average salary paid by the company, the owners want to make it sound as large as possible.

    Which average would they choose and how much is it?

Courtesy of Irv Phillips; by permission of Publishers-Hall Syndicate.

17. How much are each of the other averages?
18. Which average do you think best represents the typical salary in the Mighty Mousetrap Company?

Set II

A new application of mathematics is to the study of "queues" (pronounced "Q's"), or lines waiting for service. Examples of queues are shoppers waiting at a checkout counter of a supermarket, airplanes waiting to land at an airport, and telephone calls arriving at a switchboard. The information gained in the study of queues has some very practical applications. For example, businesses can determine how many clerks they need to service their customers and at what time of day the clerks are needed most.

Here is a simple example showing how the average number of customers in a line can be estimated.

Suppose 6 customers arrive at a counter during a period of 10 minutes and that it takes 2 minutes for each one to be served. We will represent the customers by the letters A, B, C, D, E, and F.

Here is a list showing at what time (the beginning of the minute) each of the customers arrives.

Time	Customer	Time	Customer
1	A	6	E
2	B	7	
3	C, D	8	
4		9	
5		10	F

To find the average number of customers in the line each minute, we make the following diagram showing the line during each minute:

		D	D		E				
	B	C	C	D	D	E	E		F
A	A	B	B	C	C	D	D	E	E
1	2	3	4	5	6	7	8	9	10

This diagram shows that during the first minute customer A is standing at the counter, during the second minute customer B is standing behind A, during the third minute B is at the counter (A is gone since it takes 2 minutes to be served) and customers C and D are waiting behind him, and so forth. From the diagram we see that the most frequent number of people in line, or the mode average, is 2. The mean average is $21/10 = 2.1$.

Suppose the arrival list of the customers looks like this:

Time	Customer	Time	Customer
1	A	6	
2		7	D
3	B	8	
4	C	9	E
5		10	F

1. Make a diagram similar to the one above to show the line during each minute.

2. What is the mode number of customers in line?

3. What is the mean number in line?

Suppose the arrival list of the customers looks like this:

Time	Customer	Time	Customer
1		6	
2	A	7	F
3		8	
4	B, C, D	9	
5	E	10	

4. Make a diagram showing the line during each minute.
5. What is the mode number of customers in line in this situation?
6. What is the mean number in line?

Set III

Experiment. Rolling a Die to Get All Six Numbers

Can you guess how many times a die must be rolled, on the average, before all 6 numbers come up at least once?

To find out, take a die and try it. A convenient way to keep track of the numbers that appear is with a frequency distribution like the one shown here. Make tally marks in a column until there is at least one for each number. Then write the total number of tosses at the bottom of the column. Use a new column to record the tally marks for the next trial and continue in the same way until you have completed 25 trials in all.

Number on die	Trial 1	Trial 2	Trial 3	
1	I	𝍸	II	
2	II	III	I	
3	III	I	II	
4	I	𝍸 I	I	
5	IIII	I	I	
6	II	IIII	II	
	13	20	9	etc.

1. Now make a frequency distribution of the results; label the first column "Number of tosses," and list the number of tosses made in each of the 25 trials.
2. Use the frequency distribution you made in exercise 1 to determine the median number of tosses necessary in order to get all 6 numbers.
3. Now calculate the mean number of tosses necessary.

"Oh! Oh!"

Lesson 4

Measures of Variability

The basketball team from Dribble High has arrived for their game with Wembly and, as you can see, their prospects of winning look pretty good. Although Wembly's coach had been told that the average height of the eleven players on the Dribble team was 5 ft. 10 in., he didn't know anything about Joe Dunkshot, their most promising player. This is the first game of the season and Dribble's coach had been keeping Joe a secret.

It is apparent that an average, by itself, does not tell everything that someone might want to know about a set of numbers. In addition to a measure of central tendency, it is often helpful to have some information about how the numbers vary. A simple "measure of variability" is called the *range*. The players on the Dribble team vary in height from Joe's extraordinary 7 ft. 3 in. (or 87 in.) to a very short 5 ft. 3 in. (or 63 in.). Their range in height is

$$87 \text{ in.} - 63 \text{ in.} = 24 \text{ in.}$$

► The **range** in a set of numbers is the difference between the largest and the smallest numbers in the set.

Here is a list of the heights, in inches, of the eleven members of the Dribble team, in order from shortest to tallest:

$$63 \quad 65 \quad 66 \quad 68 \quad 68 \quad 70 \quad 70 \quad 70 \quad 70 \quad 73 \quad 87.$$

Notice that the range, 24, is determined by only 2 numbers, and does not give us any information about how the other numbers in the set vary.

Another "measure of variability" is the *standard deviation.* To find the standard deviation in heights of the members of the Dribble team, we first need to know the mean average of the numbers. It is 70. Next, we find the difference between each of the heights and the mean average. The differences are squared and their mean average found. Finally, we take the square root of this average. This number, 6, in the case of the basketball players, is called the standard deviation.

Height	Difference from mean	Square of difference
63	7	49
65	5	25
66	4	16
68	2	4
68	2	4
70	0	0
70	0	0
70	0	0
70	0	0
73	3	9
+ 87	17	+ 289
770		396

$$\frac{770}{11} = 70 \qquad \frac{396}{11} = 36 \quad \sqrt{36} = 6$$

► The **standard deviation** in a set of numbers is determined by finding:

1) the mean average of the numbers,
2) the difference between each number in the set and the mean average,
3) the squares of these differences,
4) the mean average of the squares,
5) the square root of this average.

Although the standard deviation may seem like an unnecessarily complicated number to figure out, it is also a very useful number to know.

EXERCISES

Set I

The heights of the ten members of Wembly High's basketball team, in inches, are:

64 66 66 69 70 70 71 71 75 78.

1. What is the range in heights of the Wembly team?

2. Find the mean average height.

To find the standard deviation in heights, do each of the following things:

3. Make a list of the differences between each height and the mean average. (There should be 10 numbers in your list; the difference between the mean average and a number that is equal to the mean average is listed as 0.)

4. Make a list of the squares of these differences. (Again you should have 10 numbers.)

5. Find the mean average of the squares.

6. The standard deviation is the square root of this average. What is it?

7. What percent of the 10 heights are 1 standard deviation or less from the average?

8. What percent are 2 standard deviations or less from the average?

Suppose the shortest player on the Wembly team were replaced with a boy 2 inches shorter and the tallest player were replaced with one 2 inches taller.

9. Now what would be the range in heights of the Wembly team?

10. What effect do the 2 replacements have on the average height?

11. Do you think this has any effect on the standard deviation from the average?
    If, so, in what way?

Set II

Here are some examples to show why the standard deviation is worth determining.

The president of the Sleepy Hollow Grandfather Clock Company wants to find out how accurately his clocks keep time. One thousand of the clocks are set to the exact time and one week later checked to see how many minutes from the correct time each one is. A graph of the clock errors is made which looks like this:

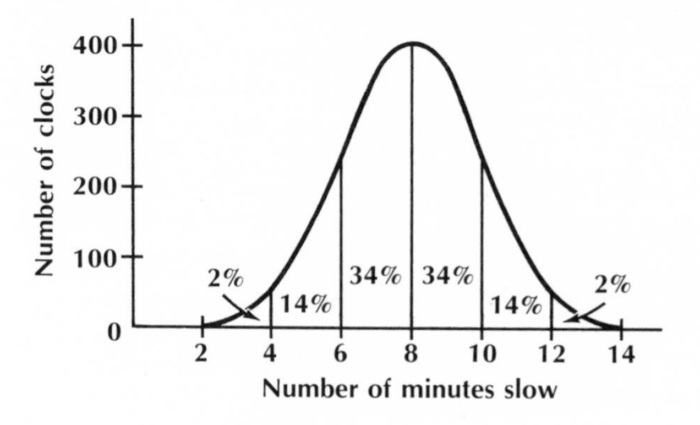

As you can see from the graph, Sleepy Hollow grandfather clocks are always slow. The graph is a bell-shaped curve and is called a **normal curve.** The vertical line in the center of the curve marks the average clock error.

1. What is the average error?

The standard deviation of the errors is 2 minutes, and, as you can see, vertical lines have been drawn across the graph at 2-minute intervals.

2. What percent of the clocks are within 1 standard deviation of the average? (In other words, what percent of the clocks are between 6 and 10 minutes slow?)
3. How many of the 1,000 clocks are within 2 minutes of the *average error*?
4. What percent of the clocks are within 2 standard deviations of the average?
5. How many of the 1,000 clocks are more than 10 minutes away from the *correct time*?

A psychologist gives a puzzle to 500 people to solve as part of an intelligence test. Each person is timed and a graph of the times looks like this:

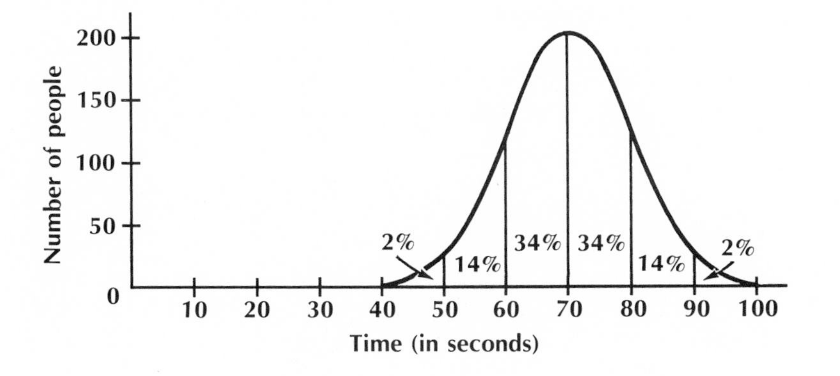

This graph is also a *normal curve.*

6. What is the average time taken to solve the puzzle?
7. Again vertical lines have been drawn across the graph to represent the standard deviations from the average. What is the standard deviation?
8. What percent of the people are within 1 standard deviation of the average?
9. How many of the 500 people is this?
10. What percent of the people are within 2 standard deviations of the average?
11. What percent of the people solved the puzzle in less than 1 minute?
12. About how many people of the 500 took more than $1\frac{1}{2}$ minutes to solve the puzzle?

Jack Beanstalk has a large farm on which he raises beans. As part of a scientific survey, the United States Department of Beans sends out some inspectors to measure each one of Jack's 10,000 plants. A graph of the heights of the beanstalks is shown at the top of the following page.

This graph has the same basic shape as the other graphs we have seen.

13. What is this curve called?

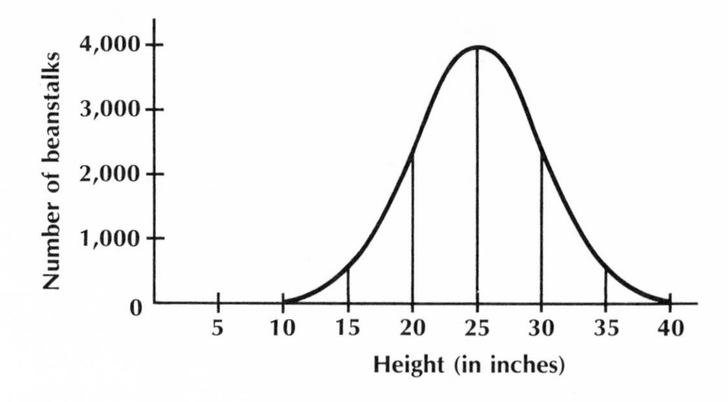

Many sets of measurements have a graph like this and that is why the standard deviation is such a useful number to know. In any set of numbers whose graph is a normal curve, 68% are within 1 standard deviation of the average, while about 96% are within 2 standard deviations of it.

14. What is the standard deviation from the average of the heights of the beanstalks?
15. What percent of the beanstalks do you think have a height between 20 and 30 inches?
16. What percent are more than 35 inches tall?

Set III

The curves below show the variations in the weights of quarters that were minted and put into circulation at the same time.

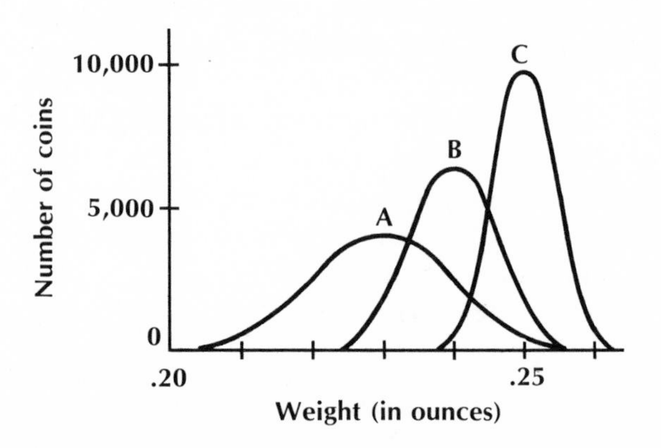

One curve shows the weight distribution when the coins were new and the other two show the distributions when they had been circulated for 5 years and for 10 years.

1. Which curve do you think shows the weights of the newly minted quarters, which curve the coins after 5 years, and which curve the coins after 10 years?
2. What happens to the average weight of the coins as time goes by?
3. What happens to the standard deviation in weight of the coins as time goes by?

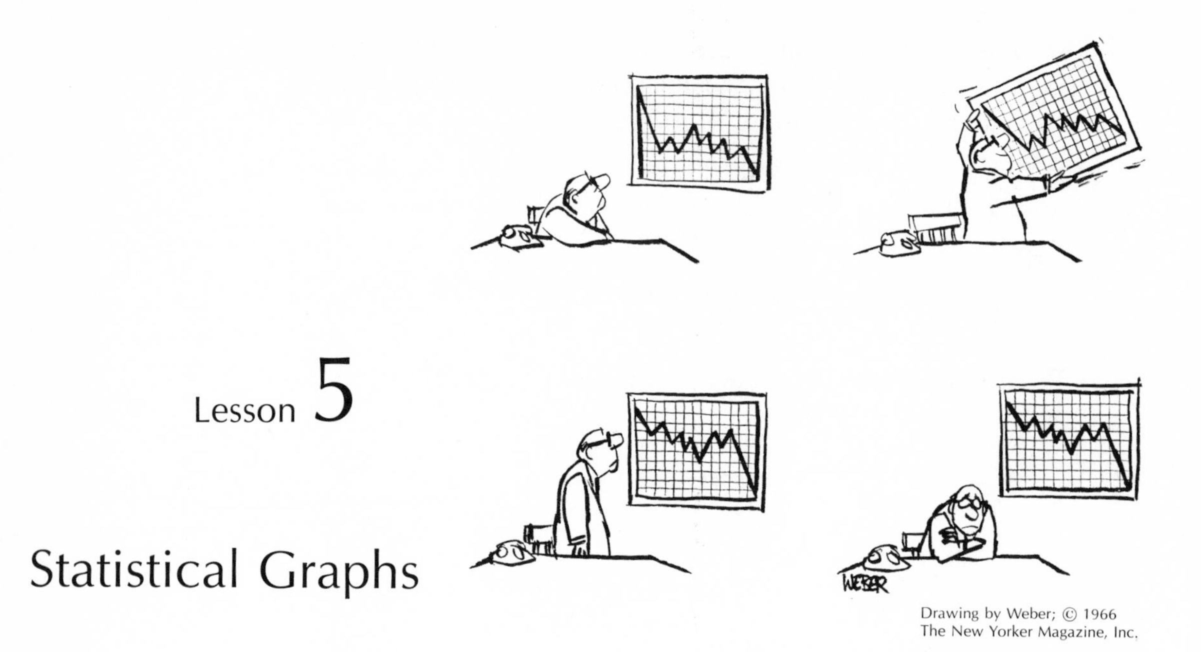

Lesson 5

Statistical Graphs

Drawing by Weber; © 1966
The New Yorker Magazine, Inc.

THIS businessman has been feeling rather discouraged by the appearance of the graph on his wall. He has decided to turn it upside-down, thinking that it would look better. Instead it looks just as bad; perhaps worse!

The advantage of a statistical graph over a table of numbers is that it can provide information at a glance. For this businessman, however, this seems to be a disadvantage instead.

There are ways to change the appearance of a graph other than by turning it upside-down, and some of them can be used to make a situation "look better" than it really is. For example, suppose that the man in the cartoon is the circulation manager of a national magazine. The magazine comes out once each month and a table showing the number of copies sold of each issue during the past year looks like this:

Month	Circulation	Month	Circulation
January	502,365	July	521,870
February	505,712	August	523,685
March	508,480	September	525,009
April	512,307	October	529,241
May	516,798	November	532,765
June	519,263	December	535,014

This table looks too complicated with so many numbers. The manager would like to show how the circulation has been growing in a simpler way, so he decides to have a member of the magazine's art department draw a graph.

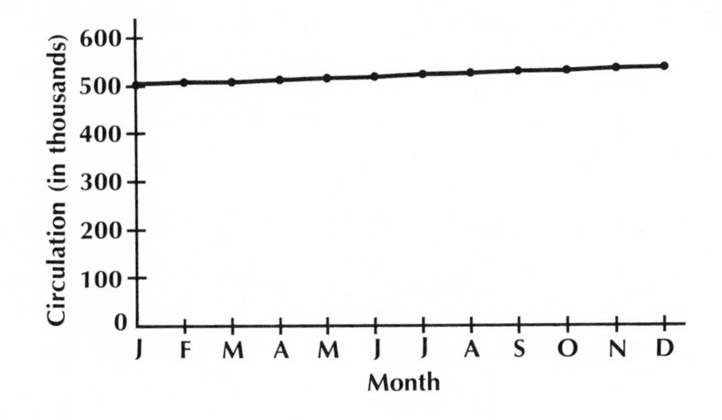

The manager had planned to send the graph to the companies that might advertise in the magazine to show them how the circulation, and thus the number of prospective customers, has been growing. Unfortunately, however, the graph does not look very impressive. He asks the artist if he can't do any better, and after several attempts, he comes up with this graph!

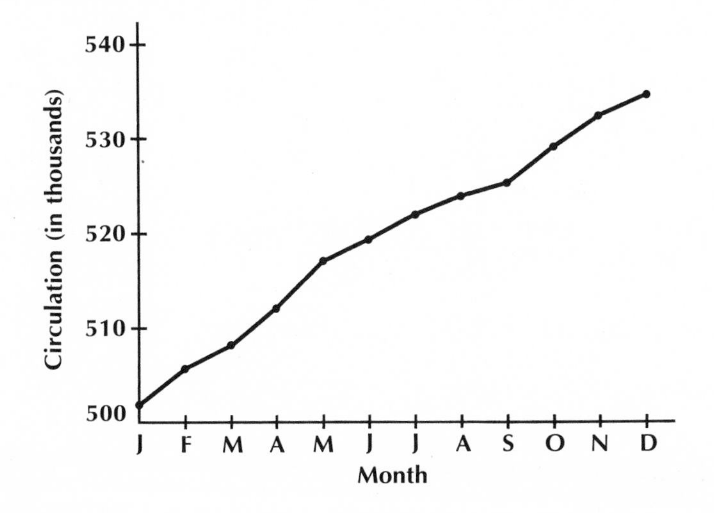

At first glance, it doesn't seem possible that both graphs could represent the same numbers! The new graph appears to show that the circulation is growing very rapidly. How could anyone help but be impressed?

Why do the graphs look so different? The scale on the side has been changed so that it no longer starts with 0 but with 500 instead, and the numbers have been spread much farther apart. The circulation manager tells the artist to erase these numbers, so that it is no longer possible to tell how the graph has been changed.

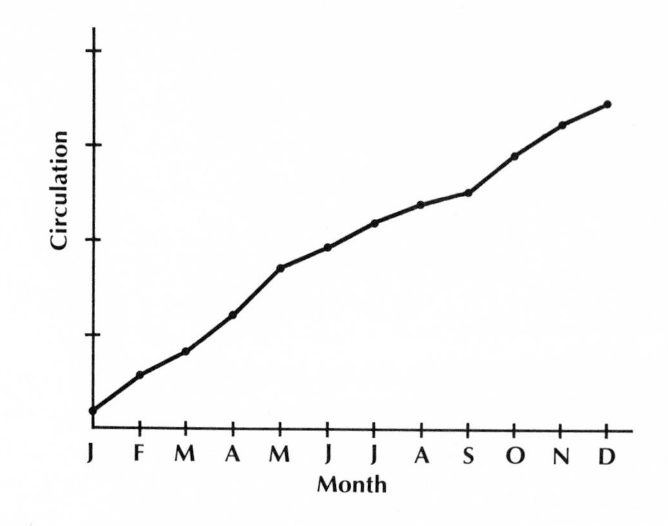

Do you see that the graph is now almost meaningless? It is still clear that the magazine's circulation has been increasing, but it is no longer possible to tell by how much. To be sure, the companies that might advertise in the magazine would not be fooled by this trick and would want to know what the missing numbers were. The moral of this story, however, is that although graphs are often used to present numerical facts, because they can provide a simple picture of the patterns in a set of numbers, they can also be deliberately drawn in a certain way to give a false impression. And even a graph that is meant to give an honest representation may be misinterpreted.

EXERCISES

Set I

The following tables give the normal low and high temperatures in San Francisco and Salt Lake City for each month of the year.

San Francisco			Salt Lake City		
	Low	High		Low	High
Jan.	46	56	Jan.	18	37
Feb.	47	59	Feb.	23	42
Mar.	49	61	Mar.	29	52
Apr.	50	62	Apr.	36	63
May	51	63	May	44	74
June	53	65	June	51	84
July	53	64	July	60	94
Aug.	54	65	Aug.	58	91
Sept.	55	69	Sept.	49	80
Oct.	54	68	Oct.	38	65
Nov.	51	64	Nov.	26	48
Dec.	47	58	Dec.	21	39

1. Make a line graph for each of these cities to show how the low and high temperatures vary during the year. Let the temperature scales range from zero to 100. The temperatures in San Francisco for the months of January, February, and March are shown on this graph.

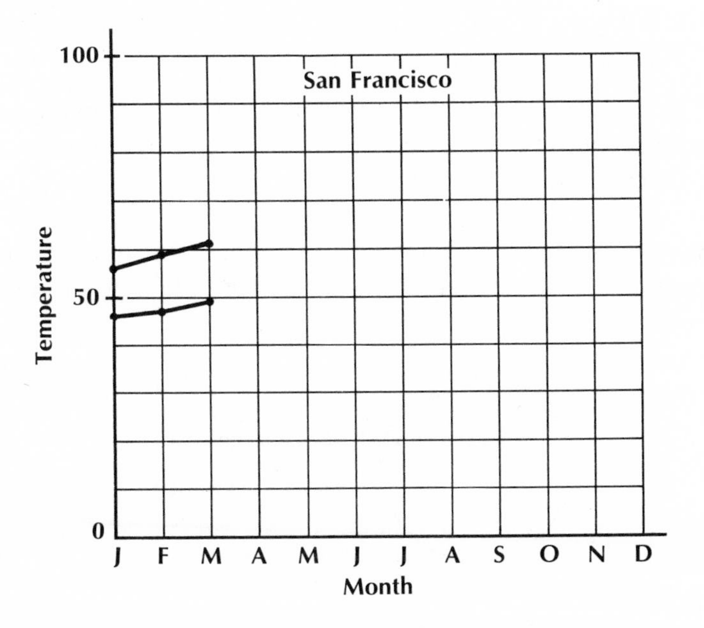

2. What advantage do your line graphs have over the tables of numbers in presenting the information about the temperature variations in the two cities?

3. Now draw the San Francisco graph again, using a temperature scale ranging from 45 to 70.

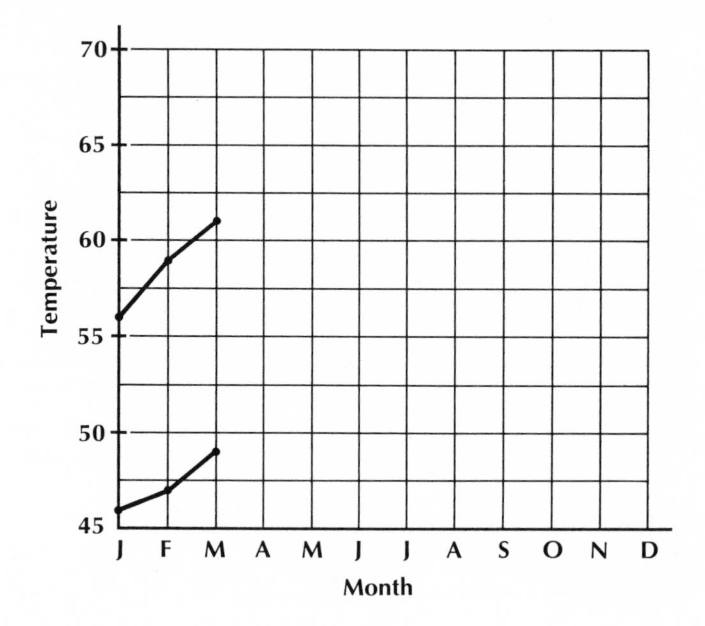

4. Compare the second San Francisco graph with the Salt Lake City graph. Which city seems to have more extreme variations in temperature now?

5. If you want to be sure of drawing correct conclusions about how data presented in two different graphs compare, what should you look at first?

Did you know that the Federal Government owns more than half of the state of Oregon? Here is a circle (or "pie") graph showing this:

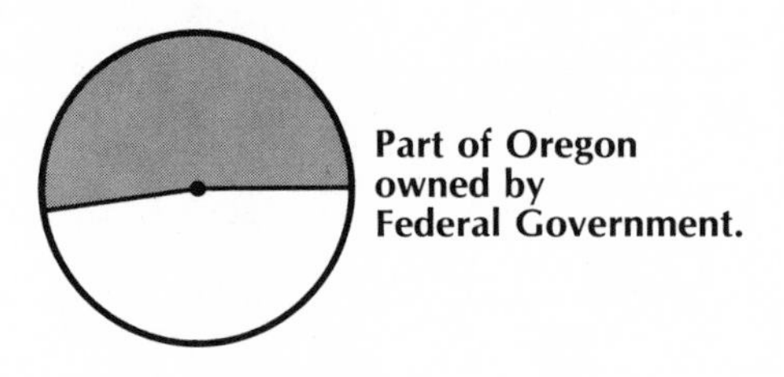

In drawing this graph, the following information was needed:

52% of Oregon is federally owned.

52% = 0.52.

The sum of the measures of all the angles surrounding a point (the center of the circle) is 360°.

$$0.52 \times 360° = 187° \text{ (rounded off).}$$

The Federal Government owns 64% of the land in Idaho and 44% of the land in California.

6. What is 64% of 360° (to the nearest degree)?
7. Draw a circle graph to show the amount of federally owned land in Idaho.
8. What is 44% of 360°?
9. Draw a circle graph to show the amount of federally owned land in California.
10. These graphs appear to show that the government owns more land in Idaho than in California, yet it is actually the other way around. Can you explain how this can be?

Set II

Bar graphs, like the one below, are often used in ads.

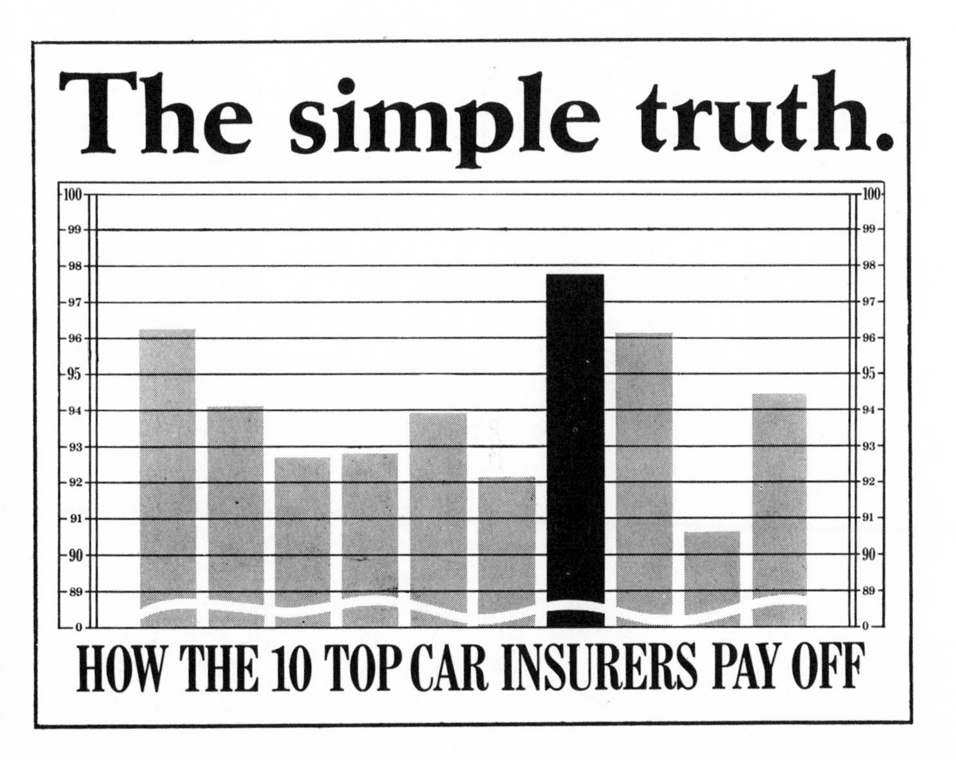

1. Does the bar graph really show the simple truth?
2. Approximately how does the shortest bar in this graph compare in height to the tallest one?
3. Make a list of the "pay-off" numbers shown for the 10 companies, rounding each ratio off to the nearest whole number. Reading from the left, the first 3 figures are: 96, 94, and 93.
4. Redraw the graph to show a more honest picture of the situation. Use graph paper and let each quarter inch represent 10 units on a scale from 0 to 100, so that the graph will have a height of $2\frac{1}{2}$ inches.
5. Do you think an insurance company that used this graph would want to use your graph?

By permission of Johnny Hart and Field Enterprises, Inc.

Set III

Peter, through a variety of means,* has twice as many clams as B.C. One way to show this is with a picture graph.

Peter's bag of clams has been drawn so that it is twice as tall as B.C.'s bag.

1. Does the larger bag look like it contains exactly twice as many clams as the smaller one?

Instead of comparing the heights of the two bags, let's compare the *areas* covered by them. Notice that the larger bag is twice as wide as the smaller one, in addition to being twice as tall.

2. How do you think their areas compare?

Hint: 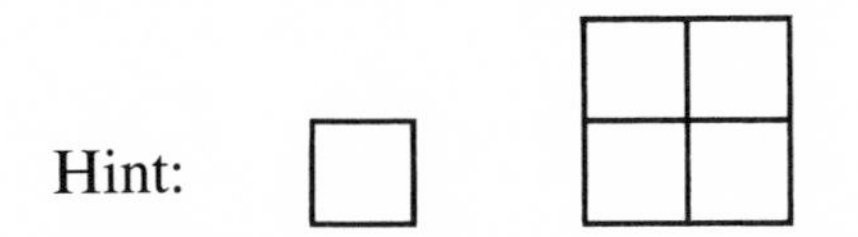

The number of clams each bag contains is not determined by the area each bag covers, but by its *volume*.

3. How do you think their volumes compare?

Hint: 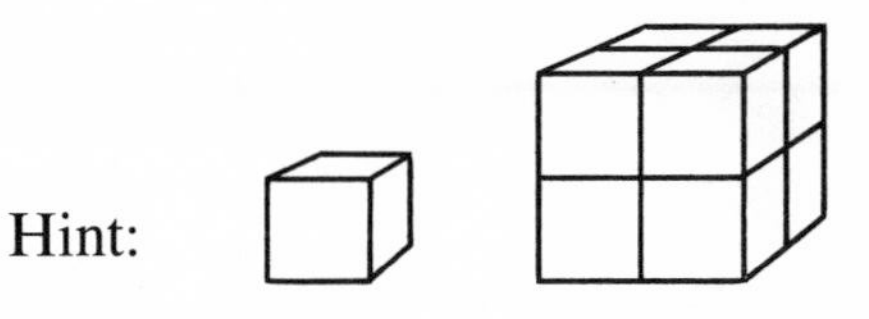

---

* See also the cartoon on page 358.

4. There is another way to draw this picture graph, also using bags of clams, which makes it perfectly clear that Peter has just twice as many clams as B.C. What is it?

5. What are the missing words in the following statement?

"Some picture graphs are easy to misinterpret because it is not clear whether heights or ________ or ________ should be compared."

Lesson 6

Sampling

HOW many stars are there in the sky? If you have ever looked at the sky on a clear moonless night, you may have the idea that there are so many stars no one could possibly count them. Actually, the number of stars that can usually be seen by the unaided eye at one time is less than 3,000. If you look through a pair of binoculars, the number increases to about 50,000, while a large telescope can extend the number to as many as 500,000,000. This number is so large that, even if you were to spend all night every night counting stars, it would take more than a lifetime to do it!

How is it possible, then, for astronomers to figure out such a large number as this? They do it by taking a *sample.* The entire sky is divided into small sections and the stars in a number of these sections in different places in the sky are counted. On the basis of these numbers, the approximate number of stars is determined.

The sampling process is the most important part of statistics.

► A **sample** is a relatively small group of items chosen to represent a larger group called the *population.*

In order for the sample to accurately represent the population, it must be *sufficiently large.* If the sample is too small, there is a good chance that the items it includes may not really be typical of the population. What would you think of the claim that 4 out of 5 doctors recommend Dr. Quack's Pep Pills if you knew that the Quack Company's sample consisted of exactly 5 doctors out of the more than 300,000 doctors that live in the United States? Statisticians have methods of determining just how large a sample should be to be reasonably dependable.

It is also important that the sample be a *random* one. In other words, the items in it should be chosen completely by chance. Unless the probability of every item in the population being included in the sample is the same, the sample may give a distorted picture of the population it is supposed to represent. Suppose Senator Fred Filibuster is wondering how likely it is that the citizens of his state will re-elect him. Would it be a good idea for him to send out questionnaires to 100 of his best friends in order to find out?

EXERCISES

Set I

There are a number of reasons for taking a sample instead of making a survey of an entire population. What do you think is the main reason for using a sample in each of these situations?

1. The Yankee Doodle Fireworks Company needs to know whether or not the firecrackers it is producing will explode. It tests one out of every 500 in order to find out.

'I can't say that I go for this kind of TV poll . . .'

2. The Nielsen television rating service determines the popularity of national programs by a sample of 1,200 homes out of the approximately 60,000,000 homes with T.V.

What do you think could cause the results in each of the following situations involving sampling to be inaccurate?

3. The Madame Whiffit Perfume Company wants to determine the average age of the women who buy its products. Madame Whiffit sends a number of representatives to various stores to ask the customers their ages.

4. In order to predict which candidate will win a presidential election, a national magazine sends ballots to all of its readers.*

5. Dr. Fetherbrane, a psychiatrist, wants to find out how many people dream in color. He asks 100 of his patients whether they do or not. (Hint: Do you dream in color?)

Set II

Two psychologists in Los Angeles recently did an experiment to find out how honest most people are. They used 375 envelopes that were addressed to one of the psychologists at his home. Of these envelopes:

75 were empty and had typed on them: "This is a research study. Drop this envelope in the nearest postbox. Thank you for your cooperation."

150 contained some blank sheets of folded paper, but did not have any message typed on the outside.

150 appeared to contain some money (2 coins and a bill that were actually fake).

The psychologists dropped each envelope on a sidewalk near a mailbox. A third of the envelopes were left in wealthy sections of the city, a third in middle income areas, and a third in poor neighborhoods. Each one was marked on the inside to show the area in which it had been left.

Of the 75 empty envelopes that were labeled "This is a research study," 68 were returned by the people who found them.

*A magazine called the *Literary Digest* did something similar to this in 1936 and predicted on the basis of their poll that Landon would defeat Roosevelt. Roosevelt was re-elected by a landslide. The magazine went out of business the following year.

Expressed as a percent, this is

$$\frac{68}{75} \times 100 = \frac{6{,}800}{75} = 91\%.$$

1. 120 of the 150 envelopes that contained blank sheets of paper were returned. What percent of these envelopes is this?
2. 102 of the 150 envelopes that appeared to contain some money were returned. What percent of these envelopes is this?

One of the purposes of the experiment was to find out how many people would do something dishonest, such as taking money that didn't belong to them, if there was very little chance of being caught.

3. What, then, do you think was the purpose of using the empty envelopes that were labeled "This is a research study" in the experiment?
4. A few of the envelopes that contained either blank sheets of paper or the fake money were opened and resealed by the people who found them before they were returned. Does this matter? How do you think these envelopes should be counted?

Here is a table showing the number of each type of envelope that were returned from each area.

	Area of City		
Kind of envelope	Poor	Middle	Wealthy
Empty envelopes labeled "research study"	22	24	22
Envelopes containing blank sheets of paper	43	37	40
Envelopes appearing to contain money	28	33	41

5. Can you draw any conclusions from the numbers shown in this table?
6. What percent of the envelopes that appeared to contain money were returned from each area? (There were

150 of these envelopes and a third of them were left in each area.)

7. Judging from the results of this experiment, do you think that it is appropriate to conclude that 56% of the people in the poor areas of the city are honest when it comes to returning envelopes that seem to contain money? Explain your answer.

Set III

Experiment. A Study in Sample Size

In this experiment, you will get some idea of how the number of items included in a sample is related to how accurately the sample represents the population.

You will be given a small box* that contains a large number of beads, some of which are black and the rest white. To determine approximately what percent of the beads are black, you will first take a sample containing 5 beads. Then you will look at 5 more to increase the size of the sample to 10 beads, then 5 more to increase it to 15 beads, and so on, until 100 beads have been observed.

In order that the results of the experiment will be similar to those for a much larger population than what the box contains, each bead will be put back with the others as soon as its color has been recorded. In fact, it will not be necessary to open the box during the experiment at all.

Shake the box and look into the window to see if the bead that appears there is black or white. Do this 4 more times so that you have observed 5 beads for your original sample.

1. Record the results in a table like the one on page 436. Let B stand for black and W stand for white. The table shows

---

*A sample box that is easy to use can be made from a small cardboard salt shaker. Remove the bottom of the shaker and pour out the salt. Put 50 small beads of two different colors (for example, 20 black and 30 white) in the shaker and replace the bottom. Rotate the top of the shaker so that when the shaker is turned upside down, a bead can be seen through the largest opening.

A paper cup can also be used as a container, with buttons or poker chips in place of beads. Cut a small window in a cardboard disk and tape the disk over the top of the cup.

that in one experiment the first 5 beads observed were black, white, white, black, and white.

Example of Table of Results

Sample	Result	Total no. of black beads obs.	Total no. of beads obs.	Percent of black beads
1	BWWBW	2	5	40%
2	WBWWW	3*	10*	30%
3	WWWWW	3	15	20%
4	WBBWW	5	20	25%

Take 20 samples, so that you have observed 100 beads altogether. Complete your table, following the example shown above.

2. Then plot the results of your samples on a graph like this one.

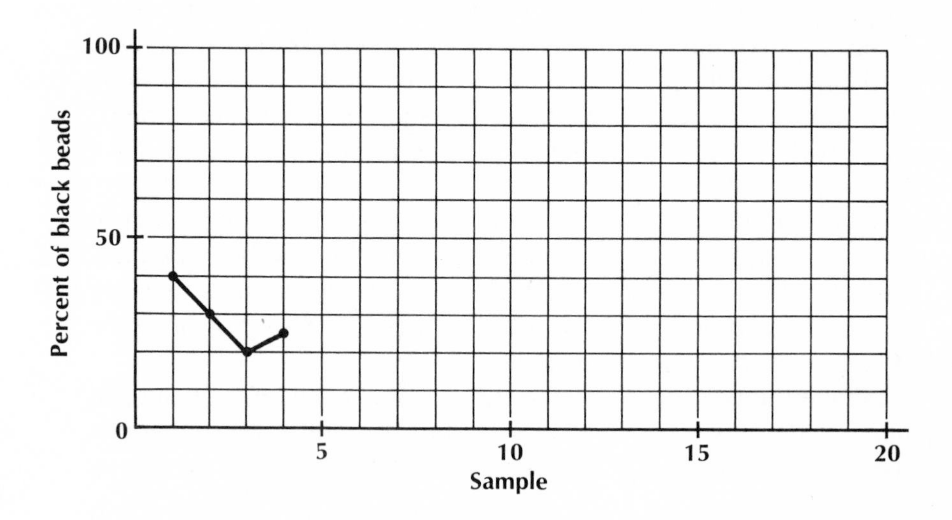

3. What do you notice about the shape of your graph as you look from left to right?

4. On the basis of your graph, make an estimate of the percent of black beads in your box.

If there is time, repeat the experiment with another box and compare the results.

---

*Each successive sample includes the results of the previous ones.

INTERESTING READING

Sampling (originally titled *Sampling in a Nutshell*), by Morris Slonim Simon and Schuster, 1960.

An interesting and witty book. Its author says concerning the people who should be able to understand it: "Those whose number sense is sufficiently keen to enable them to count all the way to ten without removing their mittens should be able to manage."

Chapter 9/Summary and Review

In this chapter we have become acquainted with the branch of mathematics that deals with the collection, organization, and interpretation of numerical facts. We have studied:

Frequency distributions *(Lessons 1 & 2)* A frequency distribution is a convenient way of organizing data to reveal what patterns it may have. It can be condensed by grouping together the data in intervals.

Measures of central tendency *(Lesson 3)* Three types of averages are commonly used:

The *mean* of a set of numbers is found by adding them and dividing the result by the number of numbers added.

The *median* is the middle number in a set of numbers arranged in order of size (or, if there is no middle number, the number halfway between the two middle numbers.)

The *mode* is the number that occurs most frequently in the set.

Measures of variability *(Lesson 4)* The *range* in a set of numbers is the difference between the largest and smallest numbers in the set.

Another measure of variability is the *standard deviation* from the mean average.

The *standard deviation* in a set of numbers is determined by finding: (1) the mean average of the numbers, (2) the difference between each number in the set and the mean average, (3) the squares of these differences, (4) the mean average of the squares, and (5) the square root of this average.

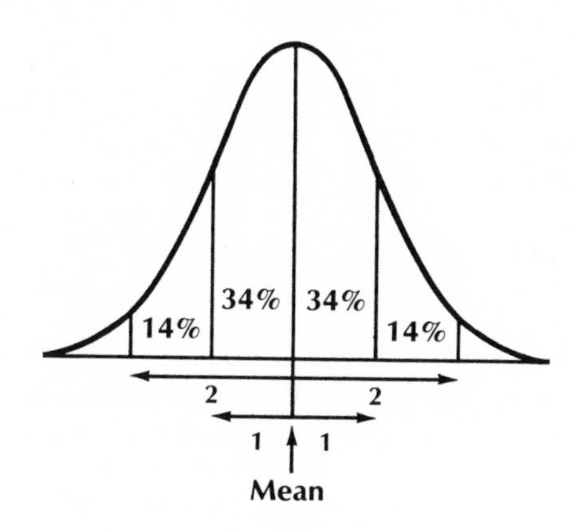

The graph of many sets of numbers is a *normal curve,* which looks like this. This graph shows that 68% of the numbers are within 1 standard deviation of the mean and that about 96% are within 2 standard deviations of it.

Statistical graphs *(Lesson 5)* Statistical graphs are sometimes drawn in a way that gives a false impression. In bar and line graphs, it is especially important to look at the numbering of the scales. Picture graphs may be misunderstood if it is not clear whether heights or areas or volumes should be compared.

Sampling *(Lesson 6)* A sample is a relatively small group of items chosen to represent a larger group called the population. For a sample to be meaningful it is important that it be sufficiently large and as random as possible.

EXERCISES

Set I

It is easy to get the impression from a map of the United States that our country is "flat," yet the average elevation of the states above sea level varies from less than 100 feet for Delaware to nearly 7,000 feet for Colorado. Here is a table listing the mean elevations of the 50 states, each to the nearest 100 feet.

1. Alabama, 500
2. Alaska, 1900
3. Arizona, 4100
4. Arkansas, 700
5. California, 2900
6. Colorado, 6800
7. Connecticut, 500
8. Delaware, 100
9. Florida, 100
10. Georgia, 600
11. Hawaii, 2000
12. Idaho, 5000
13. Illinois, 600
14. Indiana, 700
15. Iowa, 1100
16. Kansas, 2000
17. Kentucky, 800
18. Louisiana, 100
19. Maine, 600
20. Maryland, 400
21. Massachusetts, 500
22. Michigan, 900
23. Minnesota, 1200
24. Mississippi, 300
25. Missouri, 800
26. Montana, 3400
27. Nebraska, 2600
28. Nevada, 5500
29. New Hampshire, 1000
30. New Jersey, 300
31. New Mexico, 5700
32. New York, 1000
33. North Carolina, 700
34. North Dakota, 1900
35. Ohio, 900
36. Oklahoma, 1300
37. Oregon, 3300
38. Pennsylvania, 1100
39. Rhode Island, 200
40. South Carolina, 400
41. South Dakota, 2200
42. Tennessee, 900
43. Texas, 1700
44. Utah, 6100
45. Vermont, 1000
46. Virginia, 1000
47. Washington, 1700
48. West Virginia, 1500
49. Wisconsin, 1100
50. Wyoming, 6700

1. Make a frequency distribution of these elevations by grouping them together in intervals of 500 feet. Number the first column in your table like this:

Elevation
0-500
600-1000
1100-1500
1600-2000,

and so on. (The last line should read 6600-7000.)

2. What percent of the states have average elevations of 500 feet or less?

3. What percent are, on the average, more than a mile high? (1 mile = 5,280 ft.)

4. It isn't necessarily correct to conclude on the basis of your frequency distribution that the average elevation in the United States is about 1,000 feet. Why not?

Photograph from the Mount Wilson and Palomar Observatories.

The planet Jupiter
and one of its moons.

Did you know that the earth is the only planet in our solar system that has exactly one moon? Here is a list of the planets and the number of moons of each.

Planet	Number of moons
Mercury	0
Venus	0
Earth	1
Mars	2
Jupiter	12
Saturn	10
Uranus	5
Neptune	2
Pluto	0

5. Find the mean average number of moons per planet.
6. Arrange the 9 numbers in order and find the median number of moons per planet.
7. What is the mode number of moons per planet?
8. Which one of these three numbers do you think best represents the typical number of moons per planet? (Name it.)

Set II

There is a legend that many years ago someone wanted to find out how tall the emperor of China was. To ask to measure the emperor was out of the question, so this person decided to take a poll of a million of his countrymen to learn what they thought the height of the emperor was.

After the poll was completed, the investigator based his guess on the mean average height named, even though not one of the people asked had ever seen the emperor! A graph of the guesses might have looked like this.

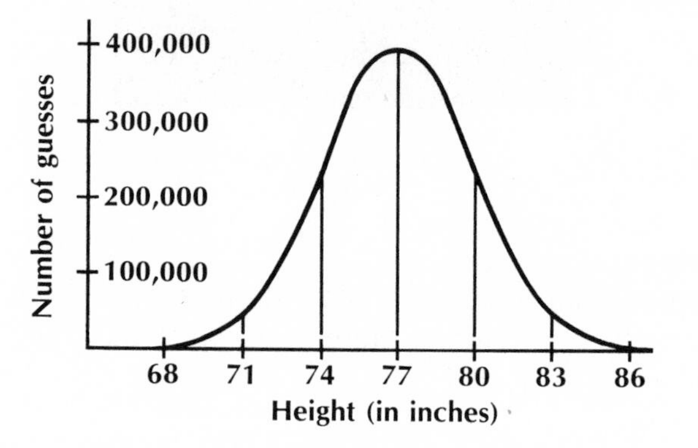

1. What kind of curve does this appear to be?

2. What is the average guess of the emperor's height, expressed in feet and inches?

3. What is the standard deviation from the average?

4. What percent of the guesses do you think are within 1 standard deviation of the average?

5. What percent of the guesses are more than 2 standard deviations from the average?

A drug company included a graph like the one shown on the facing page in one of their ads to show that their product acts two times as fast as aspirin.

6. What is missing from the graph?

7. Do you think this matters with respect to the comparison the graph is intended to show between the relative effectiveness of the drug and aspirin?

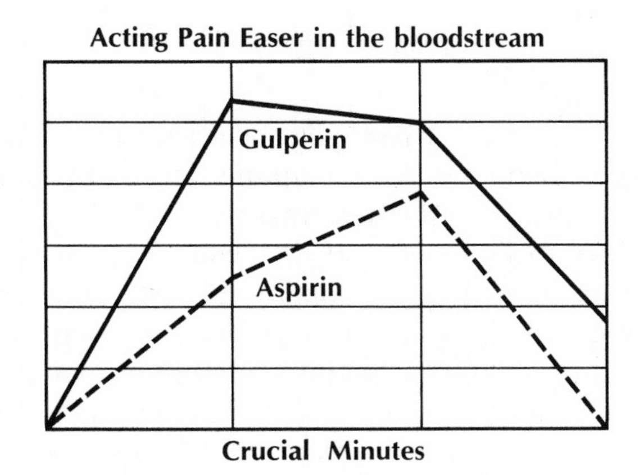

8. Can you tell from the graph how long after a person takes the drug it is until the drug has its strongest effect?
9. Is it possible to tell from the graph how long the effects of the drug will last?
10. Is there anything strange about the *shape* of the "curves" in this graph?
11. One problem in public opinion polls is that a number of people are usually "undecided." Suppose a poll is taken just before a local election. If the people who are "undecided" are *not counted* in the results of the survey, something is being assumed when the poll is used to predict the way the people of the community will vote. Do you think it is that the "undecided" people will either vote

Drawing by Dennis Renault; © 1966 Saturday Review, Inc.

in the same proportions as the decided people do or not vote at all?

12. Many airline passengers take trips that involve more than one company. For example, in taking a trip from Los Angeles to Miami, someone might fly on American Airlines to Houston and on Delta Airlines from Houston to Miami. Just one company may be involved in more than 100,000 of these transfers in a single month.

 Instead of deciding the amount of money that should be exchanged between each company on the basis of all the coupons from these transfers, the airlines take a sample of only about 10% of them and throw the rest away.

 Since the information from these samples is not as exact as that which could be obtained from all of the coupons, why do you think the airline companies use them?

By permission of Johnny Hart and Field Enterprises, Inc.

Set III

Suppose a track star has a race with a champion roller skater. How do you think their speeds would compare?

 The table at the top of the facing page shows the world record times for three different distances.

Distance	Roller skater	Runner
440 yds.	35 sec.	45 sec.
880 yds.	1 min. 13 sec.	1 min. 45 sec.
1,760 yds. (1 mi.)	2 min. 22 sec.	3 min. 51 sec.

Below are 3 bar graphs comparing the times for the distances.

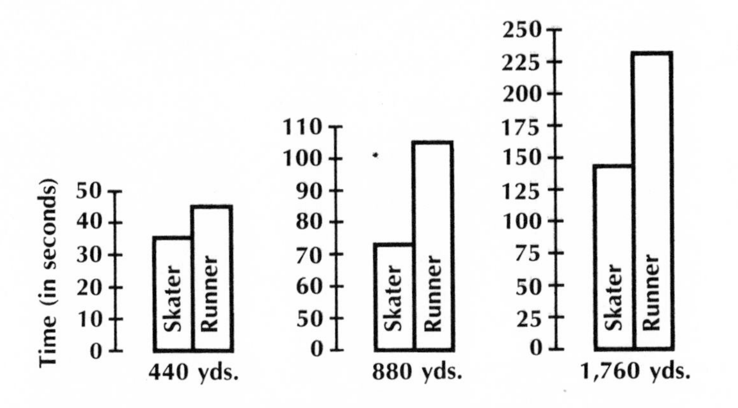

1. Which one of these graphs is misleading? Why?
2. What should be noticed in comparing the graphs for 440 yards and 1,760 yards?
3. Redraw the 3 bar graphs so that they give a clearer picture of how the times for the different distances compare.

INTERESTING READING

How to Lie with Statistics by Darrell Huff, Norton, 1954.

Chapter 10

SOME TOPICS IN TOPOLOGY

Lesson 1

The Mathematics of Distortion

HAVE you ever seen someone doing shadow tricks? The bird in this cartoon seems to be very good at stretching and bending itself into different shapes, but has suffered a crash landing after forming a pretzel. It will probably surprise you that the bird's maneuvers would be of any interest to a mathematician, yet a new and rapidly growing branch of mathematics deals with this very topic: the ways in which figures can be bent and stretched without changing certain basic properties. This kind of mathematics, sometimes called "the mathematics of distortion," is known as **topology.**

Here is an example of a property that is of interest in topology.

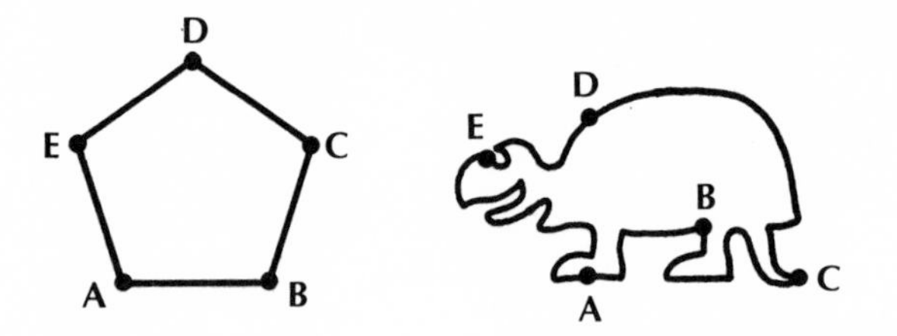

What do these drawings of a regular pentagon and a tortoise have in common? They are both *simple closed curves,* since you can start at any point of either figure and travel over every other point of the figure exactly once before returning to where you began.

Imagine that the pentagon is formed by a large rubber band

By permission of Johnny Hart and Field Enterprises, Inc.

that is stretched around five pegs and that there is a knot in the band at each peg. If the band is removed from the pegs and stretched into some other shape, say that of the tortoise, the distances between the knots may change, the area inside the band may become larger or smaller, and angles may even disappear completely. These properties involving measurements that change when the figure is distorted are not important in topology.

Another property does *not* change, however, and this is the *order* of the points along the figure. As long as the band is not broken or tied together in some new point, the order of the knots will remain unchanged—for example, knot B will always be between knots A and C—and this property is said to be a *topological* one.

EXERCISES

Set I

► If two figures can be twisted and stretched into the same shape *without* connecting or disconnecting any points, they are said to be **topologically equivalent.**

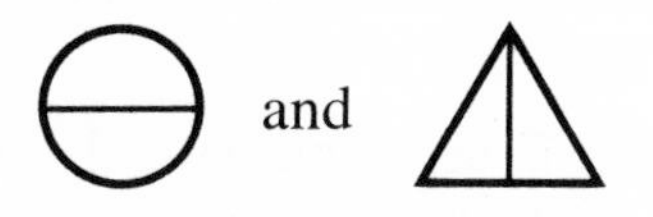

are topologically equivalent, since

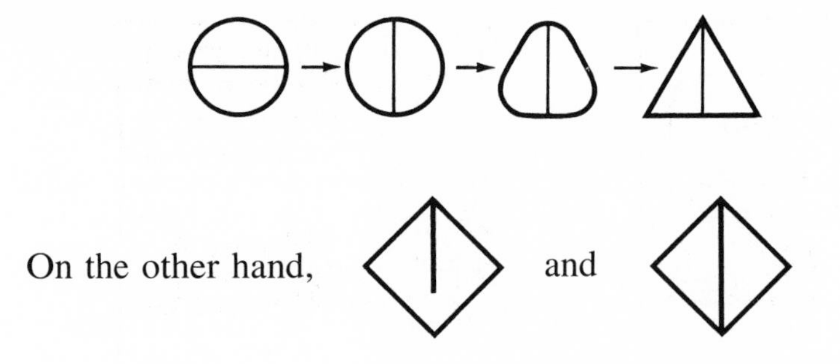

On the other hand, and

are *not* topologically equivalent, because two points of the first figure would have to be joined in order to get the second figure.

Use the figures shown here to answer the following questions.

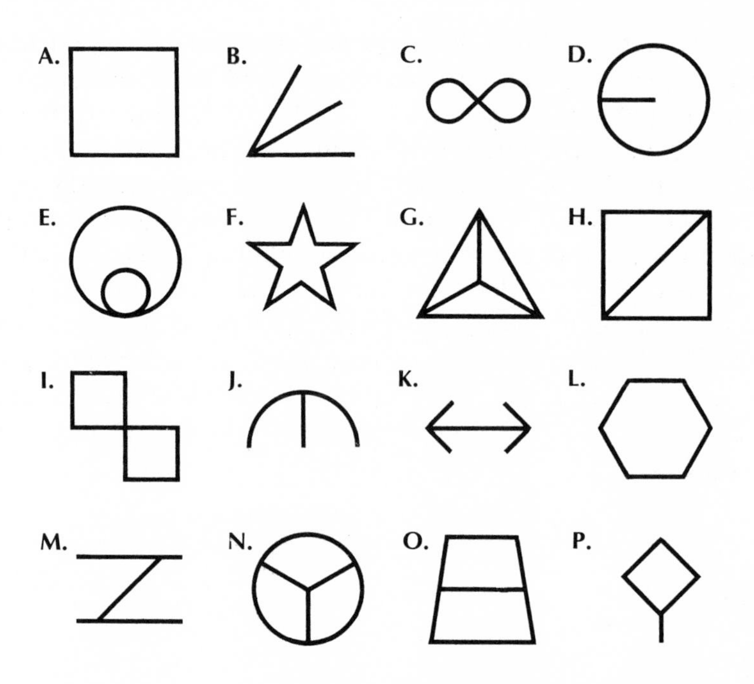

1. Figure A is a square, but the fact that it has 4 corners is not important in topology. Think of it as being made of a rubber band and you can see that if it is twisted and stretched into the shape of a circle, it wouldn't have any corners at all. Why is figure A topologically equivalent to figure F?
2. Figures A and F are topologically equivalent to a third figure shown. Which one is it?
3. Figure B consists of 3 lines that meet at one point. The

fact that the lines are straight is not important in topology. It can be twisted and stretched into another figure shown. Which one?

4. Figure C is topologically equivalent to two other figures shown. Which are they? (Hint: Notice that figure C consists of 2 loops that are joined at one point.)

5. To which figure is figure D equivalent? (Hint: Figure D consists of a loop joined to a line at one point.)

6. Which figure is equivalent to figure G?

7. Which figure is equivalent to figure H?

8. Is figure K equivalent to any other figure?

Several letters of the alphabet have more than one shape when written in capital form. For example, the seventh letter can be printed as G, in which case it is topologically equivalent to the letter C; or as G, in which case it is equivalent to the letter T. Do you see why?

Use the forms of the letters shown here to answer the following questions.

ABCDEFGHIJKLMN
OPQRSTUVWXYZ

9. Perhaps the simplest letter is the ninth, since it is merely a line. A line can be twisted into all sorts of shapes, and this letter is topologically equivalent to quite a number of other letters; in fact, 10 in all. Which letters are they?

10. Only one letter is equivalent to the letter O. Which one?

11. The letter E is equivalent to the letter G, since

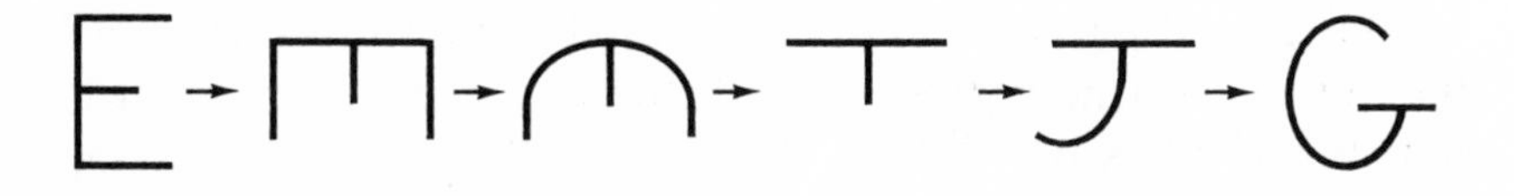

Three other letters are equivalent to the letters E and G. Which are they?

12. Which letter is equivalent to the letter K?

13. Can you find a letter that is equivalent to the letter Q? (Hint: It consists of a loop with 2 lines meeting at one of its points.)

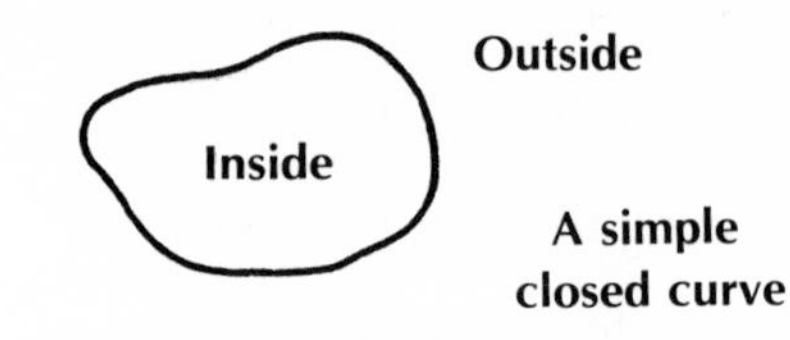

Set II

One of the fundamental ideas in topology is called the *Jordan Curve Theorem.* This theorem is about simple closed curves. A simple closed curve does not cross itself, and the theorem says that such a curve in a plane divides the plane into two regions: an inside and an outside.

1. Is this figure a simple closed curve? Explain.

2. What about this figure?

An easy way of finding out whether a point is inside or outside a simple closed curve is to draw a straight line from the point to the outside of (the region beyond) the curve and then count the number of times the straight line crosses the curve.

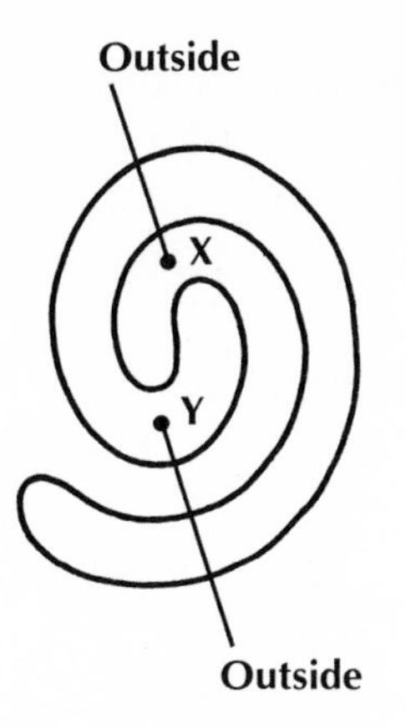

3. Is point X inside or outside the curve above? Does the straight line from point X to the outside cross the curve an even or an odd number of times?

4. What about point Y? Is the number of times that the line joining it to the outside crosses the curve even or odd?

5. Complete the following rule:
   "If a straight line joining a point to the outside of a curve crosses the curve an *even* number of times, the point is ▒▒▒▒▒▒ the curve; if it crosses an *odd* number of times, the point is ▒▒▒▒▒▒ the curve."

6. Use the rule to determine whether each of the points A, B, and C is inside or outside the curve below.

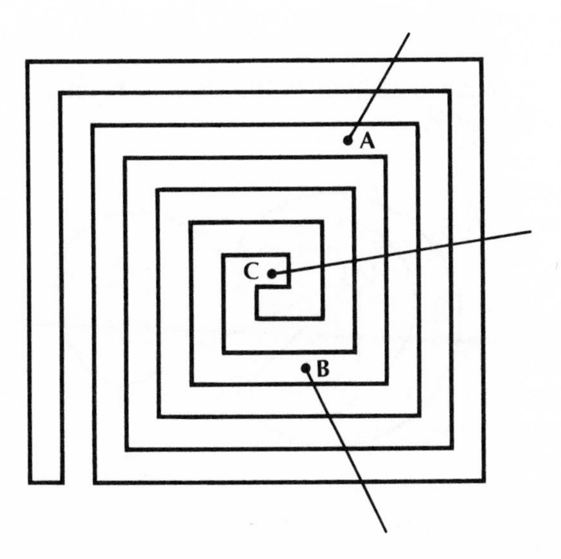

Here is a practical application of the Jordan Curve Theorem.

Suppose a printed electric circuit is to be designed so that one set of points can be connected to another set of points in every possible way, but without having any of the connections cross each other.

Such a circuit for two pairs of points is easy to design: point A is joined to both points X and Y and so is point B.

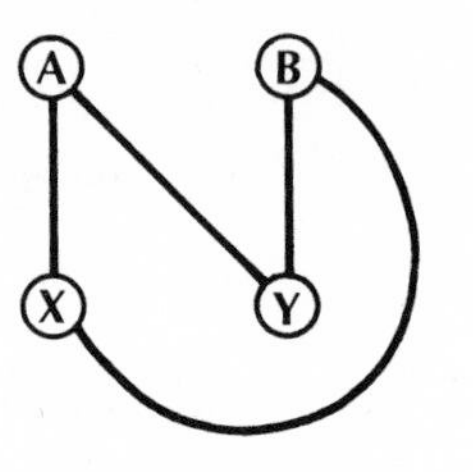

Is it possible to design a printed circuit for two sets of 3 points each? In other words, can each of the points A, B, and C be joined to each of the points X, Y, and Z without any of the connections crossing each other?

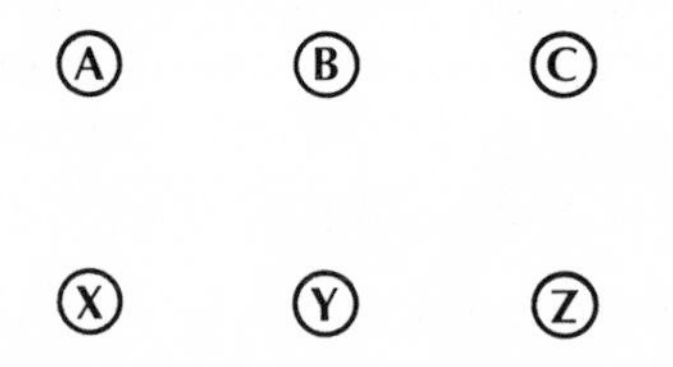

7. Copy the figure above showing the 6 points and see if you can join them as indicated.

Here is a drawing in which points A and B have been joined to points X, Y, and Z.

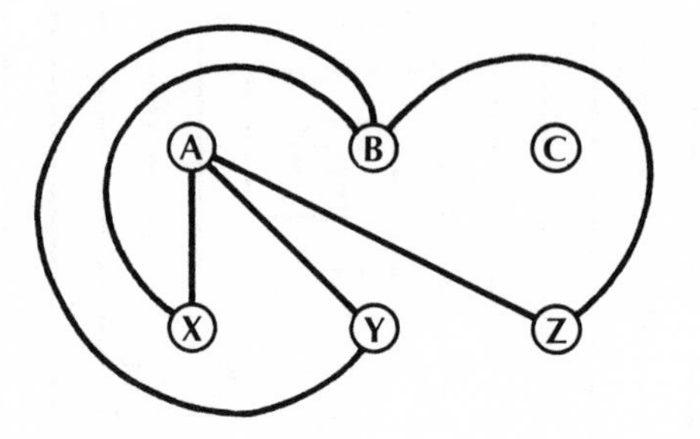

8. Copy this drawing and then shade in the inside of the

simple closed curve that passes through points X, B, Z, A, and X again.

9. Notice that point C is inside this curve and that it has not yet been joined to points X, Y, and Z. Are any of those points outside the curve?

10. Since the Jordan Curve Theorem says that a simple closed curve divides a plane into two regions, it is impossible to join a point inside the curve to a point outside without crossing it. Do you think it is possible to design a circuit for this set of points that will not short-circuit?

Set III

Experiment. A Set of Three Linked Rings,
No Two of Which Are Linked Together

It is easy to see that one of the two pairs of rings shown below is linked together and cannot be separated unless one of the rings is cut, while the other pair is not linked at all.

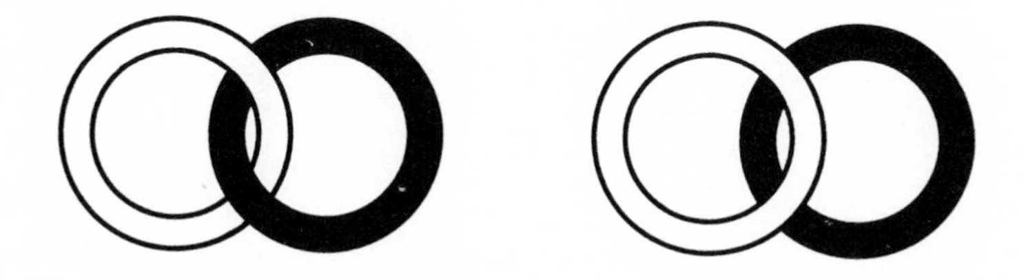

Use a compass to draw three identical rings of about the size shown below at the left on a 4 × 6 card. Cut the rings out and then cut one of them as shown on the ring below at the right.

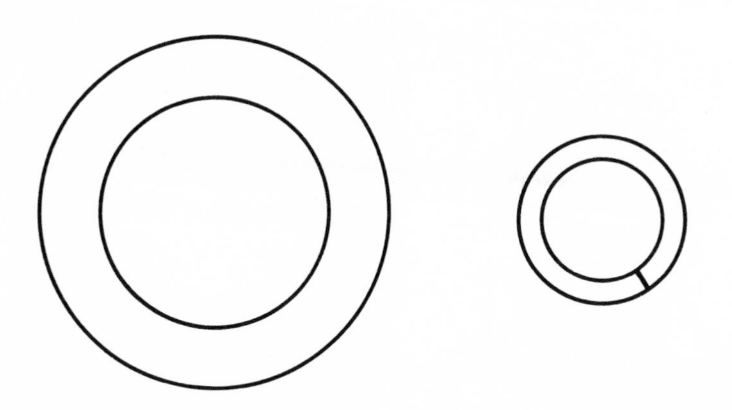

In the diagram below, the three rings are linked together in such a way that the shaded ring is not linked to either ring below it. The two unshaded rings, however, are linked to each other so that if the shaded ring were cut and removed, they still would not come apart.

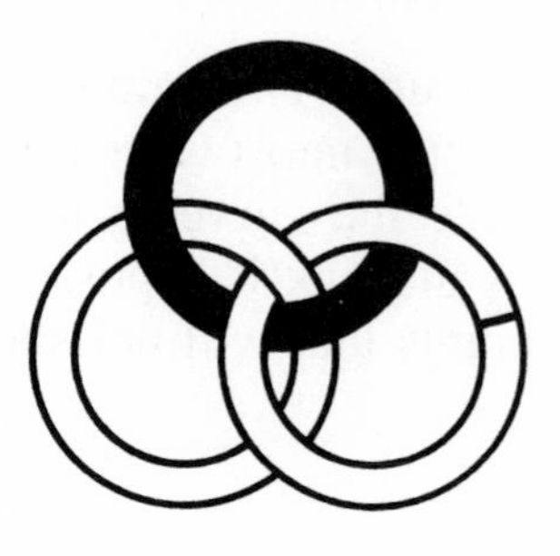

Can you link the three rings together so that no pair of them is linked like these unshaded rings?

Lesson 2

The Seven Bridges of Königsberg

ONE of the most famous problems in topology is about seven bridges in a small city, called Königsberg, in old Germany. The center of this city was on an island in the middle of a river. The island was connected by four bridges to the banks of the river and by a fifth bridge to another island, which was joined to the rest of the city by two more bridges.

The people of Königsberg wondered: is it possible to travel through the city and cross each of the seven bridges without recrossing any one of them? Whenever anyone tried it, they either

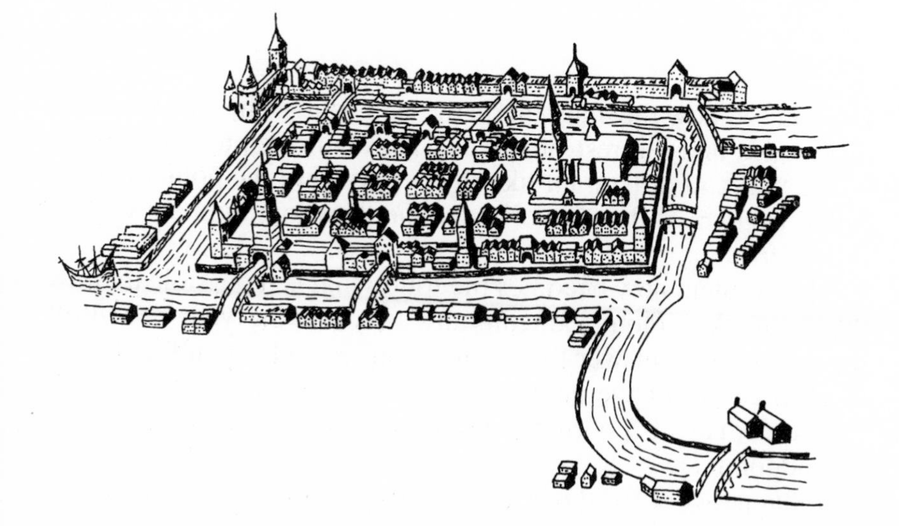

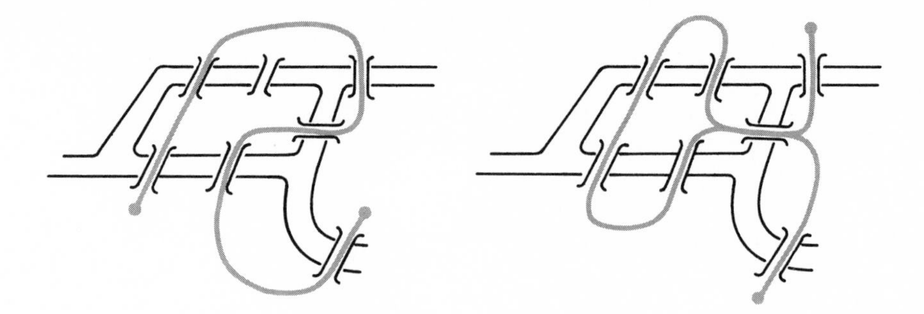

ended up skipping a bridge or else crossing one more than once. Most people were certain that it was impossible to cross each bridge exactly once, but they didn't know why.

Eventually, a great Swiss mathematician named Leonhard Euler heard about the problem of the Königsberg bridges. He found it to be an interesting puzzle and was able to prove mathematically that it couldn't be solved.

In making his analysis of the problem, Euler began by redrawing the map so that the four areas of land were represented by four points and the seven bridges by seven lines joining the points.

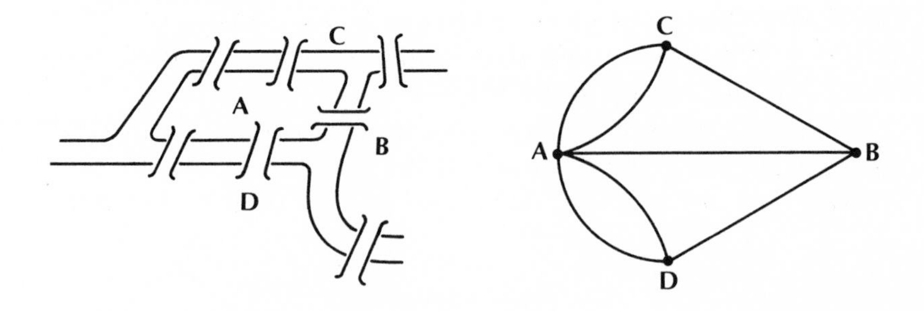

Do you see how the diagram on the right shows the way in which the islands and the rest of the city are connected to each other?

► We will call a diagram like this a **network;** the points will be called *corners* of the network, and the lines joining them will be called *arcs.*

The problem now becomes the following: is it possible to draw this network without retracing any arc or taking your pencil off the paper? The answer to this question depends upon the *degree* of each corner of the network.

► The **degree** of a corner is the *number of arcs* that have it for an endpoint.

In the network for the bridge problem, corner A is of degree 5 and corners B, C, and D are each of degree 3. All four corners, then, are of an odd degree. Euler was able to prove that a network that has four corners of an odd degree cannot be drawn with one continuous stroke of a pencil.

EXERCISES

Set I

Place a sheet of tracing paper over these networks and try to draw (travel) each of them; that is, try to travel each arc without retracing any one of them. You may start at any corner and pass through the arcs in any order. You will find, however, that certain networks can be traveled only if you start from certain corners. (Do not write in this book.) There are 2 more networks on the following page.

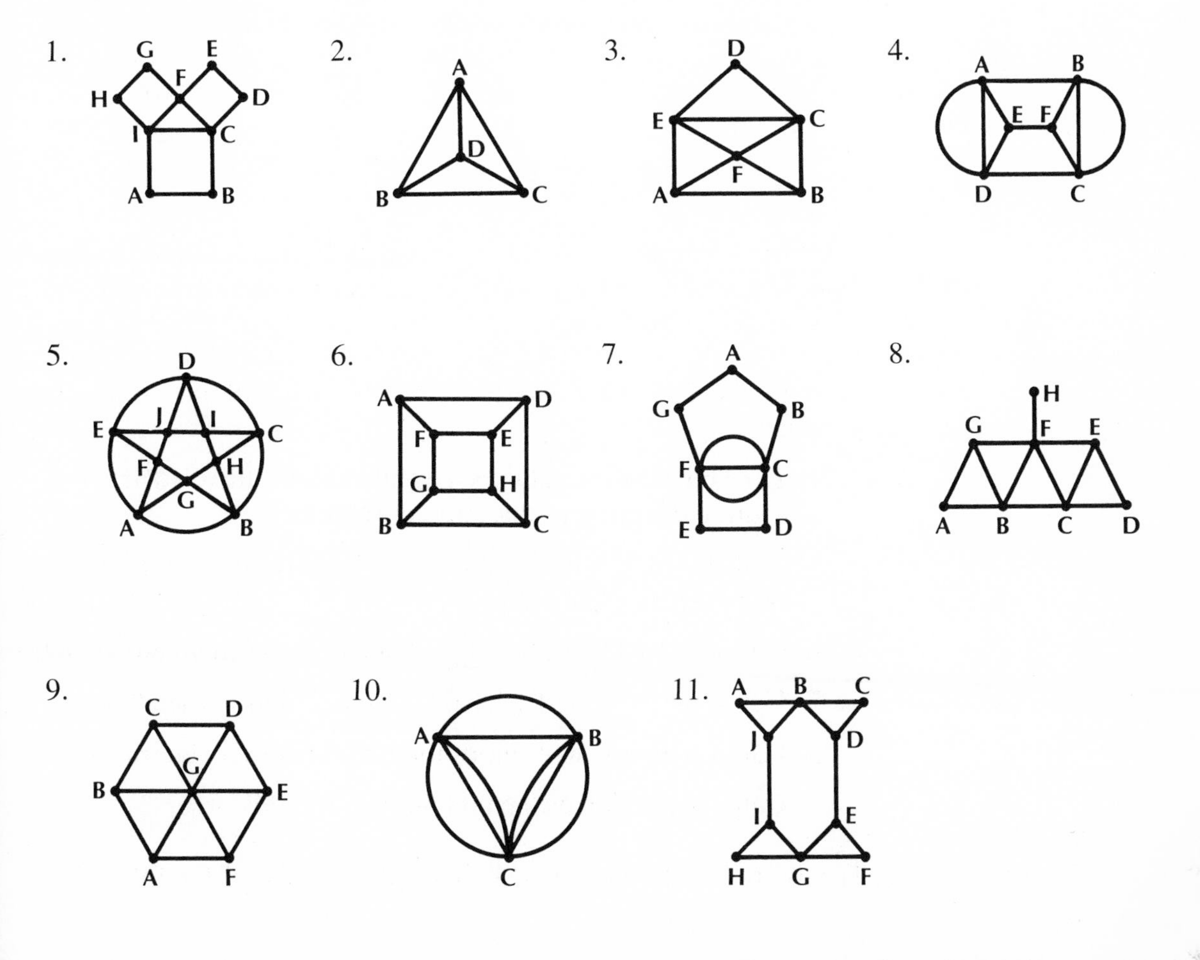

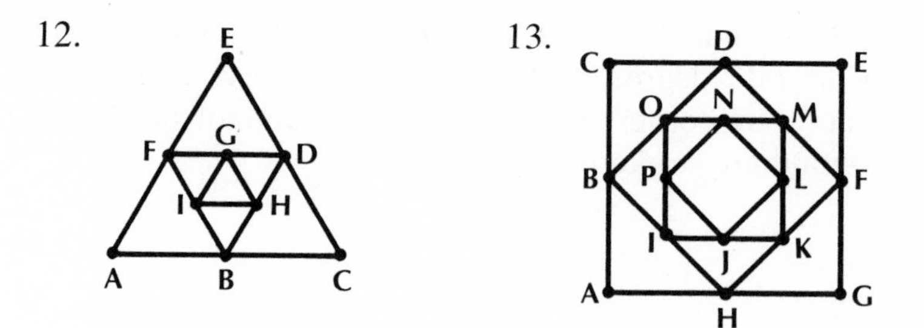

Set II

1. Count the numbers of even and odd corners in each of the networks in Set I and record the results in a table like the one below. Include a column showing whether or not the network can be traveled.

Network	Total no. of corners	No. of even corners	No. of odd corners	Can the network be traveled?
1	9	9	0	Yes
2				
3				
etc.				

Of the 13 networks, 8 can be traveled. If your table shows less than 8, go back and try drawing each of the networks that you have marked "no" again.

Answer the following questions on the basis of the results in your table.

2. Do you think a network can always be traveled if it has more even corners than odd corners?
3. Is it true that a network can *not* be traveled if it has more odd corners than even?
4. Can a network be traveled if all of its corners are even?
5. Can a network be traveled if it has 2 odd corners?
6. Can a network with more than 2 odd corners be traveled?
7. Can a network have an odd number of odd corners?

Try to draw a simple network that satisfies each of the following conditions. (If you think any of them cannot exist, write "impossible.")

8. A network with 3 even corners that can be traveled.
9. A network with no even corners that can be traveled.
10. A network with an odd number of odd corners.

Set III

An interesting new game invented by two mathematicians at Cambridge University in England is called "Sprouts."* The game is played between two people and involves drawing a network.

First, several points, which serve as the original corners of the network, are marked on a piece of paper. The players then take turns drawing arcs, following these rules.

1. Each arc must join two corners or else one corner to itself.
2. When an arc is drawn, a new corner must be chosen somewhere on it.
3. No arc may cross itself, cross another arc, or pass through any corner.
4. No corner may have a degree of more than 3.

The last person able to play wins the game. Here is a diagram showing one series of plays in a game that starts with 2 points.

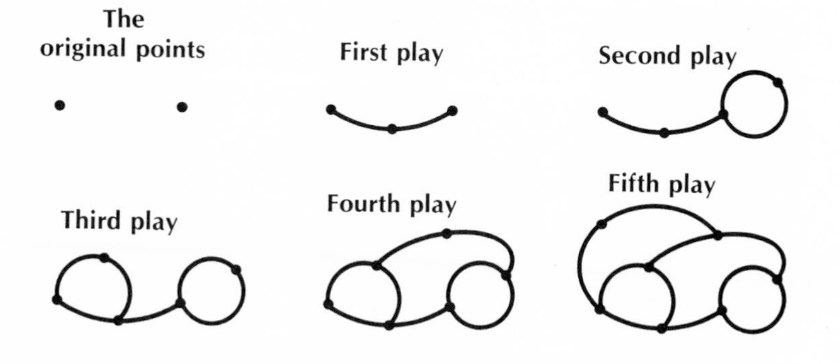

*Martin Gardner, "Mathematical Games" *Scientific American,* July 1967, pp. 112–115.

Do you see why this game ended on the fifth play?

Is there any limit to the number of moves that a "Sprouts" game can last, if the players want to continue it as long as possible? To find out, try playing it with yourself or with someone else. Play several games starting with just 2 points and see if you can draw any conclusions. Then try playing some games that start with 3 or 4 points.

Lesson 3

More on Networks

MANY years after Leonard Euler proved that it was impossible to walk through the city of Königsberg and cross each of its seven bridges exactly once, an eighth bridge was built. It isn't difficult to show that, as a result of the new bridge, a walk over the Königsberg bridges is now possible.

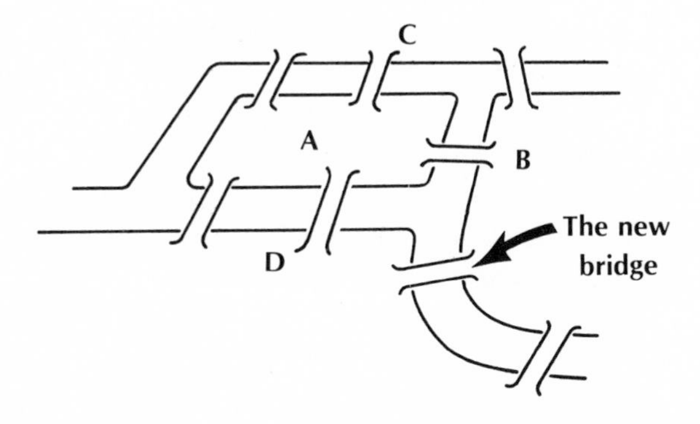

A network representing the new map of the city is shown at the top of page 464.

If you compare this network with the bottom right one on page 458, you will see that it has an additional arc between corners B and D to represent the new bridge. So corners B and D are now both of an even degree. Of course, corners A and C are still odd. One way to travel the new network is to start at corner C and

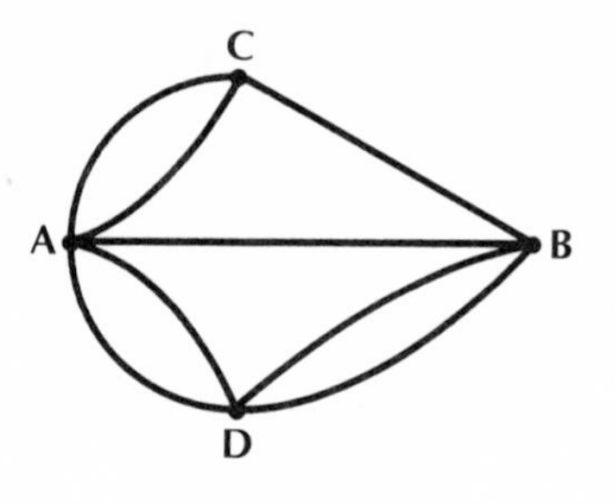

follow the path shown below, ending at corner A. This path begins at one of the *odd* corners and ends at the other one. Is it possible to travel this network by beginning at one of the *even* corners instead? The answer is no!

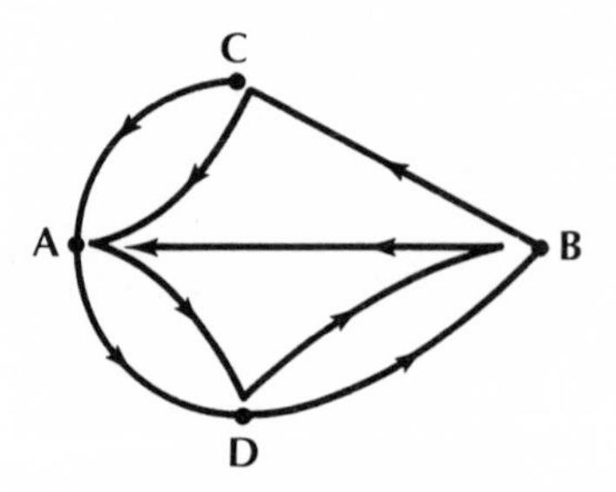

To see why, look at what would happen if we didn't start at corner C. In traveling the network we must go over each arc, including the three that end at C, exactly once. The first time we come to C, it will be along one of these three arcs and we can leave along either of the other two. That leaves one arc untraveled and when we cover it we will be "stuck" at C since there are no arcs remaining along which we can leave. In other words, if our trip doesn't *begin* at corner C, it must *end* there.

Using the same kind of reasoning, we can show that the path must also either begin or end at the other odd corner, A. Since a path has exactly one beginning and one ending, it must, therefore, begin at one of the odd corners and end at the other. It also means that a network with more than two odd corners cannot be traveled in a single trip.

EXERCISES

Set I

Which of the following networks can be traveled in a single trip? Instead of drawing them, decide by looking at their corners.

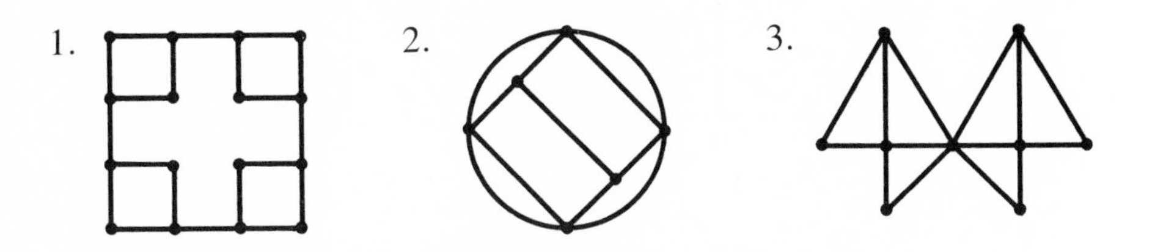

4. It is possible to travel the network shown below in a single trip. At which corners can the trip begin?

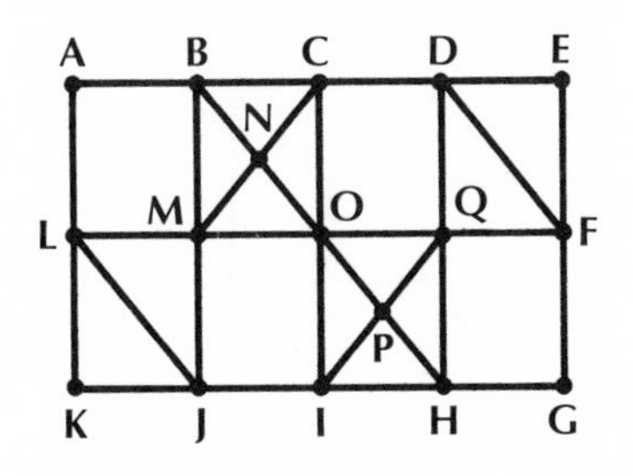

5. Try drawing the network.

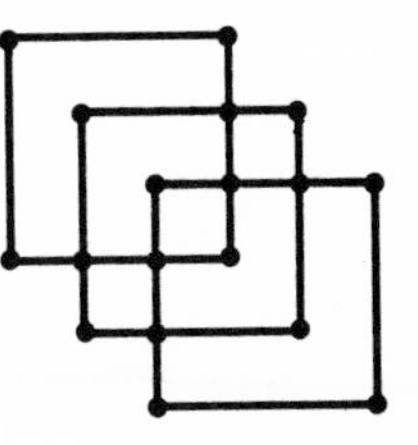

The network shown at the right was a favorite puzzle of Lewis Carroll. It is easily traveled in a single trip.

6. Try drawing it several times, starting from a different corner each time.

Use your results to help answer the following questions:

7. The network has 18 corners. Are any of them odd?
8. Do you think that it is always possible to travel the network if you choose any corner at random to start from?

9. Do you think that in traveling this network you will always end up at the corner where you started?

The network below cannot be traveled in a single trip because it has more than two odd corners. It can be traveled, however, in two trips. (In other words, it is possible to draw this figure without retracing any arc if you are allowed to remove your pencil from the paper once.) One way of doing this is shown at the right.

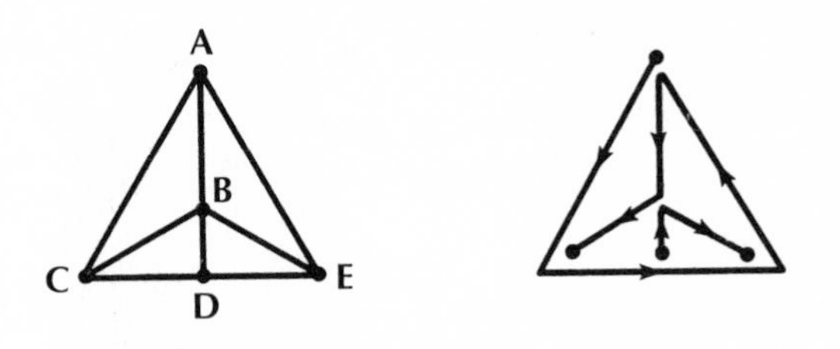

Count the number of odd corners in each of the following networks and then try to find out what is the least number of trips it takes to travel each one.

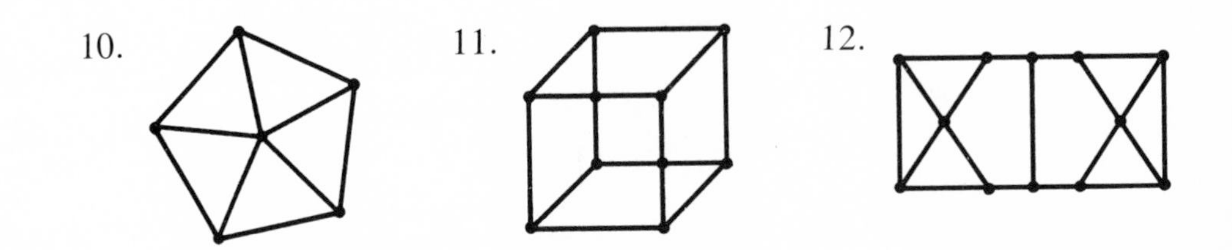

13. Can you make up a rule for determining the number of trips necessary to travel a network on the basis of the number of odd corners it contains?

14. How many trips would it take to travel a giant network that has a million corners if half of them are even and half are odd?

Set II

The network at the left here divides the paper into 3 regions: 2 are inside the network and the other is the region outside. We can think of the network at the right as having 1 region: the one outside.

There is a simple relationship between the number of regions of a network and the numbers of corners and arcs that it contains. To find out what it is, count the numbers of regions, corners, and arcs in each of the following networks.

Record your results in a table like the one below:

Network	No. of regions	No. of corners	No. of arcs
1	2		
2			
etc.			

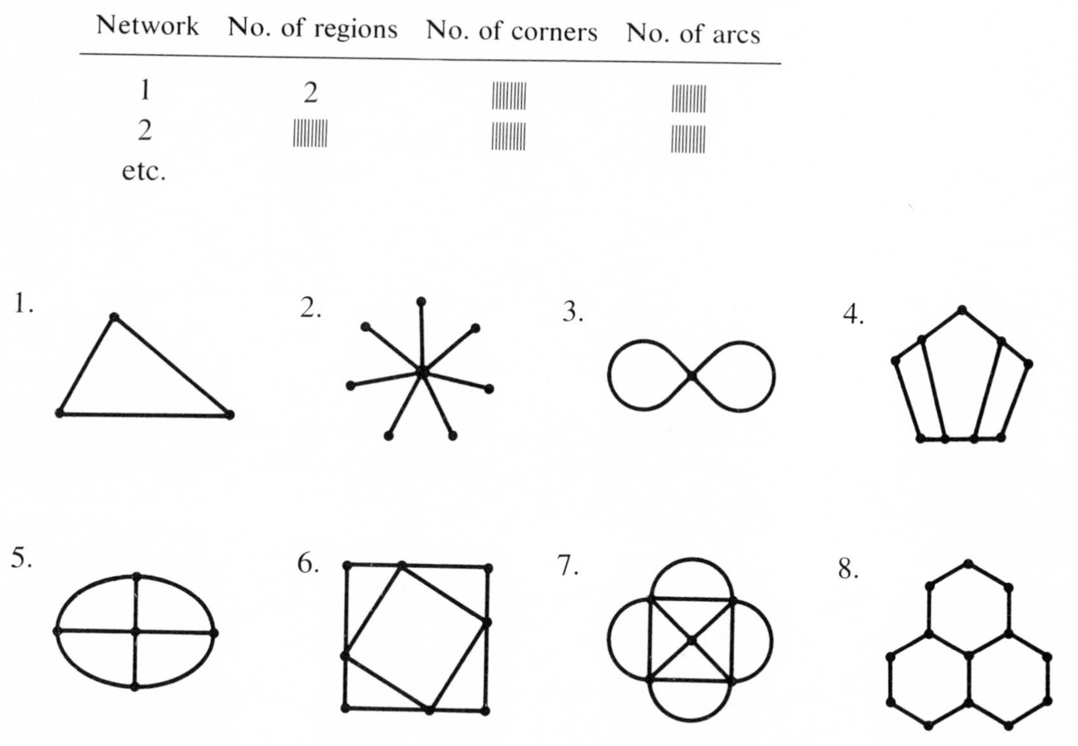

9. Can you write a formula relating the number of regions, R, in a network, the number of corners, C, and the number of arcs, A?

10. This should remind you of something else you have studied. What is it? (Hint: See pp. 235, 236.)

Set III

Another kind of network problem concerns the shortest path on a network that goes through every *corner* exactly once and ends at its starting point. A recent ad of the Bell Telephone Laboratories

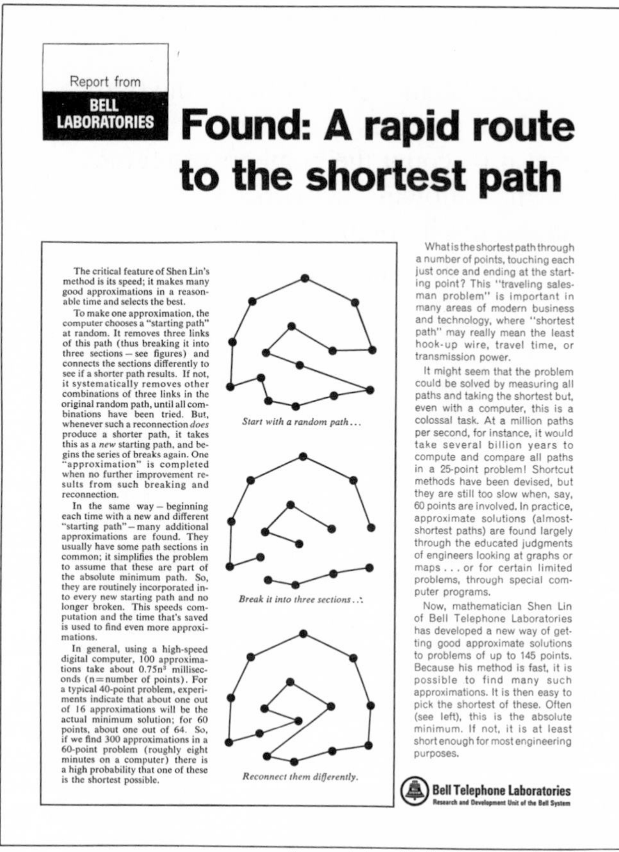

Report from
BELL LABORATORIES

Found: A rapid route to the shortest path

The critical feature of Shen Lin's method is its speed; it makes many good approximations in a reasonable time and selects the best.

To make one approximation, the computer chooses a "starting path" at random. It removes three links of this path (thus breaking it into three sections – see figures) and connects the sections differently to see if a shorter path results. If not, it systematically removes other combinations of three links in the original random path, until all combinations have been tried. But, whenever such a reconnection *does* produce a shorter path, it takes this as a *new* starting path, and begins the series of breaks again. One "approximation" is completed when no further improvement results from such breaking and reconnection.

In the same way – beginning each time with a new and different "starting path" – many additional approximations are found. They usually have some path sections in common; it simplifies the problem to assume that these are part of the absolute minimum path. So, they are routinely incorporated into every new starting path and no longer broken. This speeds computation and the time that's saved is used to find even more approximations.

In general, using a high-speed digital computer, 100 approximations take about $0.75n^3$ milliseconds (n=number of points). For a typical 40-point problem, experiments indicate that about one out of 16 approximations will be the actual minimum solution; for 60 points, about one out of 64. So, if we find 300 approximations in a 60-point problem (roughly eight minutes on a computer) there is a high probability that one of these is the shortest possible.

Start with a random path . . .

Break it into three sections . . .

Reconnect them differently.

What is the shortest path through a number of points, touching each just once and ending at the starting point? This "traveling salesman problem" is important in many areas of modern business and technology, where "shortest path" may really mean the least hook-up wire, travel time, or transmission power.

It might seem that the problem could be solved by measuring all paths and taking the shortest but, even with a computer, this is a colossal task. At a million paths per second, for instance, it would take several billion years to compute and compare all paths in a 25-point problem! Shortcut methods have been devised, but they are still too slow when, say, 60 points are involved. In practice, approximate solutions (almost-shortest paths) are found largely through the educated judgments of engineers looking at graphs or maps . . . or for certain limited problems, through special computer programs.

Now, mathematician Shen Lin of Bell Telephone Laboratories has developed a new way of getting good approximate solutions to problems of up to 145 points. Because his method is fast, it is possible to find many such approximations. It is then easy to pick the shortest of these. Often (see left), this is the absolute minimum. If not, it is at least short enough for most engineering purposes.

Bell Telephone Laboratories
Research and Development Unit of the Bell System

says that this problem is "important in many areas of modern business and technology, where 'shortest path' may really mean the least hook-up wire, travel time, or transmission power."

You know that some networks do not have a path that covers every *arc* exactly once. This figure, which contains 4 odd corners, is an example of such a network.

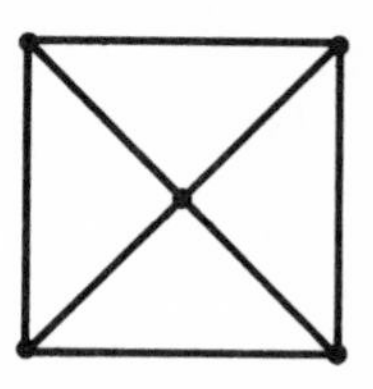

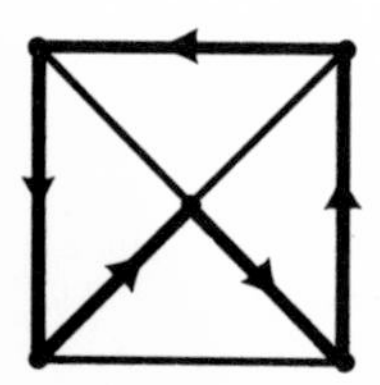

There are, however, a number of paths that go through every *corner* exactly once and end where they start. Here is an example of such a path in the figure at the left.

Are there any networks that do not have a path like this?

Place a sheet of tracing paper over these networks and, if possible, draw a path on each that goes through every *corner* once and ends where it starts.

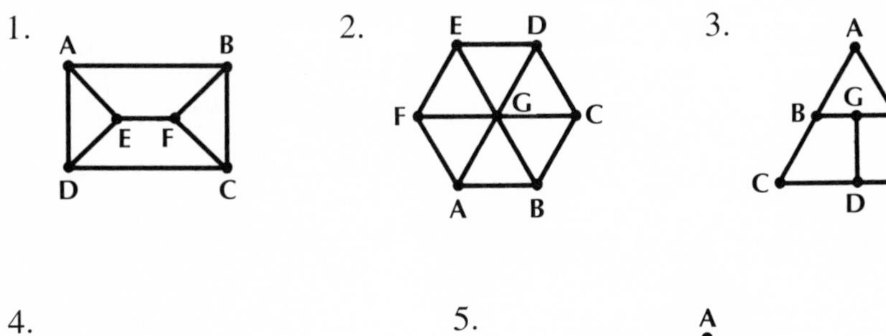

4.

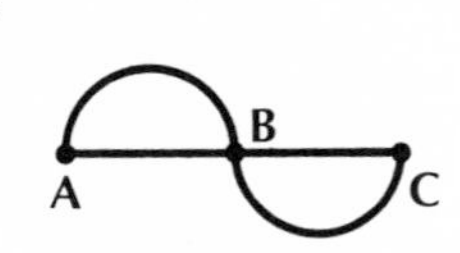

5.

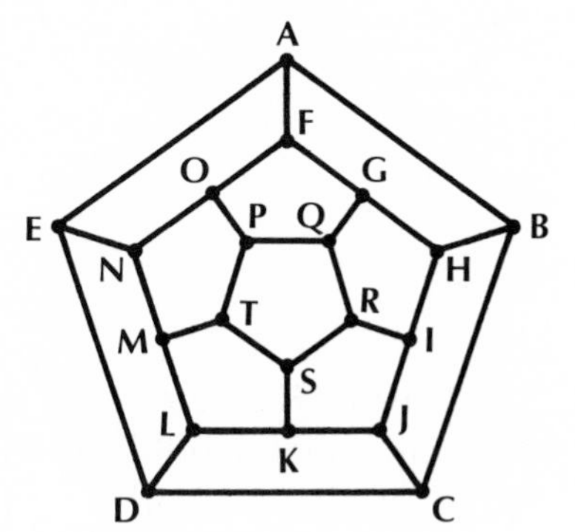

Lesson 4

Trees

ALTHOUGH Peter says that a tree is "a bush that made it," a mathematician would say that a tree is a special kind of network. Most of the networks you have seen contain "loops," which make it possible to begin at a corner, travel over part (or all) of the network, and return to the original corner without passing over any arc twice.

► A **tree network** is one that does not have any "loops."

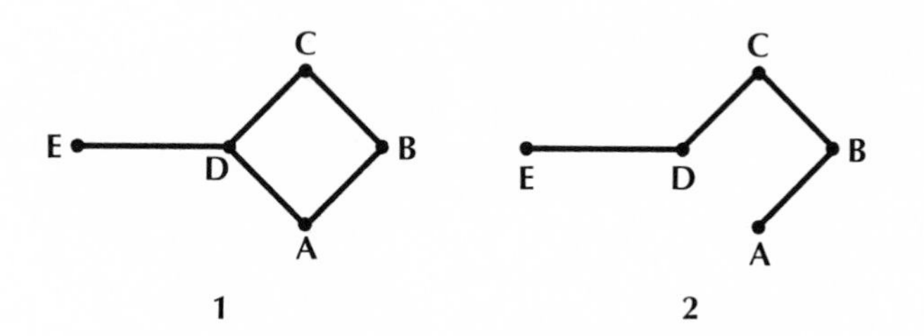

Network 1 shown here contains a loop, so that if you start, for example, at corner A, there is a path through corners B, C, and D that leads back to A without retracing itself. A path like this on network 2, on the other hand, is impossible. There is no way of starting at any corner and returning to that corner over a different route. So network 2 is an example of a tree.

A real tree, the kind Peter and B.C. are looking at, is like a topological tree in that it consists of a trunk, branches, and twigs that ordinarily do not grow back together. A squirrel on one

By permission of Johnny Hart and Field Enterprises, Inc.

branch cannot take a trip over other branches of the tree and come back to where he began without returning along the same branches (assuming, of course, that he does not jump from one branch to another!)

Tree networks appear in a wide variety of subjects. Map makers draw complicated trees to represent rivers and their tributaries. The map here shows the Mississippi River and some of the other rivers that flow into it. A genealogist uses trees to show the relationships between members of different generations of a

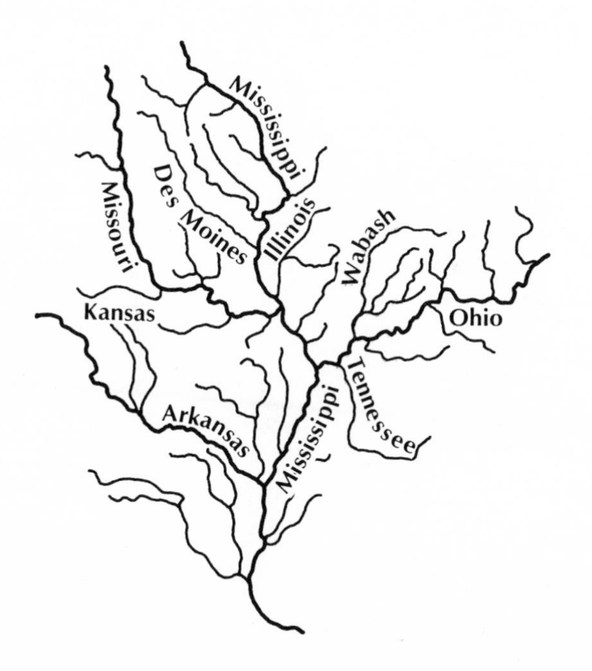

family. The process that a postal clerk uses to sort mail can be represented by a tree; the zip code identifies its main branches. Tree diagrams are drawn by chemists to show the arrangements of atoms in molecules. And you have already seen trees put to use in solving problems involving probability (see pp. 359 and 361).

EXERCISES

Set I

Count the numbers of corners and arcs in each of the following trees.

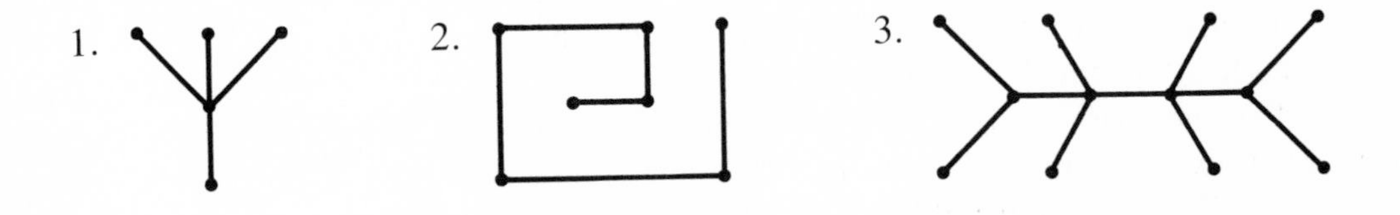

4. Write a formula relating the number of corners, C, in a tree to the number of arcs, A.

Does your formula hold true for these networks? Explain your answers.

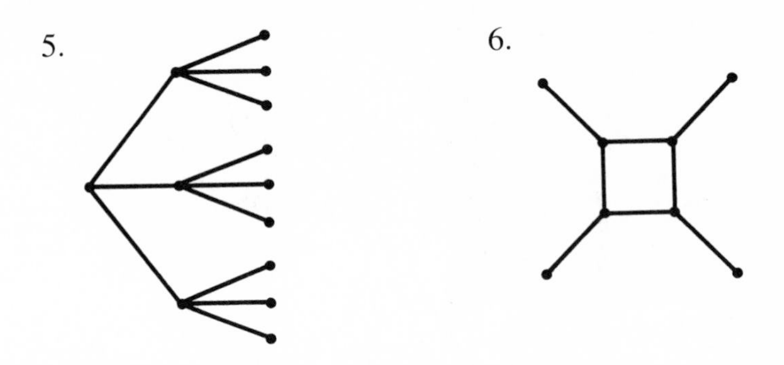

Try to draw a tree that satisfies each of the following conditions. (If you think any of them cannot exist, write "impossible.")

7. A tree with all odd corners.

8. A tree with all even corners.
9. A tree whose arcs can be traveled in a single trip.
10. A tree with a path that goes through every corner once and ends where it starts.

There is just one path from any corner of a tree to any other. In the tree below, the path between corners B and E goes over 1 arc, while the path from A to C goes over 2 arcs and the path from A to F covers 3 arcs.

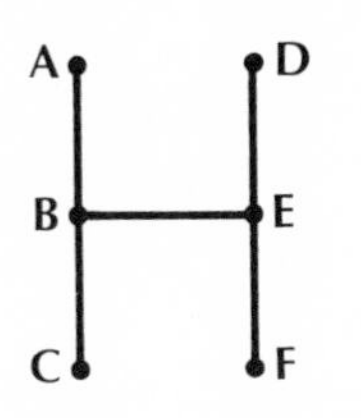

► The number of arcs in the longest possible path on a tree is called its **diameter.**

The diameter of the tree above is 3.

11. What are the diameters of the trees shown in problems 1, 2, 3, and 5?
12. Many snowflakes have the shapes of trees. Here are a photograph of one taken through a microscope and a tree diagram of its structure. What is the diameter of this tree?

Set II

There is only one tree possible that contains just 2 corners, •—•, and only one that contains 3 corners, •—•—•

You will recall that two figures that can be twisted and stretched into the same shape are said to be topologically equivalent. This means that such trees as and are equivalent to the tree,

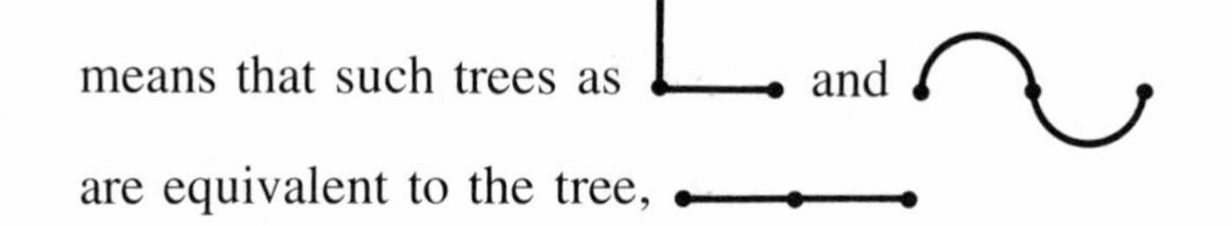

There are 2 different trees that contain 4 corners.

Notice that the first has a diameter of 3 and the second has a diameter of only 2.

To which of these two trees are the following trees containing 4 corners each equivalent?

3. There are 3 different trees that contain 5 corners. Make a set of drawings to show them. (Hint: One has a diameter of 4, one a diameter of 3, and one a diameter of 2.)

The number of different trees containing 6 corners is 6. One is

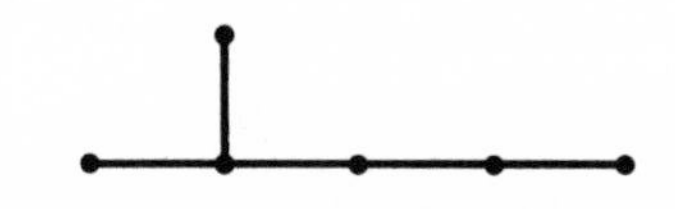

4. Look at the trees below. Except for one, all of them are topologically equivalent to the 6-cornered tree shown above. Which one is *different*?

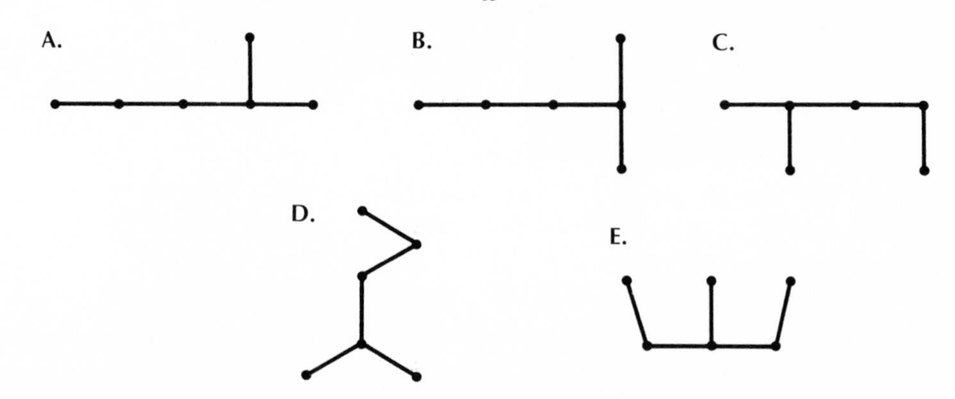

5. Here is another tree that contains 6 corners.

To which one of the following trees is it equivalent?

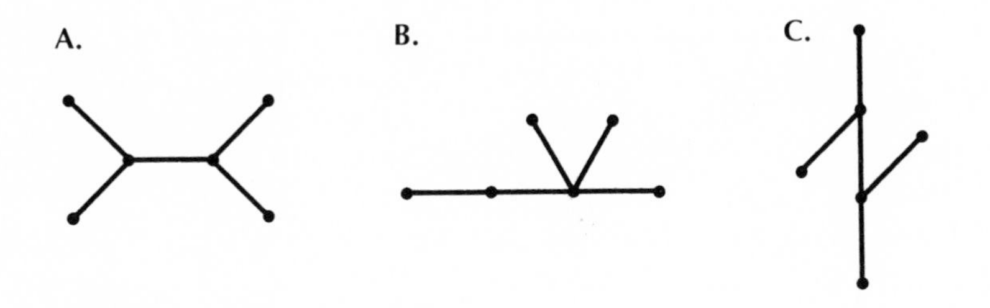

6. Here is a set of drawings of 5 of the 6 different trees that contain 6 corners. Can you draw the sixth one?

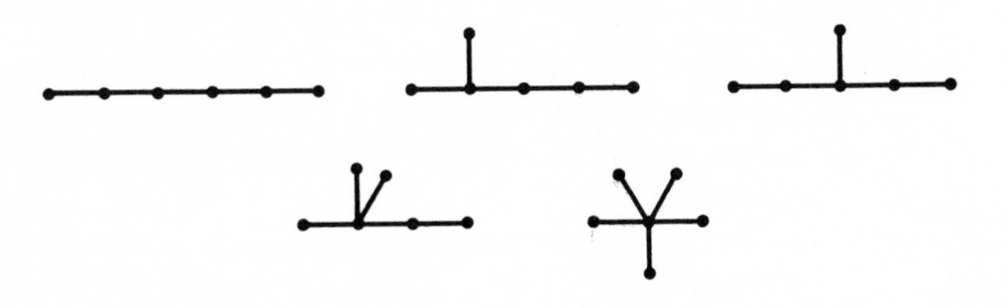

7. There are 11 different trees that contain 7 corners. Make drawings of as many of them as you can think of. (Hint: There is 1 tree with a diameter of 6, 2 trees with a diameter of 5, 5 trees with a diameter of 4, 2 trees with a diameter of 3, and 1 tree with a diameter of 2.)

Set III

Chemists use tree diagrams to show the ways in which atoms are linked together in molecules. On page 476 are some photographs of models of several hydrocarbon molecules, along with trees showing their structure.

The corners of the trees labeled C represent carbon atoms and the corners labeled H represent hydrogen atoms. Notice that the degree of each H corner is 1.

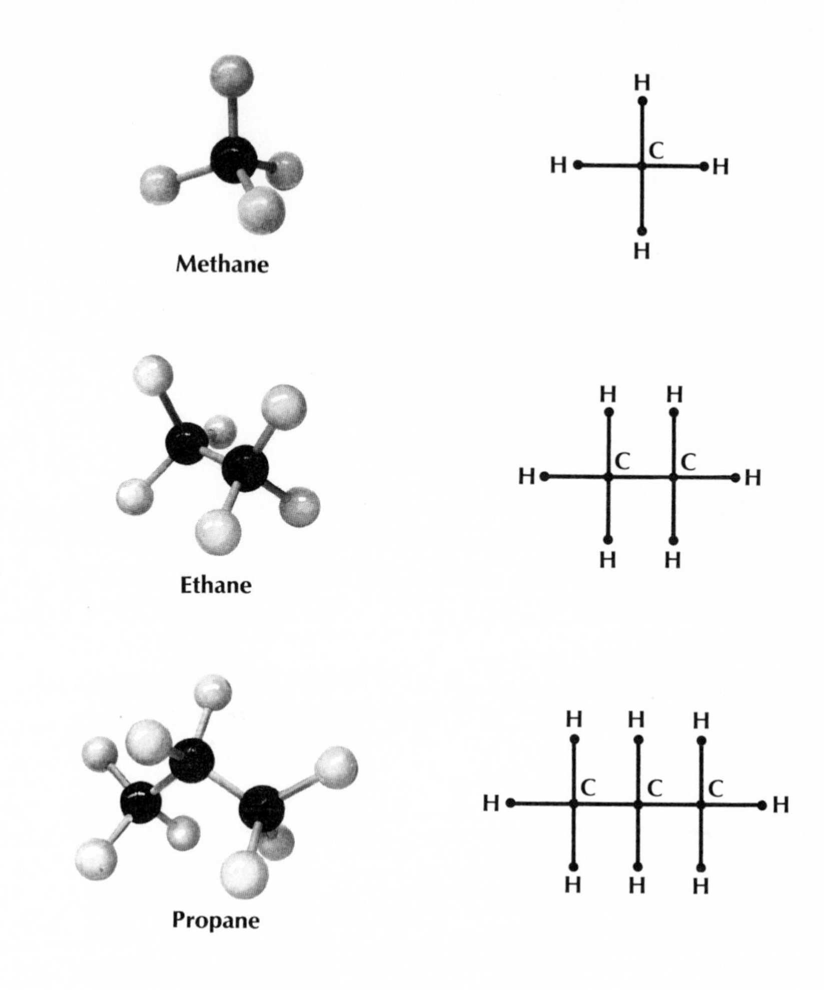

1. What is the degree of each C corner?

The "skeleton" of each of these molecules consists of its carbon atoms and the bonds between them. Here is the skeleton of the propane molecule:

There are 2 different butane molecules containing 4 carbon atoms each. Their skeletons are:

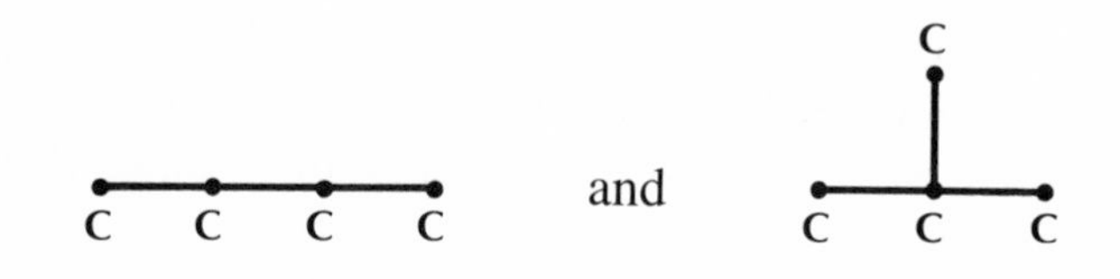

2. Copy these figures and then add enough arcs and corners representing hydrogen atoms to them so that the degree of each C corner is the number you gave as the answer to the first question. (Each figure should have 10 hydrogen atoms.)

3. There are 3 different pentane molecules containing 5 carbon atoms each. First, draw tree diagrams of their skeletons, and then complete each tree to show the rest of the molecule.

4. Does each of your three drawings contain the same number of hydrogen atoms?

5. Notice that a pentane molecule contains 5 carbon atoms and that a pentagon contains 5 corners. What, do you suppose, is the name of the hydrocarbon with 8 carbon atoms in each molecule?

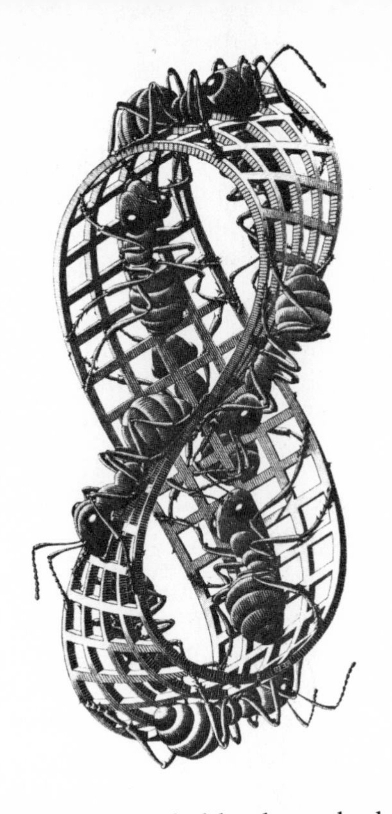

Drawing by Maurits Escher; courtesy of G. W. Breughel, Zwolle, Netherlands.

Lesson 5

The Moebius Strip and Other Surfaces

THERE is something rather remarkable about the band pictured in this drawing by Maurits Escher. Do you see what it is?

The artist says: "An endless ring-shaped band usually has two distinct surfaces, one inside and one outside. Yet on this strip nine red ants crawl after each other and travel the front side as well as the reverse side. Therefore the strip has only one surface."*

This band, called a **Moebius strip,** is named after one of the pioneers in the subject of topology, a German mathematician who discovered it in the middle of the 19th century and wrote a paper about its properties. To get a clear idea of just what a Moebius strip is, look at these photographs of two bands. The first is an ordinary "belt-shaped" loop without any twists in it.

* Maurits Escher, in *The Graphic Work of M. C. Escher,* Meredith, 1967, p. 17.

It is easy to see that it has 2 sides and 2 edges; it is impossible to travel on this loop from one side to the other without crossing over an edge. The second loop is a Moebius strip.

Notice that it contains a twist. It is as a result of this twist that the strip has only 1 side and 1 edge! If this seems hard to believe, take another close look at the drawing of the Moebius strip with the ants crawling on it. The strip's onesidedness gives it a number of strange properties, which you will discover by experiment.

Some of these properties have been put to practical use. The B. F. Goodrich Company has patented a rubber conveyor belt in the shape of a Moebius strip—the belt lasts longer since it has only one side instead of two. A continuous loop recording tape sealed in a cartridge will play twice as long if it has a twist in it. The ad below reveals that the Moebius strip is even put to use in the design of electronic resistors!

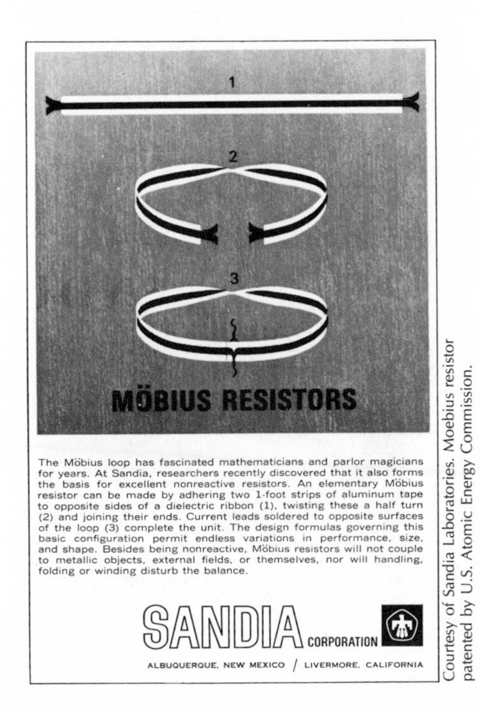

Courtesy of Sandia Laboratories. Moebius resistor patented by U.S. Atomic Energy Commission.

EXERCISES

Set I

Experiment. The Moebius Strip and Other Surfaces

Some of the properties of the Moebius strip are so unusual that they are hard to predict or even imagine unless you have discovered them yourself.

Cut about 10 strips, each 11 inches long and 1 inch wide, from graph paper ruled 4 units per inch.

Take one of the strips, make it into a loop and turn one end over before taping the two ends together. This is called giving the loop a "half-twist" since the end is turned through an angle of 180°, half of 360°. Do the same thing to 2 more strips so that you have 3 Moebius strips altogether.

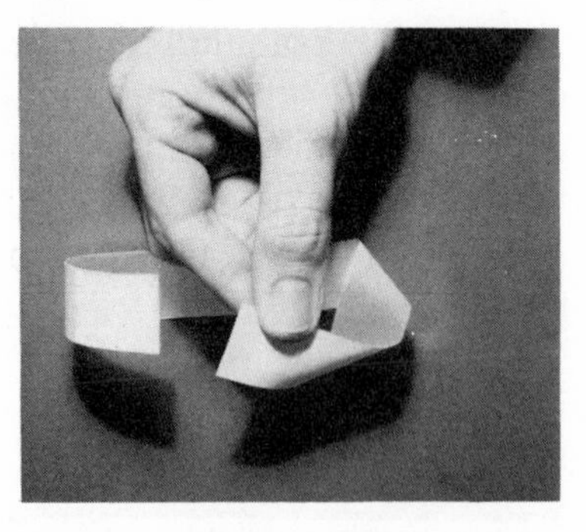

Put your pencil down midway between the edges of one of the Moebius strips and draw a line down its center, continuing until you come back to the same point from which you started.

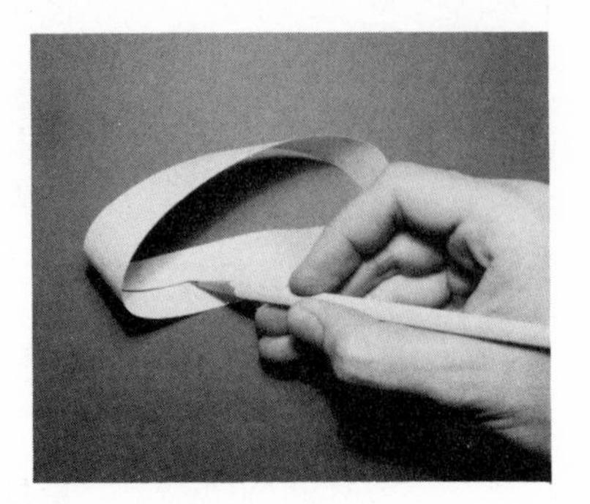

1. What happens?

This proves that the strip has only 1 side since in drawing the line you never crossed over the edge.

Now cut the band along the line you have drawn.

2. What is the result?

Take another Moebius strip and cut along it parallel to an edge and about one-third of the way from the edge. When you have cut all the way around the loop you will find that you are across from the point where you started. Continue cutting, staying the same distance from the edge as before, until you come back to where you began.

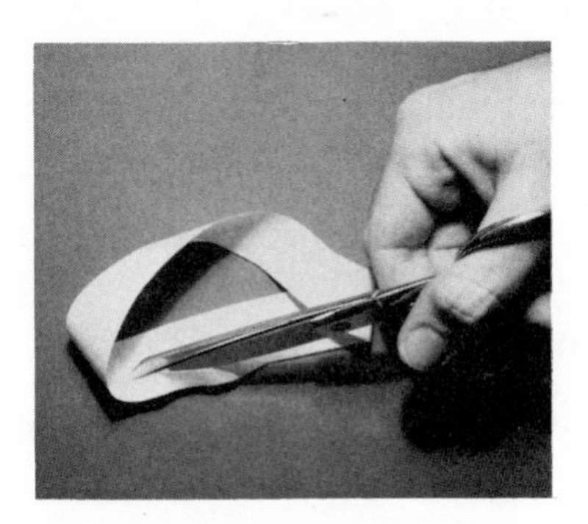

The result is 2 interlocking loops.

3. How do the loops compare in length?
4. How do they compare in width?

Take your third Moebius strip and this time cut along it one-fourth of the way from an edge.

5. In what way is the result similar to the previous one?
6. In what way is it different?
7. Without trying it out, can you guess what the result would be if you cut around a Moebius strip one-fifth of the way from an edge?

Set II

Take two more of your strips and make each one into a band having 2 half-twists.

Draw a line down the center of one of the bands.

1. How many sides does a band with 2 half-twists in it have?

Cut the band along the line.

2. Describe the result as specifically as you can (so that someone who couldn't see it could imagine what it is.)

Cut the other band one-third of the way from an edge.

3. Describe the result.
4. Without trying it out, what do you think would be the result if you cut around a band containing 2 half-twists one-fourth of the way from an edge?

Take three strips and make them into bands having 3 half-twists. Draw a line down the center of one of the bands.

5. How many sides does a band with 3 half-twists in it have?

Cut the band along the line.

6. The result this time is unlike any of the previous ones. What is it?

Cut the second band one-third of the way from an edge. The result will be so tangled up that you will have to pull at it some and study it closely in order to figure out what it is.

7. Describe the result as specifically as you can.

Can you guess what the result of cutting a band containing 3 half-twists one-fourth of the way from an edge would be? Try it out with the third band and see.

8. What is the result?

Take another strip and make it into a band having 4 half-twists.

9. How many sides do you think it has? Check to see if your guess is correct.
10. What is the relationship between the number of half-twists in a band and the number of sides it has?
11. Cut the band down the center. What is the result?
12. What do you think the result would be if you took a band with 100 half-twists in it and cut it down the center?

Set III

Cut out a strip of paper about $1\frac{1}{2}$ inches wide and 11 inches long. Fold the strip in half and cut 2 slits into both ends as shown in the drawing below.

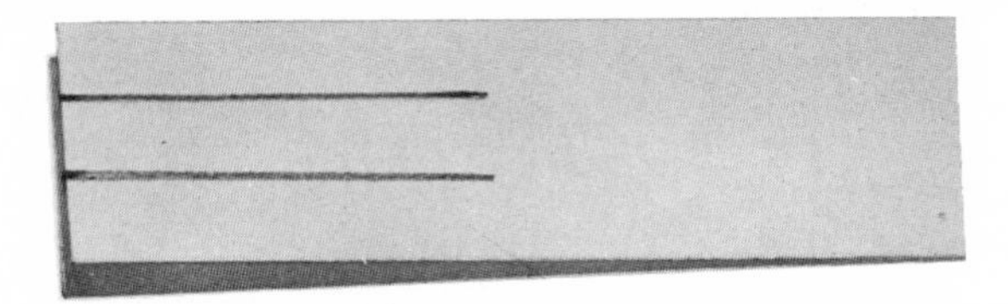

Unfold the strip and number it like this:

Put 3 short pieces of tape on ends 4, 5, and 6, and then make a loop like this:

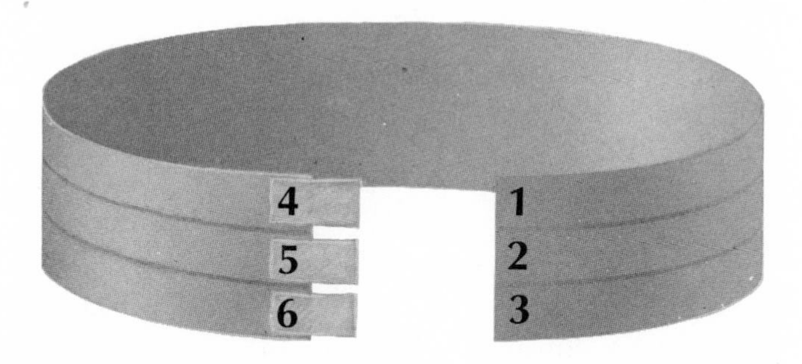

Now tape end 4 to end 1.

Pass end 5 *over* end 4, and end 2 *under* end 1 and tape together.

Pass end 6 *between* ends 4 and 5; pass end 3 *over* end 1 and tape together.

Then finish both cuts so that they go all the way around the band.

1. What is the result?

Prepare another strip in exactly the same way as described in the first paragraph of this exercise.

Then turn end 6 over (give it a half-twist) and tape it to end 1.
Turn end 4 over and tape it to end 2.
Turn end 5 over and tape it to end 3.
Finish both cuts so that they go all the way around the band.

2. What is the result?

INTERESTING READING

Mathematics, Magic and Mystery, by Martin Gardner, Dover, 1956: "Topological Tomfoolery," pp. 69–94.

Fantasia Mathematica, edited by Clifton Fadiman, Simon & Schuster, 1958: "A. Botts and the Moebius Strip," pp. 155–170.

The Mathematical Magpie, edited by Clifton Fadiman, Simon & Schuster, 1962: "Paul Bunyan versus the Conveyor Belt," pp. 33–35.

Chapter 10/Summary and Review

In this chapter we have studied some ideas in topology.

The mathematics of distortion *(Lesson 1)* Two figures that can be twisted and stretched into the same shape are said to be topologically equivalent.

A simple closed curve is a figure in which it is possible to start at any point and travel over every other point of the figure exactly once before returning to the starting point.

The Jordan Curve Theorem says that a simple closed curve in a plane divides it into two regions: an inside and an outside.

Networks *(Lessons 2, 3, & 4)* A network is a figure consisting of points, called corners, joined by lines, called arcs.

The degree of a corner of a network is the number of arcs that have it for an endpoint.

A network with more than two odd corners cannot be traveled in a single trip.

A tree is a network that does not contain any "loops."

The diameter of a tree is the number of arcs in the longest possible path on the tree.

The Moebius strip and other surfaces *(Lesson 5)* A Moebius strip is a band that contains a half-twist. It has only one side and one edge.

A band containing an odd number of half-twists has only one side; a band with an even number of half-twists has two sides.

EXERCISES

Set I

Each of these networks represents the structure of a molecule.

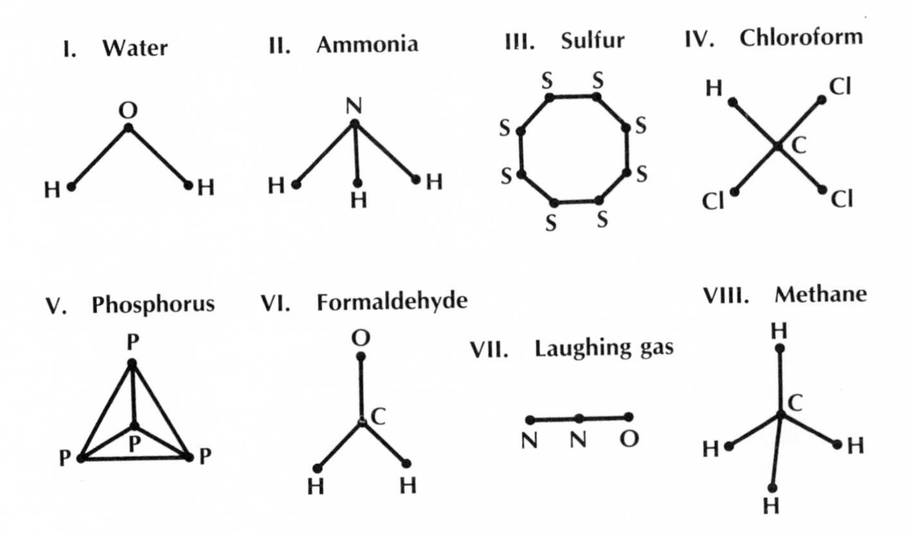

1. Are any of these networks simple closed curves?
2. Which networks are trees?
3. Which networks are topologically equivalent to each other?

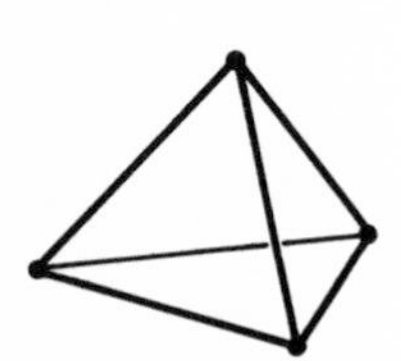
Tetrahedron

The corners and edges of the five regular polyhedra can be considered to be three-dimensional networks.

Cube

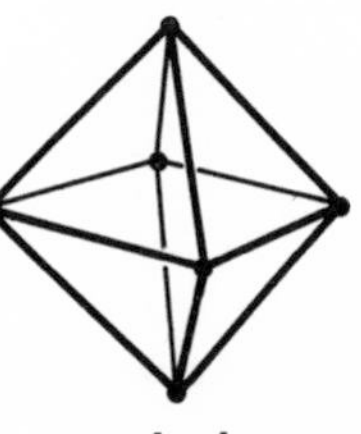
Octahedron

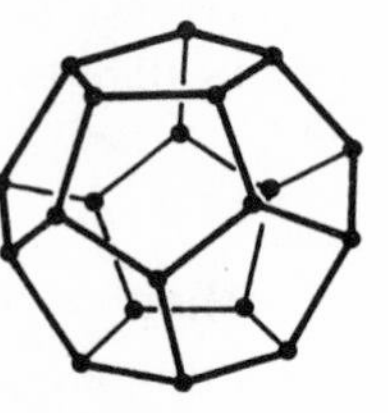
Dodecahedron

Icosahedron

4. It is possible to travel the edges of only one of them in a single trip. Which one is it?
5. Why is it impossible to travel any of the other four figures in a single trip?
6. What is pictured in this postage stamp issued in 1967 in Brazil?
7. How many surfaces does it have?
8. What happens when a band like this is cut down the center?

In Lesson 3 of this chapter, you discovered a formula relating the number of regions, R, of a network, to its number of corners, C, and number of arcs, A. The formula is:

$$R + C = A + 2.$$

Remember that one of the regions of a network is the region outside it.

Imagine that the peculiar network from the cover of *Mad* magazine shown below is flat instead of solid.

Copyright © 1964 by E. C. Publications, Inc.

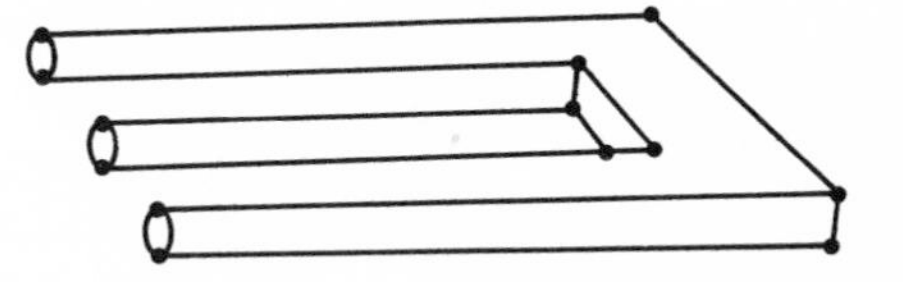

9. How many regions does it have?
10. How many corners?
11. How many arcs?
12. Do these numbers agree with the formula?

Set II

Although the network at the left below is not a tree, it can easily be changed into one by removing some of its arcs.

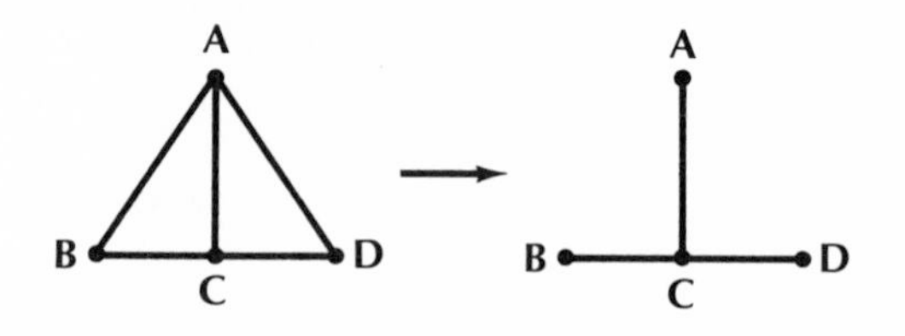

Is there any relationship between the numbers of arcs and corners in a network and the number of arcs that must be removed to change it into a tree? To find out, count the numbers of arcs and corners in each of the following networks. Then make a light drawing of each network in pencil and erase arcs until you have changed it into a tree. Count the number of arcs erased and record your results in a table like this:

Network	No. of arcs	No. of corners	No. of arcs removed
1			1
2			
etc.			

1.

2.

3.

4.

5.

6. Can you write a formula for finding the number of arcs, X, that must be removed from a network containing A arcs and C corners to change it into a tree?

Courtesy of Irv Phillips; by permission of Publishers-Hall Syndicate.

If you think of the ladder in this cartoon as a network, it is easy to see that it is quickly being changed into two separate trees.

7. Suppose the ladder has 10 steps. Draw a network to represent its structure, showing each step as a single arc.

If your diagram is correctly drawn and you have marked all of its corners, you should count 32 arcs.

8. How many corners does it contain?
9. How many arcs must be removed to change your ladder network into a tree?
10. Do these numbers agree with your formula?

Set III

A popular puzzle involves drawing a continuous line that crosses each arc of the network below exactly once, without going through any of its corners.

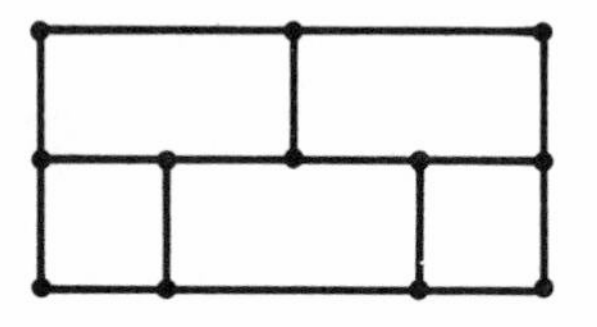

Here are a couple of unsuccessful attempts to solve this puzzle. In the first one, one of the arcs has not been crossed, and in the second one an arc has been crossed more than once.

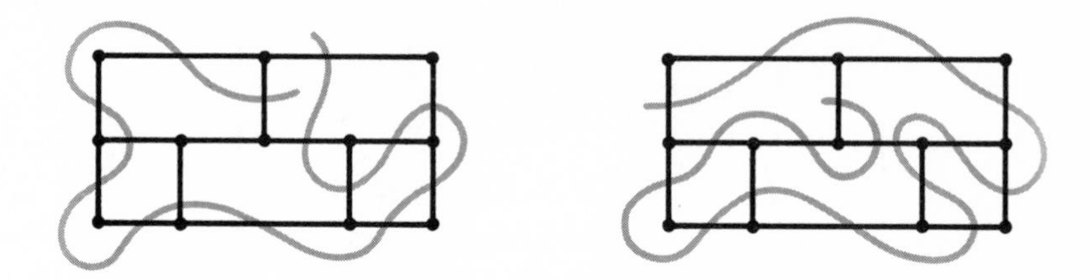

Can the puzzle be solved, or is it impossible? To find out, place a sheet of tracing paper over the networks below and try to draw such a line on each. It should cross each arc exactly once and should not pass through any corner.

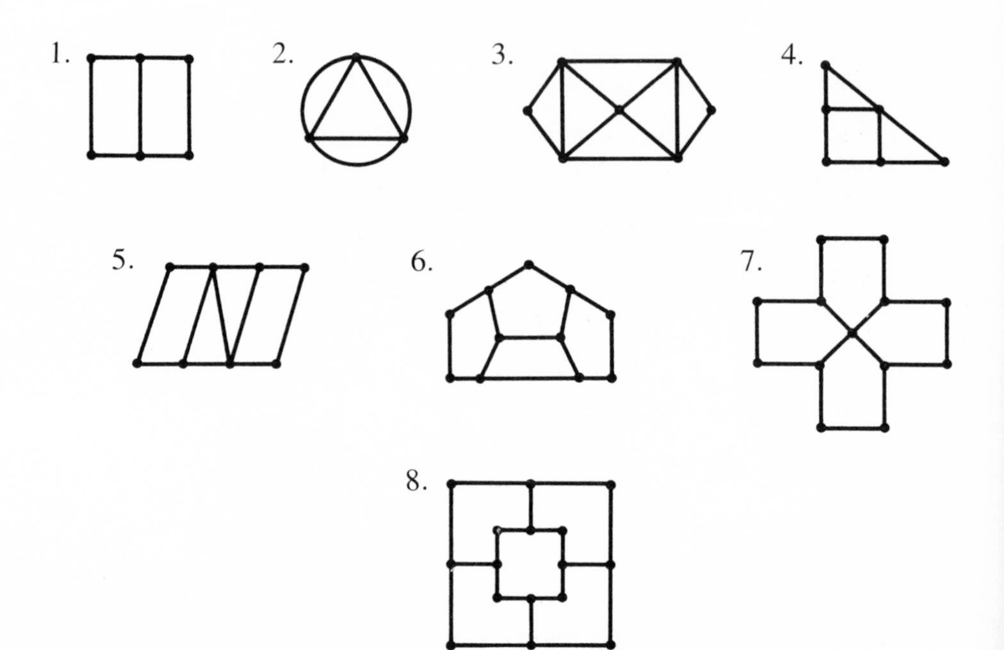

Each of these networks contains 2 or more regions. We will consider each region to be of an even or odd degree according to whether it is surrounded by an even or an odd number of arcs. For example, the first network contains 2 even regions, since each is surrounded by 4 arcs. (We will ignore the region outside.)

9. Count the numbers of even and odd regions in each of the other networks and record the results in a table like the one below. Include a column showing whether or not you think the network has a path.

Network	No. of even regions	No. of odd regions	Does the network have a path?
1	2	0	Yes
2	▥	▥	▥
etc.			

10. Can you make up a rule for determining whether or not it is possible to draw a continuous line that crosses each arc of a network exactly once?

11. On the basis of the rule you have written, does the puzzle have a solution? Explain your answer.

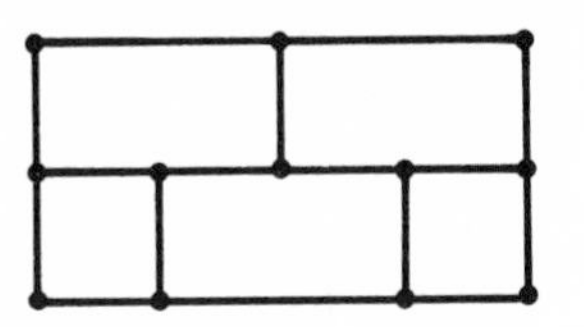

INTERESTING READING

Mathematics by David Bergamini, a book in Time-Life's "Life Science Library" series, 1963: "Topology: The Mathematics of Distortion," pp. 176–191.

Mathematical Snapshots, by Hugo Steinhaus, Oxford, 1969: Chapter 12, "Platonic Bodies, Crossing Bridges, Tying Knots, Coloring Maps, and Combing Hair," pp. 252–281.

APPENDIX

Some Fundamental Ideas

1 Angles and Their Measurement

Imagine a large pie that has been divided by a very sharp knife into 360 equal slices.* Viewed from above, one of the slices looks like the one below. The slice's two straight edges form an *angle* and the point where they meet is called the *vertex*. The edges lie along the *sides* of the angle.

The *measure* of an angle represents the "amount of opening" between its sides. The angle of this skimpy slice is said to have a measure of one degree, which we write as 1°. If all 360 slices are left together, they completely surround the center of the pie; hence, the number of degrees about a point is 360.

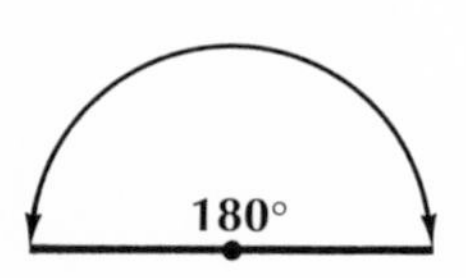

The measures of other angles can be conveniently expressed in terms of this basic unit, the degree. Here are some examples. Suppose we consider a more generous piece of the pie, half of it in fact. What is the measure of the angle formed by its sides? Since this piece is equal to 180 of the small slices, its angle has

* If you've ever cut up a pie, this may be very hard to imagine.

a measure of 180°. You probably know that such an angle is called a *straight angle,* since its sides lie along a straight line.

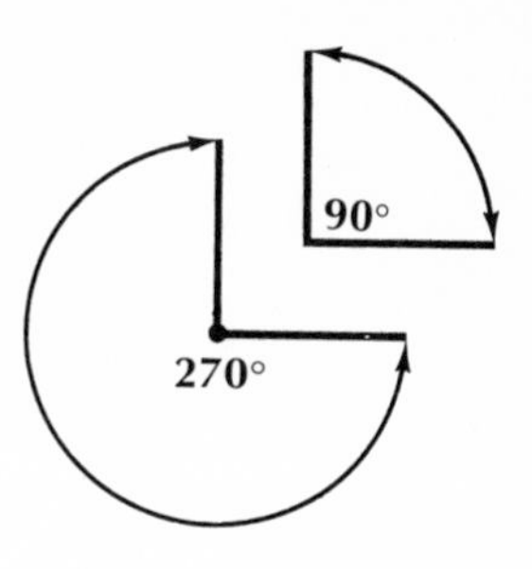

Now imagine a pie from which a quarter slice has been cut. The angle of the quarter slice has a measure of 90° and is called a *right angle.* We can think of the rest of the pie as another slice, and if we measure the "amount of opening" between its sides by looking around the "pie side," the angle formed has a measure of 270°. (360° − 90° = 270°.)

Angles are measured with an instrument called a protractor.

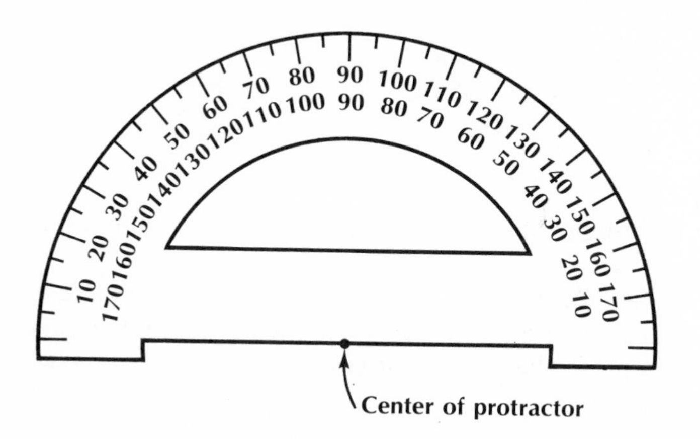

Most protractors have two scales, so that you can measure an angle in either a clockwise or a counterclockwise direction. Here is an example of how an angle can be measured using a protractor.

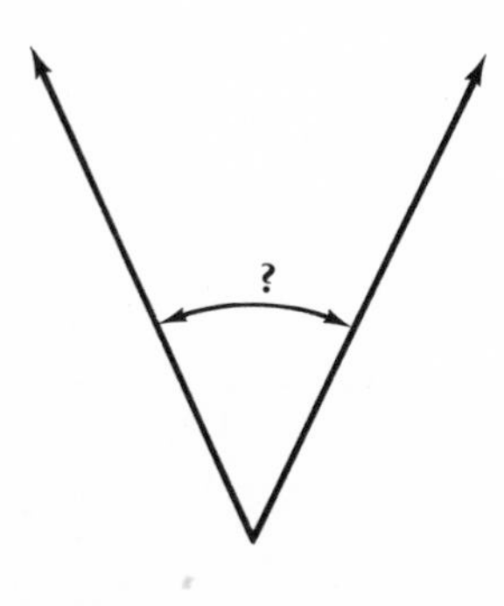

First, place the center of the protractor on the vertex of the angle. (Notice that the center of the protractor is on its *outer* edge.) Next, line up the edge of the protractor with one side of the angle. (It doesn't matter which side; both possibilities are shown on page 496.) Finally, read the measure of the angle by looking at where its other side falls underneath the scale.

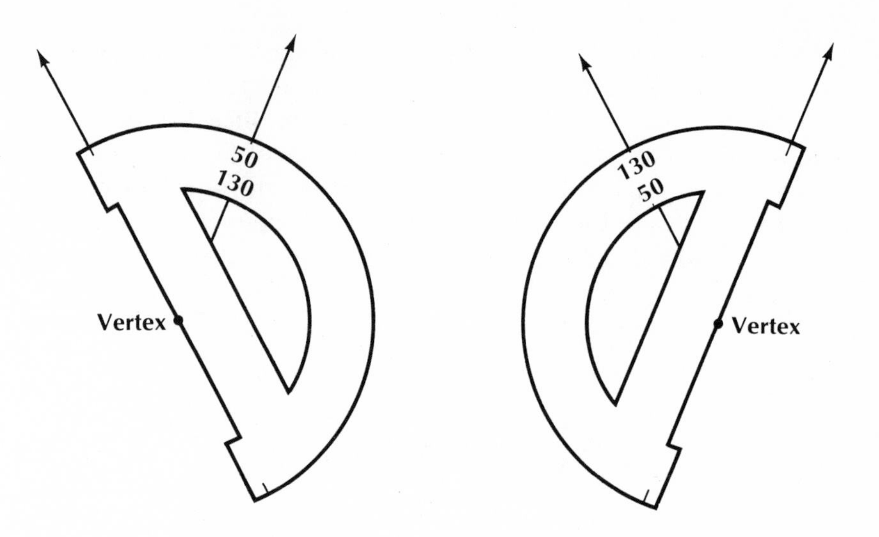

Since this angle is smaller than a right angle, its measure must be 50° and not 130°.

For practice, measure the angles below and on page 497. The measures of the angles are given at the bottom of the page so that you will know whether or not you are measuring them correctly.

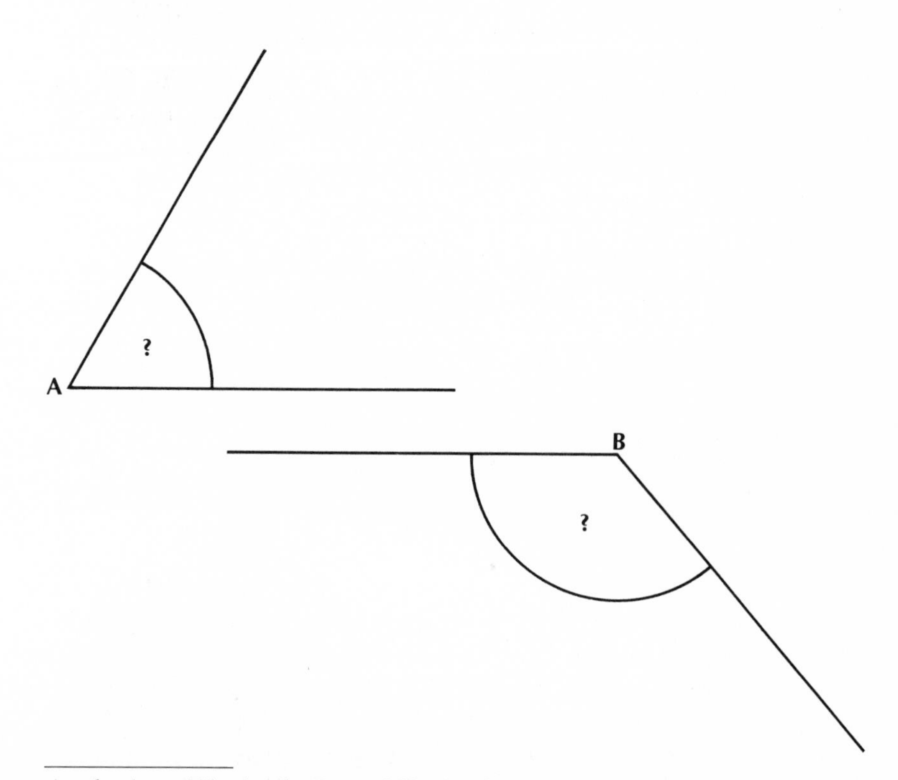

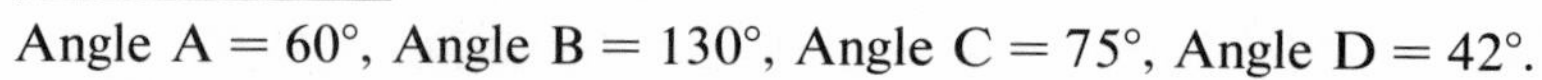

Angle A = 60°, Angle B = 130°, Angle C = 75°, Angle D = 42°.

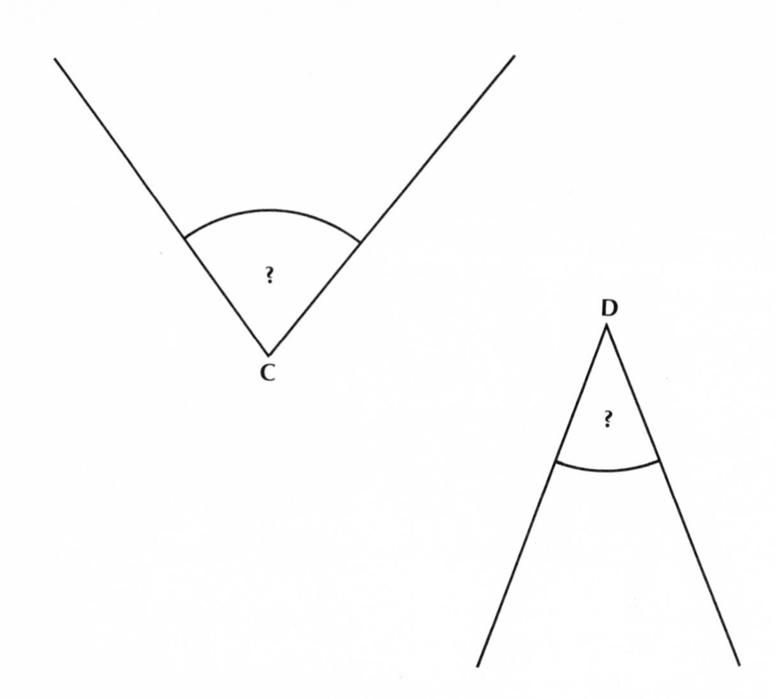

2 The Distributive Law

The distributive law relates multiplication to addition. It says that if you want to multiply the sum of two numbers by a third number, you can multiply each of the two numbers by the third number and then add the results.

For example,

$$3 \times (2 + 4) = 3 \times 2 + 3 \times 4$$
$$= 6 + 12$$
$$= 18.$$

This is correct, since

$$3 \times 6 = 18.$$

Let's see how this applies to the number puzzle.

Choose a number.	n
Add five.	$n + 5$
Double the result.	$2 \times (n + 5) = 2 \times n + 2 \times 5$ $= 2n + 10$

Before we go any further with the puzzle, we need to know that the distributive law relates division to addition in the same way. If you want to divide the sum of two numbers by a third number, you can divide each of the two numbers by the third number and then add the results.

For example,

$$\begin{aligned}\frac{9+12}{3} &= \frac{9}{3} + \frac{12}{3} \\ &= 3 + 4 \\ &= 7.\end{aligned}$$

This is correct, since

$$\frac{21}{3} = 7.$$

Now back to the number puzzle. We left off with:

	$2n + 10$
Subtract four.	$2n + 6$
Divide by two.	$\frac{2n+6}{2} = \frac{2n}{2} + \frac{6}{2} = n + 3$

Then, when you subtract the original number, n, it is easy to see that the result is three.

Check your understanding of the distributive law by trying these problems. The answers are given on page 502.

1. Multiply by 2: $n + 3$.
2. Multiply by 3: $n + 5$.
3. Divide by 2: $2n + 10$.
4. Divide by 4: $4n + 12$.

3 Signed Numbers

Everyone knows that the whole numbers have a definite order. After a child has learned to count he is able to say, for example, what number comes after 12 or what number comes before 8. We can show the order of the whole numbers by representing them as evenly spaced points along a line.

0 1 2 3 4 5 6 7

Suppose we extend the line beyond 0 in the opposite direction. How shall we number the points on the other side? The customary way of doing it looks like this:

−4 −3 −2 −1 0 1 2 3 4

The numbers of the points on the line left of the zero are called *negative* and the numbers to the right of the zero are called *positive*. Since we identify them with plus and minus signs, they are often called *signed numbers*. (A number without a sign is assumed to be positive.)

In an algebra course, it is necessary to learn all of the rules for making calculations with signed numbers. Only those that you need to know for the exercises in this book are explained here.

Addition To add two signed numbers, think in terms of gains and losses. You might use a football field to help in picturing the situation. For example, what is the sum of 2 and −5? A gain of 2 yards followed by a loss of 5 yards is the same as a loss of 3 yards.

$$2 + -5 = -3.$$

What is the sum of −3 and 7? A loss of 3 yards followed by a gain of 7 yards is equivalent to a net gain of 4 yards.

$$-3 + 7 = 4.$$

And what is the sum of −8 and −4? A loss of 8 yards followed by a loss of 4 yards is the same as a net loss of 12 yards.

$$-8 + -4 = -12.$$

Try the following exercises. The answers are on page 502.

5. $-6 + 5$
6. $9 + -2$
7. $-3 + -15$
8. $12 + -27$
9. $-64 + 10$

Subtraction To subtract a positive number, think of it as a loss. For example, what is 10 subtracted from 3? If we have made a

gain of 3 yards and then lose 10 yards, this is the same as a net loss of 7 yards.

$$3 - 10 = -7.$$

What is 6 subtracted from 0? A gain of 0 yards followed by a loss of 6 yards is the same as a loss of 6 yards.

$$0 - 6 = -6.$$

Try these exercises. The answers are on page 502.

10. $8 - 9$

11. $1 - 12$

12. $5 - 20$

Multiplication Multiplication of signed numbers is easily illustrated by a multiplication table.

×	3	2	1	0	−1	−2	−3
3	9	6	3	0	−3	−6	−9
2	6	4	2	0	−2	−4	−6
1	3	2	1	0	−1	−2	−3
0	0	0	0	0	0	0	0
−1	−3	−2	−1	0	1	2	3
−2	−6	−4	−2	0	2	4	6
−3	−9	−6	−3	0	3	6	9

This table shows, for example, that

$$3 \times -2 = -6$$

and that

$$-1 \times -3 = 3.$$

In general, the product of two numbers having *opposite* signs is *negative,* and the product of two numbers having the *same* sign is *positive.*

Try these exercises. The answers are on page 502.

13. -4×3

14. -5×-5

15. 7×-2

Division The rules for division are the same as those for multiplication. The quotient of two numbers having *opposite* signs is *negative;* the quotient of two numbers having the *same* sign is *positive.*

For example,

$$\frac{12}{-3} = -4 \quad \text{and} \quad \frac{-20}{-2} = 10.$$

Try these exercises. The answers are on page 502.

16. $\frac{-35}{5}$

17. $\frac{-24}{-8}$

18. $\frac{10}{-4}$

4 Percent

The word "percent" literally means "per hundred." For example, if you are told that there is a 17% chance of getting a 6 when a die is rolled once, this means that it will turn up about 17 times per 100 rolls. If we change 17% to a fraction, we get

$$\frac{17}{100} = 0.17.$$

To change a fraction to a percent, simply *multiply it by 100.*

$$0.17 \times 100 = 17\%$$

What is 1/38 expressed as a percent? Here are the steps involved in finding out:

Step 1. $\frac{1}{38} \times 100 = \frac{100}{38}$

Step 2.

$$\begin{array}{r} 2.6 \\ 38\overline{)100.0} \\ \underline{76} \\ 24\ 0 \\ \underline{22\ 8} \\ 1\ 2 \end{array}$$

Step 3. 2.6 rounded off to the nearest whole number is 3.

So $\frac{1}{38} = 3\%$ (rounded off).

Check your understanding of percent by changing each of the following numbers to percents. (Round each answer to the nearest whole number.) The answers are given below.

19. $\frac{3}{4}$

20. 1

21. $\frac{2}{25}$

22. $\frac{5}{36}$

Answers to Exercises in Appendix.

1.	$2n + 6$	9.	−54	16.	−7
2.	$3n + 15$	10.	−1	17.	3
3.	$n + 5$	11.	−11	18.	−2.5
4.	$n + 3$	12.	−15	19.	75%
5.	−1	13.	−12	20.	100%
6.	7	14.	25	21.	8%
7.	−18	15.	−14	22.	14%
8.	−15				

Answers to Selected Exercises

CHAPTER 1

Lesson 1

Set I

2.

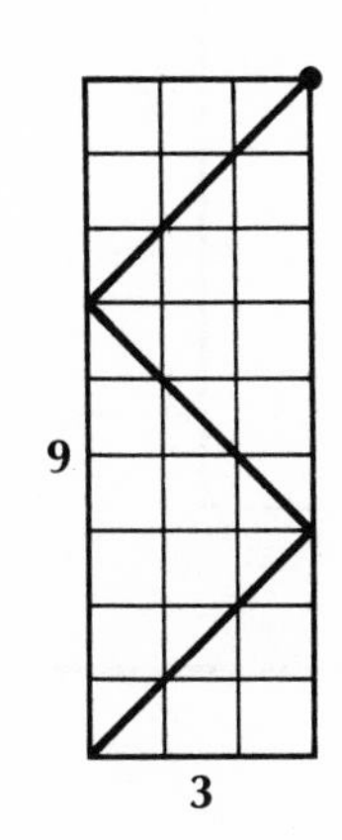

5.

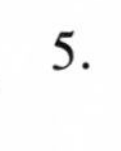

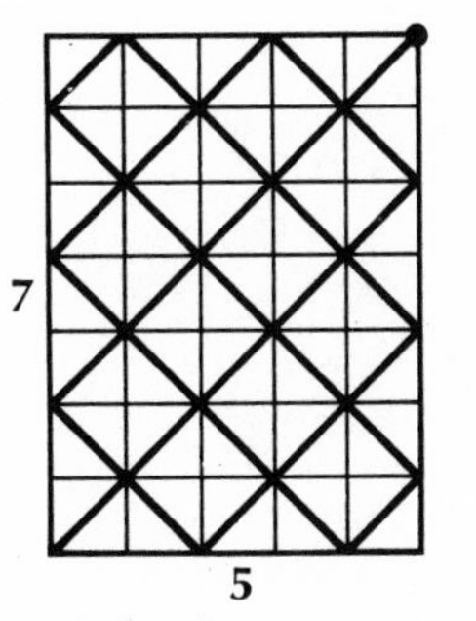

Set II

1.

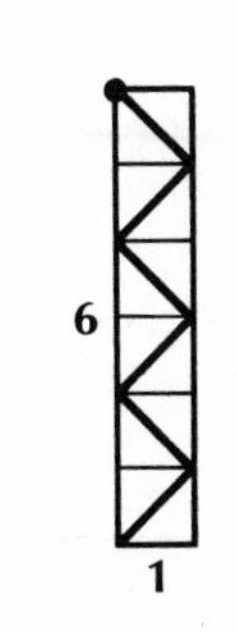

Lesson 2

Set I

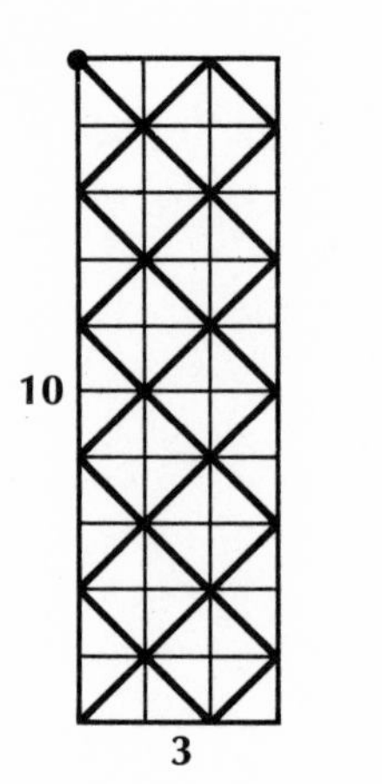

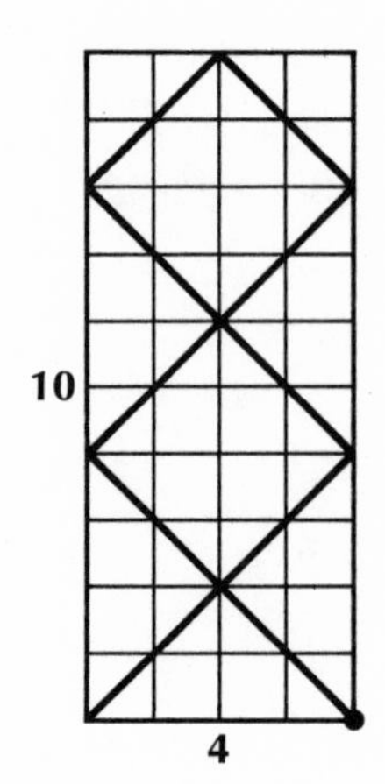

Set II

1.

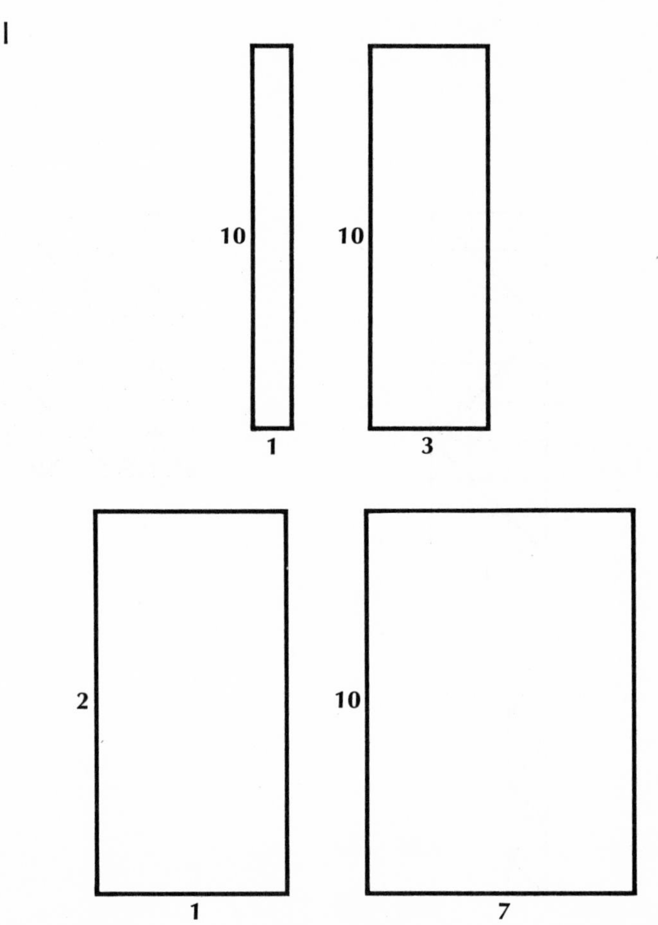

Lesson 3

Set I

2. 25 units.
5. 28.

Lesson 4

Set I

1. The figure: Two perpendicular line segments; segment AB is horizontal and segment CD is vertical.
   The illusion: CD appears to be longer than AB; it is actually shorter!
2. The figure: A set of parallel line segments, equal in length and equidistant from each other.
   The illusion: The figure seems to be taller than it is wide, yet the two dimensions are the same.
8. The figure: A set of cubes.
   The illusion: There are either 12 cubes (if you picture the black squares as their bottoms), or 9 cubes (if the 4 black squares in the middle are pictured as spaces between cubes).

Set II

2. 64 square units.

Lesson 5

Set I

1. 8.
2. 12.
7. Along the edges between the corner cubes.
9. In the centers of the faces.
13. 8.
14. 24.

Set II

5. No.

Lesson 6

Set I

2. Choose a number. ■ n

   Double it. ■ ■ $2n$

   Add nine. ■ ■ ::::. $2n + 9$

Add your original number.	■ ■ ■ ::::.	$3n + 9$
Divide by three.	■ •••	$n + 3$
Add four.	■ :::.	$n + 7$
Subtract your original number. Your result is seven.	:::.	7

Set II.

8. $\frac{143n}{11} = 13n.$

CHAPTER 2

Lesson 1

Set I

2. 1 4 7 10 [13] [16]
8. 7 □ □ □. Many sets of numbers will fit here; one number does not ordinarily determine a number sequence.
10. Yes. The number added each time is 0.

Set II

1. $4 + 9 \cdot 5 = 49.$
2. $3 + 19 \cdot 4 = 79.$

Set III

3. $$\begin{array}{l} \;\;1 + \;\;2 + \;\;3 + \ldots \\ \underline{20 + 19 + 18 + \ldots} \\ 21 + 21 + 21 + \ldots = 10 \cdot 21 \doteq 210. \end{array}$$

Lesson 2

Set I

1. 1 5 25 [125] [625]
4. [2] [6] 18 54 162
8. 2 4 8 16
9. No. The numbers are not being multiplied by the same number each time.

Set II

4. $5 \cdot 2^9$.

Set III

1. $312.50.

Lesson 3

Set I

1.

Number	16	8	4	2	1
1					1
2				1	0
3				1	1
4			1	0	0
5			1	0	1
6			1	1	0
7			1	1	1
8		1	0	0	0
9		1	0	0	1
10		1	0	1	0
11		1	0	1	1
12		1	1	0	0

etc.

Set II

1. 2^{15}.
3. $2^{16} - 1$.

Lesson 4

Set I

3. 1 4 9 6 5 6 9 4 1 0. This pattern repeats.
5. $2 + 5 + 6 = 13$; $1 + 3 = 4$.

Lesson 5

Set I

4. The digital roots of the first four cube numbers are 1, 8, 9, and 1.

Lesson 6

Set I

2. $1 + 1 + 2 + 3 + 5 = 12$.

CHAPTER 3

Lesson 1

Set I

1.

x	0	1	2	3	4
y	5	6	7	8	9

3.

x	0	1	2	3	4
y	1	3	5	7	9

5.

x	0	1	2	3	4
y	0	1	4	9	16

Set II

1. $y = 2x$.
3. $y = x - 3$.
7. $y = 10x + 1$.

Lesson 2

Set I

1. A(4, 3); B(3, 4); C(0, 5); D(−3, 4); E(−5, 0); F(−4, −3); G(0, −5); H(3, −4); I(5, 0).

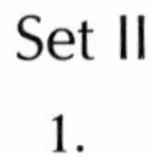

Set II

1.

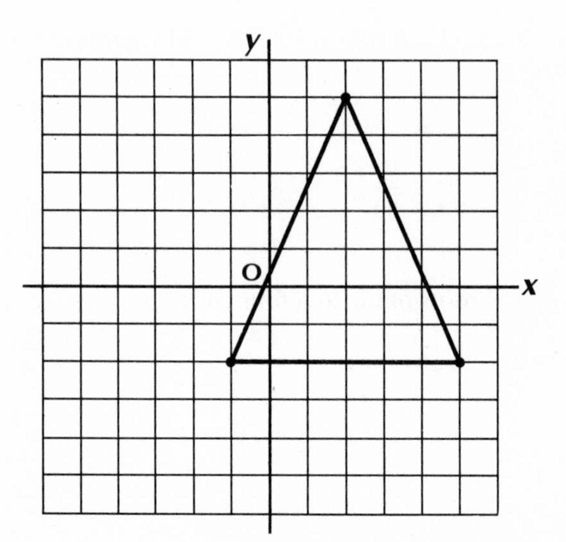

Lesson 3

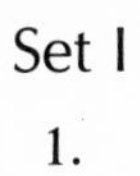

Set I

1.

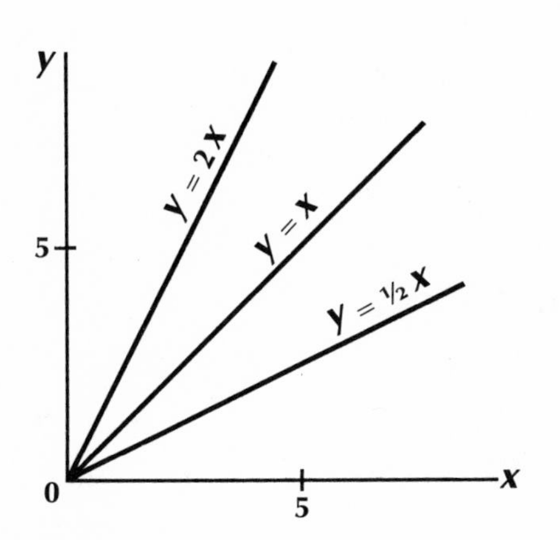

Set II

1.

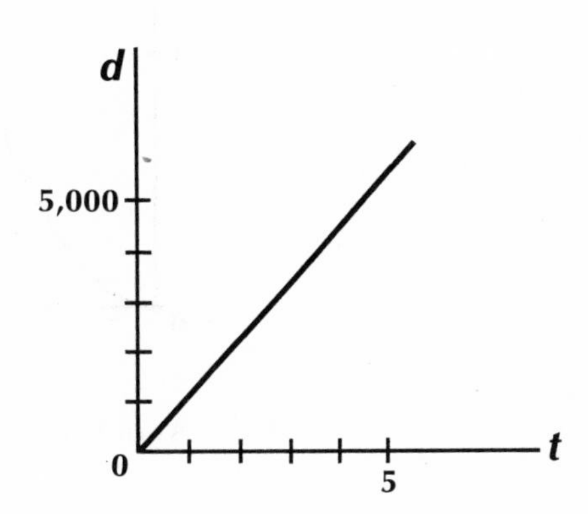

Set III

4. About 82 degrees.

Lesson 4

Set I

1.

x	−3	−2	−1	0	1	2	3
y	11	6	3	2	3	6	11

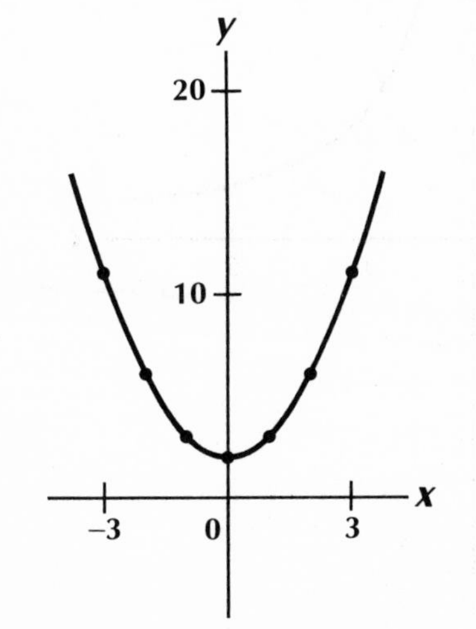

5.

x	−3	−2	−1	0	1	2	3
y	18	8	2	0	2	8	18

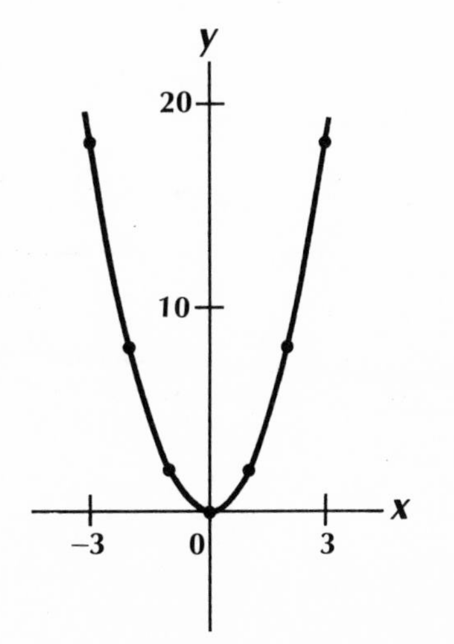

Set II

1.

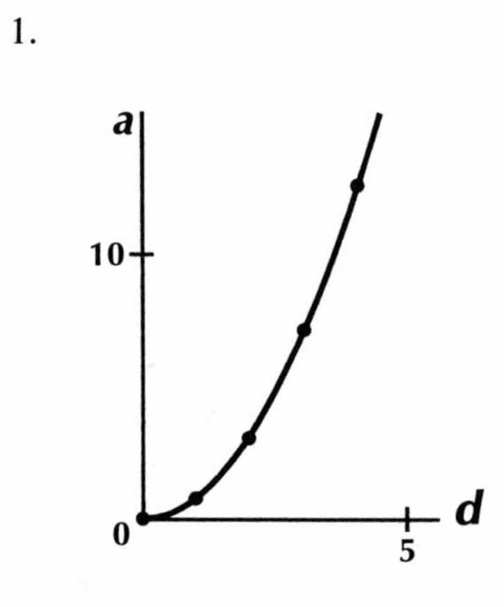

4. About 67 feet.

Set III

5. $2\frac{1}{2}$ seconds.

Lesson 5

Set I

1.

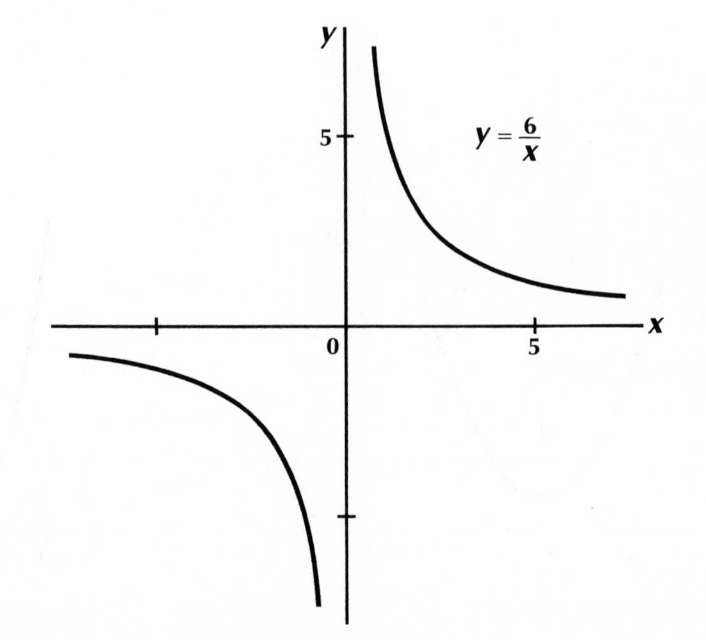

4. $y = \dfrac{4}{x}$

5.

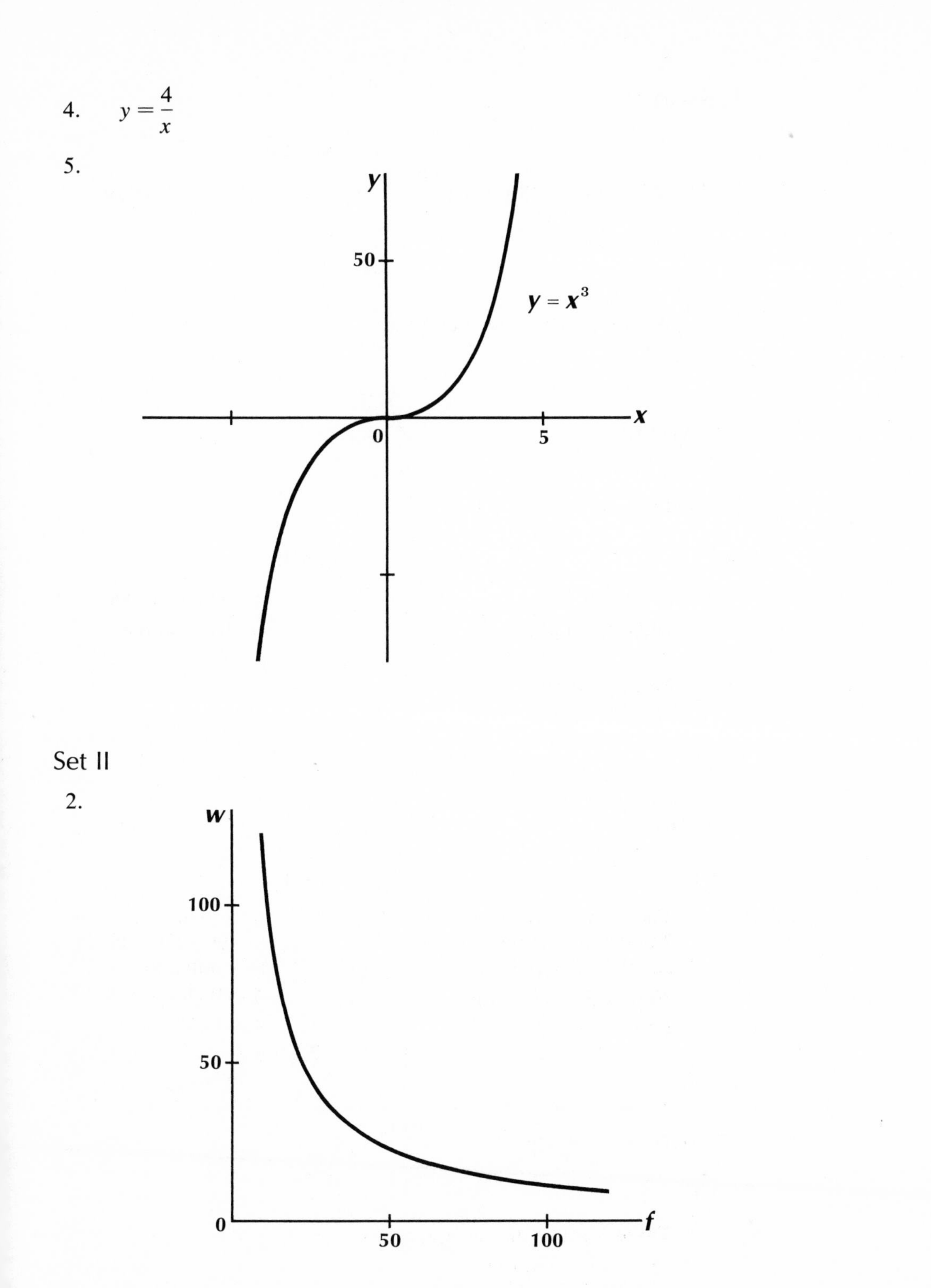

Set II

2.

4. 0, 33.

7. About 11 pounds.

Lesson 6

Set I

4. $\sqrt{20} = 4.5$; $\sqrt{30} = 5.5$; $\sqrt{40} = 6.3$; etc.

Set II

2. About 1.9 billion.

Set III

4. About 26,000 miles per hour.

CHAPTER 4

Lesson 1

Set I

2. a) 10^7; b) 10^{26}
6. 10,000,000,000; ten billion.
10. More than 10^5 or 100,000 years.

Set II

2. $10{,}000 \times 10{,}000$ or 100,000,000.

Lesson 2

Set I

1. There are approximately 2,000 birds in the picture; $2{,}000 = 2 \times 10^3$.
2. 26,000,000; 26 million.
5. 1.58×10^{18}.

Set II

2. $(4 \times 10^2) \times (2 \times 10^4) = 400 \times 20{,}000 = 8{,}000{,}000$; $8{,}000{,}000 = 8 \times 10^6$.
4. 4×10^9.
7. 5.8×10^{11} miles.

Lesson 3

Set I

1. Three lines of the table are:

Number	Logarithm
32,768	15
1,048,576	20
33,554,432	25

2. 8,192. $(6 + 7 = 13)$
6. 16,384. $(7 + 7 = 2 \cdot 7 = 14)$
9. 256. $(16 \div 2 = 8)$
12. 16,777,216. $(4 \cdot 6 = 24)$

Set II

3. $4 \times 8 = 2^2 \times 2^3 = 2 \times 2 \times \boxed{2 \times 2 \times 2} = 2^{\boxed{5}}$.
5. $2^1 \times 2^4 = 2^{\boxed{1+4}} = 2^{\boxed{5}}$.

Set III

1. 1.6×10^4.

Lesson 4

Set I

5. $\log 20 = \log 4 + \log \boxed{5}$; 1.301.
8. $\log \boxed{72} = \log 8 + \log 9$; 1.857.
13. 1.690. $(\log 49 = 2 \times \log 7)$

Set II

3. 2.301. $(2 + 0.301)$
6. It is the whole number part of the logarithm.
9. The fraction part.
10. $\log (3 \times 10^5) = 5.477$.

Lesson 5

Set I

1. 1.43×10^3; 3.155.
2. 4×10^1; 1.602.
5. 8.4.
7. $$\begin{aligned} \log (1.34 \times 10^5) &= 5.127 \\ + \log (4.1 \times 10^2) &= 2.613 \\ \hline \log \text{ of answer} &= 7.740 \\ \text{answer} &= 5.5 \times 10^7 \\ &= 55{,}000{,}000. \end{aligned}$$
9. 6.2.
11. 2.4×10^{12}.

Set II

2. 5.270.
5. 3.5×10^{10}.
9. 3.556.

Set III

3. 1.380.
5. 4.0×10^8.

Lesson 6

Set I

2. It is an arithmetic sequence.

Set III

2. The keys have a logarithmic arrangement with respect to their frequencies.
5.

Relative loudness	1	10	100	etc.
Decibels	0	10	20	

6. 10,000 times louder.
8. It is 1/10 of it.

CHAPTER 5

Lesson 1

Set I

1.

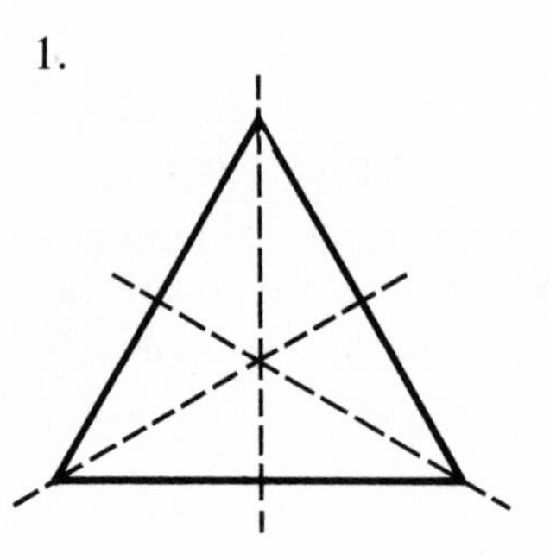

6.

B C A B

A D D C

etc.

Set II

6. 4°.

Lesson 2

Set I

1.

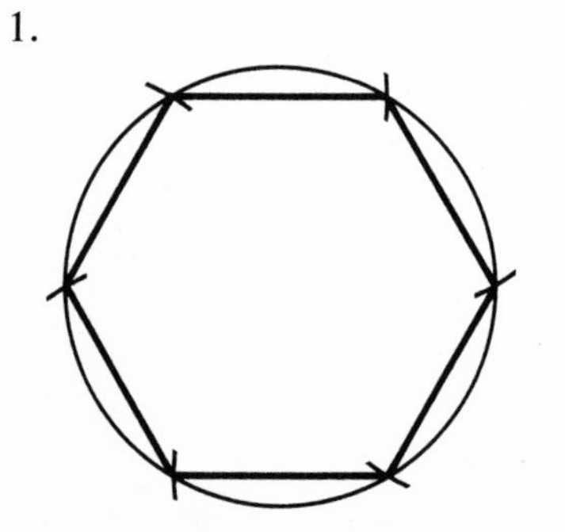

Lesson 3

Set II

1. n 3 4 5
   a 60° $\boxed{90°}$ $\boxed{108°}$ etc.

Lesson 4

Set I

1. Cube: 4-4-4.
4. 5.
5. 60.
11. 5.
12. 12.
14. 2.

Lesson 5

Set I

1. 3-6-6.
6. 3-4-3-4.

Set II

1. 24 corners and 36 edges.
   $\left(\frac{6 \times 4 + 8 \times 6}{3} = \frac{72}{3}; \frac{72}{2}\right)$
3. 48 corners and 72 edges.
   $\left(\frac{12 \times 4 + 8 \times 6 + 6 \times 8}{3} = \frac{144}{3}; \frac{144}{2}\right)$
8. 60 edges.

Lesson 6

Set I

3. A circle.

Set II

1.

Pyramid	No. of sides in base	No. of faces	No. of corners	No. of edges
Pentagonal	5	6	6	10

4. $2n$.
6. $F + C = E + 2$
   $(n + 1) + (n + 1) = 2n + 2$
   $2n + 2 = 2n + 2.$

7.

Prism	No. of sides in a base	No. of faces	No. of corners	No. of edges
Pentagonal	5	7	10	15

CHAPTER 6

Lesson 1

Set I

1. 16 units.
3. The major and minor axes of ellipse B are both longer than those of ellipse A.
8. (12, 0) and (−12, 0).

Set II

6. A circle.

Lesson 2

Set I

1.

x	0	1	2	3	4	5	6	7	8	9	10
y	16	9	4	1	0	1	4	9	16	25	36

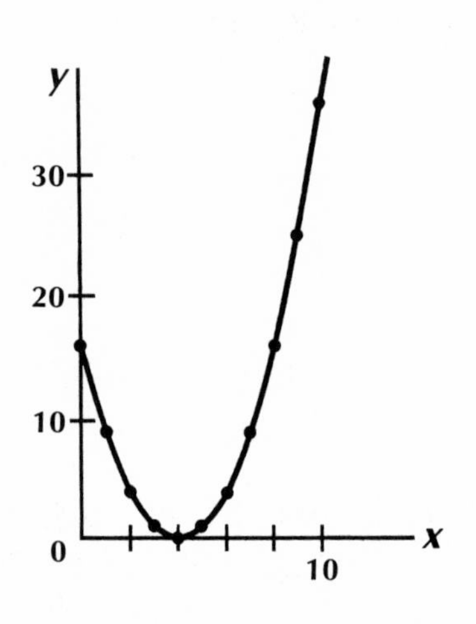

Set II

2. 204.
3. 361.

Lesson 3

Set I

3. Infinite.
7. A hyperbola.
10. Yes. It can be rotated 180° to coincide with its original position.

Lesson 4

Set I

3. 1 and −1.
6. 0°, 180°, 360°.
8. 175°.

Set II

1. Amplitude: 2; period: 360°.
6. 5.

Lesson 5

Set I

2. 200°.
9. 135 revolutions.
11. It slows down.

Set II

3. A geometric sequence.

Lesson 6

Set II

2. No.
7. 5 units.
8. 78.5 square units.

CHAPTER 7

Lesson 1

Set I

2. 54 ways.
4. 25 combinations.
8. 100 phrases.
11. 64 groups.

Set II

2. 3 bells: 9 ways;
   3 plums: 25 ways.
3. 150 combinations.
5. 64 key designs.
9. 1,352 sets.

Set III

2. 24 positions.

Lesson 2

Set I

1. Two lines of the table are:
   $7! = 5{,}040$
   $10! = 3{,}628{,}800$
6. True.
7. 3,628,800 ways.

Set II

1. 3,540 ways.
3. 64 squares.
6. 86,400 ways.
7. 6 ways.

Set III

2. 288 ways.

Lesson 3

Set I

1. Two more paths are:

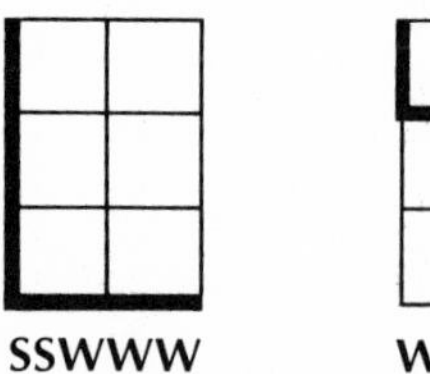

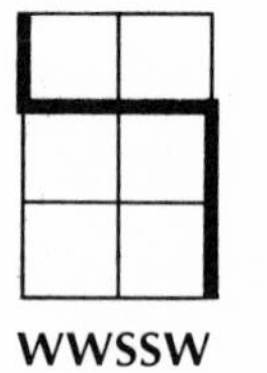

2. 110 ways.
6. Digits for his house numbe
7. 2,520 stacks.
8. 70 orders.

Set II

3. 1 way.
8. Each line reads the same in reverse.

Set III

1. 10 orders.
5. 35.

Lesson 4

Set I

1. 10 "doubles" dominoes.
2. 45 "singles" dominoes.
7. 190 handshakes.
9. 20 combinations.

Set II

1. 161,700 committees.
3. 792.
6. 35.

CHAPTER 8

Lesson 1

Set I

3. 1/3.
5. 10 times.
11. 8/8 = 1.
12. 3/8.
13. 2/3.
16. 1/8.

Set II

2. 32%.
5. 47%.
7. 1/2.
10. 5/1.

Lesson 2

Set I

1-2.

Sum of 2 dice:	2	3	4	
Number of ways:	1	2	3	
Probability:	$\frac{1}{36}$	$\frac{1}{18}$	$\frac{1}{12}$	etc.
Percent probability:	3	6	8	

3. 4 ways.

Set II

1. 216 ways.
5. There are 5 times as many ways of getting a sum of 7 as of getting a sum of 17.
10. 1/36.

Lesson 3

Set I

2. 1/4.
6. 7/8.
9. 3/8.
12. 1/8.

Set II

1. 1, or 100%.
4. 1/2, or 50%.

Lesson 4

Set I

4. The 6th row.
8. 7 heads.
9. 1/256.
14. 63/256, or approximately 25%.

Lesson 5

Set I

1. 9/8,000.
4. 169/400.
7. 1/36.
9. 1/231.

Set II

1. 1/221.
2. 1/2,652.
4. 2/17.
7. 12.
9. 1%.
11. $52/52 \times 12/51 \times 11/50 \times 10/49 \times 9/48$.

Lesson 6

Set I

1. $1 - 364/365 \times 363/365 \times 362/365 \times 361/365$.
5. About 90%.
8. 17/18.
10. 25/36.
13. 31/32.

Set II

2. 5,764,801/100,000,000.

CHAPTER 9

Lesson 1

Set I

3. 21 presidents.
5. 8%.
7. 40%.

Set II

1. 10 times.

Lesson 2

Set I

1. The symbols seem to be chosen at random.
4. 26.
8. 6%.

Set II

5. Z represents H.
7. H represents O.

Lesson 3

Set I

4. 12.4.
12. 4.5 letters.
16. The owners would choose the mean, which is $6,000.

Set II

1.

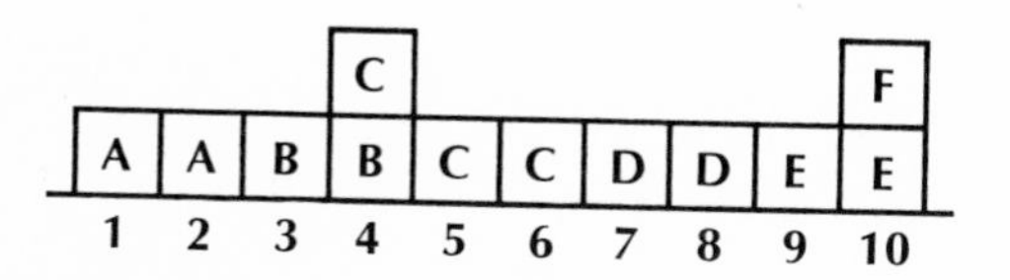

3. 1.2 customers.

Lesson 4

Set I

1. 14 inches.
5. 16.
7. 70%.

Set II

1. 8 minutes slow.
3. 680 clocks.
4. 96%.
5. 160 clocks.
7. 10 seconds.
12. 10 people.

Lesson 5

Set I

6. 230°.
7.

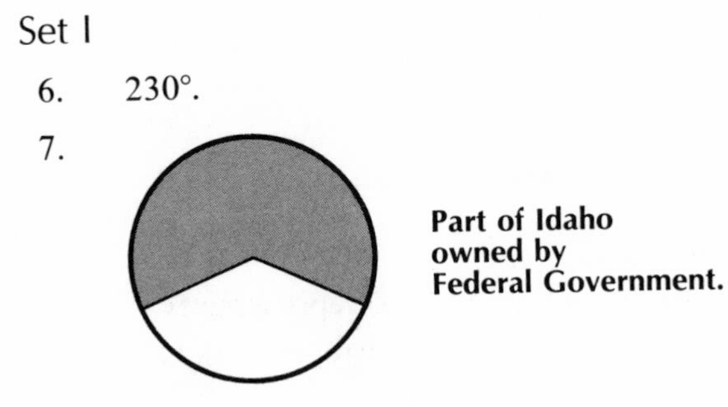

Set II

2. It is about one-fourth as tall.

Lesson 6

Set I

3. Many of the customers would probably claim to be younger than they really were.

Set II

1. 80%.

CHAPTER 10

Lesson 1

Set I

1. Because it can be twisted and stretched into the same shape.
3. J.
9. Three of the letters are L, M, and S.
12. X.

Set II

3. Outside; even.

Lesson 2

Set I

3. Start at either A or B.
4. Start at either E or F.
7. Start at either C or F.

Set II

2. No.
10. Impossible.

Lesson 3

Set I

1. No. (It has 8 odd corners.)
10. 6 odd corners; 3 trips.

Set II

Network	No. of regions	No. of corners	No. of arcs
1	2	3	3
2	1	8	7
etc.			

Lesson 4

Set I

2. 7 corners; 6 arcs.
6. No. It has 8 corners and 8 arcs, but this network is not a tree.
10. Impossible. (This is contrary to the definition of a tree.)
11, #1 has a diameter of 2, #2 has a diameter of 6.

Set II

3.

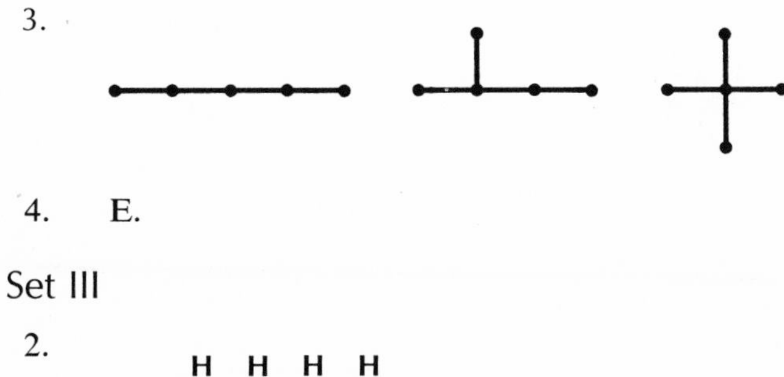

4. E.

Set III

2.

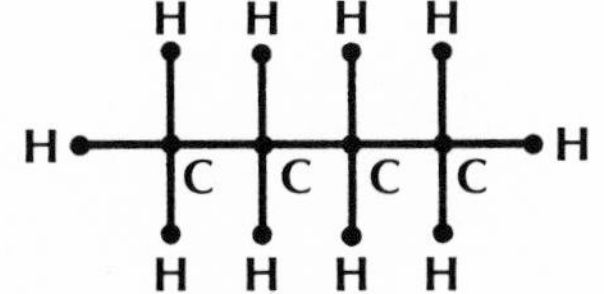

Lesson 5

Set I

2. The result is a twisted band twice as long and half as wide.
5. Two interlocking loops are produced, one of which is twice as long as the other.

Set II

2. Two interlocking loops of the same length and width. (Both loops are twisted.)
5. One side.
6. A single loop with a knot in it is produced.
11. Two interlocking loops of the same length and width.

Index